D1064809

Lecture Notes in Statistics

Edited by J. Berger, S. Fienberg, J. Gani, K. Krickeberg, and B. Singer

51

J. Hüsler R.-D. Reiss (Eds.)

Extreme Value Theory

Proceedings of a Conference held in Oberwolfach, Dec. 6–12, 1987

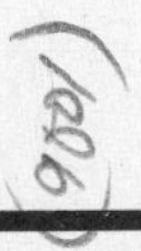

Springer-Verlag
New York Berlin Heidelberg London Paris Tokyo

Editors

Jürg Hüsler
Universität Bern, Institut für mathematische Statistik
Sidlerstrasse 5, 3012 Bern, Switzerland

Rolf-Dieter Reiss
Universität Gesamthochschule Siegen
Hölderlinstr. 3, 5900 Siegen, Federal Republic of Germany

Mathematics Subject Classification (1980): 60G55, 60F05, 62E20, 62G30

ISBN 0-387-96954-3 Springer-Verlag New York Heidelberg Berlin
ISBN 3-540-96954-3 Springer-Verlag Berlin Heidelberg New York

Printed in Germany

Printing and binding: Druckhaus Beltz, Hemsbach/Bergstr.
2847/3140-543210

PREFACE

The urgent need to describe and to solve certain problems connected to extreme phenomena in various areas of applications has been of decisive influence on the vital development of extreme value theory. After the pioneering work of M. Fréchet (1927) and of R.A. Fisher and L.H.C. Tippett (1928), who discovered the limiting distributions of extremes, the importance of mathematical concepts of extreme behavior in applications was impressively demonstrated by statisticians like E.J. Gumbel and W. Weibull. The predominant role of applied aspects in that early period may be highlighted by the fact that two of the "Fisher-Tippett asymptotes" also carry the names of Gumbel and Weibull.

In the last years, the complexity of problems and their tractability by mathematical methods stimulated a rapid development of mathematical theory that substantially helped to improve our understanding of extreme behavior. Due to the depth and richness of mathematical ideas, extreme value theory has become more and more of interest for mathematically oriented research workers. This was one of the reasons to organize a conference on extreme value theory which was held at the Mathematische Forschungsinstitut at Oberwolfach (FRG) in December 1987.

The theory of extreme values has become a broad subject which is difficult to cover by a few authors. It is the purpose of this book to lay out in an expository way the broad spectrum of extremes by contributions, some of which are reviewing recent developments and some are including original ideas and results. It is evident that, by the nature of a conference proceedings, the coverage must be selective. From the main subjects of the conference and of this book we mention

extremes of special stochastic processes,
rate of convergence and finite expansions,
probabilistic aspects concerning records,
extremes of dependent random variables,
the role of point processes in extreme value theory,
statistical inference about the type of the domain of attraction,
standard and non-standard statistical models for
univariate and multivariate extremes.

The book is split up into three parts with a total number of nine sections. Part I, that concerns the probabilistic theory, is arranged in five sections on limit laws and expansions, strong laws, records, exceedances, and characterizations.

Section 1 deals with the convergence of distributions of extremes in the classical case. M. Falk verifies that a convergence rate that holds in case of generalized Pareto random variables is the best attainable convergence rate with respect to the joint behavior of largest order statistics. T.J. Sweeting discusses asymptotic expansions of distributions of extremes by using an expansion of an integral of a regular varying function.

Strong laws are studied in Section 2. The present results clarify much of the asymptotic representation and characterization of extremes in the classical case. P. Deheuvels gives bounds and characterizations for the k-th largest extremes where k may slowly tend to infinity as the sample size increases. The almost sure behavior of sums of upper extremes is discussed by P. Révész for the case that the underlying distribution has a very heavy tail.

Three papers in the book are concerned with records. Section 3 treats topics from univariate theory. D. Pfeifer and Y. Zhang prove strong approximations for records and record times under particular distributions. Records and their time points may be studied from the point of view of the theory of self-similar random measures. U. Zähle discusses this relation and studies the Hausdorff dimension of such measures related to records.

In the last decades the classical theory of extreme values has been extended to the more general framework of random sequences and stochastical processes. The present state-of-the-art is well developed in textbooks. In Section 4, the very powerful point process approach is used by M.R. Leadbetter and S. Nandagopalan in their investigation of clustering of extreme exceedances in case of stationary random sequences satisfying weak dependence conditions. S.M. Berman explores the question of extreme sojourn times for time-continuous, stationary Gaussian processes; a central limit theorem is proved for the sojourn time above a certain function which increases sufficiently slowly. I. Rychlik reviews recent results on the distribution of waves and cycles which are related to local maxima and minima of particular continuous processes.

Section 5 deals with the important question of characterizing distributions by means of distributional properties of extremes and order statistics. This aspect is studied by L. Gajek and U. Gather in the particular case of the exponential distribution. Z. Buczolich and G.J. Székely characterize location models of uniform distributions by the property that the midrange is the maximum likelihood estimator.

Part II is devoted to the statistical theory of extreme values. The three sections concern the estimation of the tail of the distribution, an estimation problem in case of dependent random variables, testing the type of extreme value model, and questions concerning the sufficiency of extremes.

Stochastic properties of extremes are determined by the tail behavior of the underlying distribution. In Section 1 of Part II, parameters are estimated that characterize the tail behavior. S. Csörgő and D.M. Mason consider the estimation of the lower endpoint in a parametric Weibull-type model; a continuous time regression method is used to derive asymptotically efficient estimators. The contribution by J. Beirlant and J.L. Teugels further specifies the normal domain of attraction of the Hill estimator. By R.-D. Reiss a first step is made towards the statistical inference in extreme value models that are built by expansions of distributions of maxima; this approach leads to adaptive estimators of the tail index.

An estimation procedure of a different nature is studied by W.P. McCormick and G. Mathew who propose an estimator of the autocorrelation coefficient for a first order autoregressive sequence based on an extreme value statistic and derive its asymptotic properties.

Test procedures in the model of actual distributions and in limiting extreme value models are treated in Section 2. E. Castillo, J. Galambos, and J.M. Sarabia are testing the domain of attraction of the actual distribution; the test is based on the curvature of the upper tail of the empirical distribution function. M.I. Gomes considers test procedures in the limiting model of samples of annual upper extremes (in other words, multidimensional GEV model).

Sections 3 consists of one paper. Sufficiency is one of the basic concepts of statistical theory. In connection with extremes this question is dealt with by A. Janssen.

Part III concerns multivariate extremes and records. This part of the book covers to some extent the recent, intensive investigations in the multivariate set-up.

In the univariate case the definition of extremes and records are obvious, but in the multivariate situation different directions are possible, as for instance the frequently used componentwise ordering or the definition of a record by the property that each of its components is larger than the corresponding components of the previous observations. In a paper by K. Kinoshita and S.I. Resnick such records are discussed and utilized to study the shape of the convex hull of the observations. The componentwise definition of extremes is taken up by J. Hüsler. He investigates dependence properties of limit distributions that occur under weak dependence conditions in general nonstationary random sequences.

Various representations and characterizations of multivariate extreme value distributions are surveyed by J. Tiago de Oliveira and by J. Pickands. Moreover, Tiago discusses bivariate, parametric extreme value models.

We take this opportunity to express our heartiest thanks (also from the side of the participants) to the Oberwolfach institute represented by Prof. M. Barner, the director of the institute, for the permission of spending a week at Oberwolfach and of making use of the splendid facilities. Thanks are also due to A. Hofmann who took the photograph. We are particularly grateful to F. Marohn for the careful preparation of the subject index and author index. Finally, we would like to thank the following colleagues for their support by acting as a referee: A. Barbour, S.M. Berman, H. Carnal, R.A. Davis, M. Falk, J. Galambos, L. de Haan, A. Janssen, M.R. Leadbetter, D.M. Mason, M.A.J. van Montfort, D. Pfeifer, J. Pickands, S.I. Resnick, P. Révész, R.L. Smith, T.J. Sweeting, J. Tiago de Oliveira.

October 1988

Jürg Hüsler, Bern
Rolf-Dieter Reiss, Siegen

IN FRONT OF LECTURE HALL

List of Participants

Anderson, C.W.	1	Department of Probability and Statistics, The University of Sheffield, Sheffield S10 2TN, Great Britain.
Berman, S.M.	2	Courant Institute of Mathematical Sciences, New York University, 251 Mercer Street, New York, NY 10012, USA.
Cooil, B.	3	Owen Graduate School of Management, Vanderbilt University, 401 21st Ave., Nashville, TN 37203, USA.
Csörgő, S.	4	Bolyai Institute, Szeged University, Aradi Vértanúk Tere 1, 6720 Szeged, Hungary.
Davis, R.A.	5	Department of Statistics, Colorado State University, Fort Collins, CO 80521, USA.
Deheuvels, P.	6	7, Avenue du Château, 92340 Bourg-la-Reine, France.
Dziubdziela, W.	7	Institute of Mathematics, University of Wroclaw, Pl. Grunwaldski 2/4, 50-384 Wroclaw, Poland.
Falk, M.	8	Department of Mathematics, University of Siegen, Hölderlinstr. 3, 5900 Siegen 21, West Germany.
Gänßler, P.	9	Institute of Mathematics, University of Munich, Theresienstr. 39, 8000 München 2, West Germany.
Gajek, L.	10	Institute of Mathematics, Politechnika Lodzka, Al. Politechniki 11, 90-924 Lodz, Poland.
Galambos, J.	11	Department of Mathematics, Computer Building 038-16, Temple University, Philadelphia, PA 19122, USA.
Gather U.	12	Department of Statistics, University of Dortmund, P.O. Box 500500, 4600 Dortmund, West Germany.
Goldie, C.M.	13	Mathematics Devision, University of Sussex, Falmer, Brighton BN1 9QH, Great Britain.
Gomes, M.I.	14	Department of Statistics, University of Lisboa, 58, Rua da Escola Politecnica, 1294 Lisboa Codex, Portugal.
Grill, K.	15	Institute of Statistics, TU Wien, Wiedner Hauptstr. 8-10, 1040 Wien, Austria.
de Haan, L.	16	Institute of Economics, Erasmus University, Postbus 1738, 3000 DR Rotterdam, The Netherlands.
Häusler, E.	17	Institute of Mathematics, University of Munich, Theresienstr. 39, 8000 München 2, West Germany.
Hsing, T.	18	Department of Statistics, Texas A&M University, College Station, TX 77843, USA.
Hüsler, J.	19	Institute of Mathematical Statistics, University of Bern, Sidlerstr. 5, 3012 Bern, Switzerland.
Janssen, A.	20	Department of Mathematics, University of Siegen, Hölderlinstr. 3, 5900 Siegen 21, West Germany.
Leadbetter, M.R.	21	Department of Statistics, University of North Carolina, 321 Phillips Hall 039 A, Chapel Hill, NC 27514, USA.
Marohn, F.	22	Department of Mathematics, University of Siegen, Hölderlinstr. 3, 5900 Siegen 21, West Germany.
Mason, D.M.	23	Department of Mathematical Sciences, University of Delaware, 501 Ewing Hall, Newark, DE 19716, USA.

Matsunawa, T.	24	Institute of Statistical Mathematics, 4-6-7 Minami-Azabu, Minato-ku, Tokyo 106, Japan.
McCormick, W.	25	Department of Statistics, University of Georgia, Athens, GA 30602, USA.
Obretenov, A.	26	Institute of Mathematics, Bulgarian Academy of Sciences, P.O. Box 373, 1090 Sofia, Bulgaria.
O'Brien, G.	27	Department of Mathematics, York University, 4700 Keele Street, North York, Ont. M3J 1P3, Canada.
Pfeifer, D.	28	Department of Mathematics, University of Oldenburg, P.O. Box 2503, 2900 Oldenburg, West Germany.
Pickands, J.	29	Department of Statistics, The Wharton School, University of Pennsylvania, 3000 Steinberg Hall - Dietrich Hall, Philadelphia, PA 19104-6302, USA.
Reiss, R.-D.	30	Department of Mathematics, University of Siegen, Hölderlinstr. 3, 5900 Siegen 21, West Germany.
Resnick, S.I.	31	Department of Operations Research, Cornell University, Upson Hall, Ithaca, NY 14853, USA.
Révész, P.	32	Institute of Statistics, TU Wien, Wiedner Hauptstr. 8-10/107, 1040 Wien, Austria.
Rootzén, H.	33	Department of Mathematical Statistics, University of Lund, Box 118, 221 00 Lund, Sweden.
Rychlik, I.	34	Department of Mathematical Statistics, University of Lund, Box 118, 221 00 Lund, Sweden.
Schüpbach, M.	35	Institute of Mathematical Statistics, University of Bern, Sidlerstr. 5, 3012 Bern, Switzerland.
Siebert, E.	36	Institute of Mathematics, University of Tübingen, Auf der Morgenstelle 10, 7400 Tübingen, West Germany.
Smith, R.L.	37	Department of Mathematics, University of Surrey, Guildford, Surrey GU2 5XH, Great Britain.
Steinebach, J.	38	Institute of Mathematical Stochastics, University of Hanover, Welfengarten 1, 3000 Hannover 1, West Germany.
Stute, W.	39	Department of Mathematics, University of Gießen, Arndtstr. 2, 6300 Gießen, West Germany.
Sweeting, T.J.	40	Department of Mathematics, University of Surrey, Guildford, Surrey GU2 5XH, Great Britain.
Székely, G.J.	41	Mathematical Institute, Eötvös University, Muzeum krt. 6-8, 1088 Budapest, Hungary.
Teugels, J.L.	42	Department Wiskunde, Cath. University of Leuven, Celestijnenlaan 200 B, 3030 Heverlee, Belgium.
Tiago de Oliveira, J.	43	Academy of Sciences of Lisbon, 19, Rua Academia das Ciências, 1200 Lisboa, Portugal.
Weissman, I.	44	Faculty of Industrial Engineering & Management, Technion, Israel Institute of Technology, Haifa 32000, Israel.
Witting, H.	45	Institute of Mathematical Stochastics, University of Freiburg, Hebelstr. 27, 7800 Freiburg, West Germany.
Zähle, U.	46	Department of Mathematics, University of Jena, Universitätshochhaus, 17. OG, 6900 Jena, GDR.
van Zwet, W.R.	47	Institute of Mathematics, University of Leiden, Postbus 9512, 2300 RA Leiden, The Netherlands.

Contents

PART I. UNIVARIATE EXTREMES: PROBABILITY THEORY

1. LIMIT LAWS AND EXPANSIONS

2. STRONG LAWS

3. RECORDS

4. EXCEEDANCES

5. CHARACTERIZATIONS

PART II. UNIVARIATE EXTREMES: STATISTICS

1. ESTIMATION

2. TEST PROCEDURES

3. SUFFICIENCY OF EXTREMES IN PARAMETRIC MODELS

PART III. MULTIVARIATE EXTREMES

BEST ATTAINABLE RATE OF JOINT CONVERGENCE OF EXTREMES

MICHAEL FALK

Department of Mathematics, University of Siegen
Hölderlinstr. 3, 5900 Siegen 21
Federal Republic of Germany

Abstract. Suppose that the underlying distribution function is ultimately continuous and strictly increasing. In this case the rate of joint convergence of the k largest order statistics, equally standardized, in a sample of n i.i.d. random variables is $O(k/n)$, uniformly in k if, and only if, the underlying distribution function is ultimately a generalized Pareto distribution. Thus, generalized Pareto distributions yield the best rate of joint convergence of extremes.

1. Introduction and main results

Let $X_1, \ldots, X_n$ be independent and identically distributed random variables ($\equiv$ rvs) with common distribution ($\equiv$ df) F and denote by $X_{1:n}, \ldots, X_{n:n}$ the corresponding order statistics arranged in the increasing order.

The central limit theorem in extreme value theory was first discovered by Fisher and Tippett (1928) and later proved in complete generality by Gnedenko (1943): If $(X_{n:n} - b_n)/a_n$ tends in distribution to some nondegenerate limiting distribution G for some choice of constants $a_n > 0$, $b_n \in R$, $n \in N$, then this limit must be one of the following types ($\alpha > 0$)

$$G_{1,\alpha}(x) := \begin{cases} 0, & x \leq 0 \\ \exp(-x^{-\alpha}), & x > 0, \end{cases}$$

$$G_{2,\alpha}(x) := \begin{cases} \exp(-(-x)^{\alpha}), & x \leq 0 \\ 1, & x > 0, \end{cases} \tag{1}$$

$$G_3(x) := \exp(-e^{-x}), \qquad x \in R.$$

Moreover, Gnedenko (1943) gave necessary and sufficient conditions for F to belong to the domain of attraction ($\equiv \mathcal{D}(G)$) of each of the above limits.

The key step in the derivation of this result is the following idea. Suppose that $(X_{n:n} - b_n)/a_n$ converges in distribution to some nondegenerate G and let x be a point of continuity of G. (In our notation we do not distinguish between distributions and their dfs). Then,

$$\begin{aligned} & F^n(a_n x + b_n) \to_{n\to\infty} G(x) \\ \Rightarrow\ & (F^n(a_{nk}x + b_{nk}))^k = F^{nk}(a_{nk}x + b_{nk}) \to_{n\to\infty} G(x) \text{ for any } k \in N \\ \Rightarrow\ & (X_{n:n} - b_{nk})/a_{nk} \text{ converges in distribution to } G^{1/k} \text{ for any } k \in N. \end{aligned}$$

Now, by a classical result of Khintchine (1938) (see Galambos (1987), Lemma 2.3) the limiting distribution of a sequence of rvs is up to a scale and location shift uniquely determined. Hence, for any $k \in N$ there exist constants $A_k > 0$, $B_k \in R$ such that $G^{1/k}(x) = G(A_k x + B_k)$, i.e.

Key words. Extreme order statistics, generalized Pareto distribution, rate of convergence.
AMS 1980 subject classifications. Primary 62G30; secondary 62H05

$$G(y) = G^k(A_k y + B_k), \qquad y \in R. \tag{2}$$

This is the fundamental property of extreme value distributions and the distributions given in (1) are the only nondegenerate ones which satisfy (2). In particular (2) shows that the maximum of n observations coming from an extreme value distribution has again an extreme value distribution. This is the max-stability of $G \in \{G_1, G_2, G_3\}$, already observed by Fisher and Tippett (1928).

While the rate of convergence of central order statistics to a normal distribution is commonly of order $O(n^{-1/2})$ which is well-known (see Reiss (1974)), the situation is completely different in the case of extremes.

Here the distance $\sup_{x \in R} |F^n(a_n x + b_n) - G(x)|$ may be zero if F is itself an extreme value df, or it may be of order $O(n^{-1})$ in the exponential case as well as of order $O(1/\log(n))$ in the normal case. Therefore, results on the rate of convergence of extremes have been sporadic for a long time and a unifying approach was given only by Cohen (1982) for the case G_3 and by Smith (1982) for the case G_1. These results were mainly derived by the concept of regularly varying functions.

An alternative and quite appealing approach to the problem of determining the rate of convergence of extremes can be based on the concept of generalized Pareto distributions. The significance of gPds in extreme value theory was first observed by Pickands (1975) who showed that, roughly speaking, $F \in \mathcal{D}(G)$ if and only if the conditional distribution $(F(u+y) - F(u))/(1 - F(u))$ is approximately given by a shifted gPd if u is large.

A simple derivation of the class of gPds is the following one. Obviously, if $0 < G(x) < 1$

$$\begin{aligned} & F^n(a_n x + b_n) && \to_{n\to\infty} && G(x) \\ \Leftrightarrow\ & n \log(F(a_n x + b_n)) && \to_{n\to\infty} && \log(G(x)) \\ \Leftrightarrow\ & n(1 - F(a_n x + b_n)) && \to_{n\to\infty} && -\log(G(x)) \\ \Leftrightarrow\ & \frac{1 - F(a_n x + b_n)}{1 - (1 + \log(G(x)^{1/n}))} && \to_{n\to\infty} && 1, \end{aligned} \tag{3}$$

where, with $\alpha > 0$, if $G = G_{1,\alpha}, G_{2,\alpha}, G_3$

$$1 + \log(G(x)) = \begin{cases} 1 - x^{-\alpha} =: W_{1,\alpha}(x), & x \geq 1, \\ 1 - (-x)^{\alpha} =: W_{2,\alpha}(x), & x \in [-1, 0], \\ 1 - \exp(-x) =: W_3(x), & x \geq 0. \end{cases}$$

$W \in \{W_1, W_2, W_3\}$ is called *generalized Pareto distribution* (≡gPd). Notice that $W_{1,\alpha}$ is the standard Pareto distribution, $W_{2,1}$ is the uniform distribution on $[-1, 0]$ and W_3 is the standard exponential distribution.

Moreover,

$$1 + \log(G(x)^{1/n}) = \begin{cases} W_{1,\alpha}(n^{1/\alpha} x), & x \geq n^{-1/\alpha}, \\ W_{2,\alpha}(n^{-1/\alpha} x), & -n^{1/\alpha} \leq x \leq 0, \\ W_3(x + \log(n), & x \geq -\log(n), \end{cases}$$

$$=: W_{(n)}(x),$$

and hence, we obtain from (3) if $0 < G(x) < 1$

$$\begin{aligned} & F^n(a_n x + b_n) \to_{n\to\infty} G(x) \\ \Leftrightarrow\ & \frac{1 - F(a_n x + b_n)}{1 - W_{(n)}(x)} \to_{n\to\infty} 1 \end{aligned} \tag{4}$$

where $W_{(n)}$ is the shifted gPd corresponding to G. Thus, $F \in \mathcal{D}(G)$ iff its upper tail can be approximated in an appropriate way by a shifted gPd.

Suppose now that F has a derivative f near the upper endpoint of its support. Under suitable regularity conditions on F the ratio of the derivatives of numerator and denominator in (4) will tend to 1 as n increases, i.e.

$$a_n f(a_n x + b_n)/w_{(n)}(x) \to_{n\to\infty} 1 \tag{5}$$

where, if $W'_{(n)}(x) > 0$

$$w_{(n)}(x) = W'_{(n)}(x) = n^{-1} w(x)$$

with $w(x) := (\log(G(x)))'$, $0 < G(x) < 1$.

Formula (5) holds for example if F satisfies one of the well-known von Mises conditions (cf. Sweeting (1985), Falk (1985), Pickands (1986)). Consequently, we obtain the representation

$$a_n f(a_n x + b_n) = w_{(n)}(x)[1 + h_n(x)] \tag{6}$$

where $h_n(x) \to_{n\to\infty} 0$, $0 < G(x) < 1$.

This representation of the closeness of $F(a_n x + b_n)$ and $W_{(n)}$ in terms of their densities was the starting point in Falk (1986), where bounds for the variational distance $\sup_{B\in\mathcal{B}} |P\{(X_{n:n} - b_n)/a_n \in B\} - G(B)|$ were established. By $\mathcal{B}$ we denote the Borel-σ-algebra of the real line. These bounds show that the variational distance is determined by the rate at which h_n converges to zero. Roughly, it is of order $O\left((\int h_n^2\, dG)^{1/2}\right)$. See also formula (11) and Proposition 12 below.

However, the particular role of gPds in extreme value theory is revealed if in place of the largest order statistic the joint distribution of the k largest order statistics is considered. This is shown in the following. To this end we remind of the following result due to Dwass (1966).

THEOREM 7. *$(X_{n:n} - b_n)/a_n$ converges in distribution to an extreme value distribution G iff for any $k \in N$ $((X_{n-i+1:n} - b_n)/a_n)_{i=1}^k$ converges in distribution to $G^{(k)}$, where $G^{(k)}$ has the k-dimensional Lebesgue-density $g^{(k)}(x_1, \ldots, x_k) = G(x_k)\Pi_{i=1}^k G'(x_i)/G(x_i)$ if $x_1 > \ldots > x_k$ and zero elsewhere.*

Hence, $G^k \in \{G_{1,\alpha}^{(k)}, G_{2,\alpha}^{(k)}, G_3^{(k)}\}$ is the only possible limit for the joint distribution of the k largest order statistics, equally standardized. A proof of this result which is stated in Dwass (1966) without proof, is given by Weissman (1975). The following short proof demonstrates the usefulness of the integral probability transformation technique.

PROOF: Denote by F^{-1} the generalized inverse of F, i.e. $F^{-1}(q) = \inf\{x \in R : F(x) \geq q\}$, $q \in (0,1)$ and let $U_1, \ldots, U_n$ be independent and on $(0,1)$ uniformly distributed rvs. Then we may choose $X_i = F^{-1}(U_i)$ and hence, $X_{i:n} = F^{-1}(U_{i:n})$, where $U_{1:n}, \ldots, U_{n:n}$ denote the order statistics pertaining to $U_1, \ldots, U_n$. Furthermore, we have for $q \in (0,1)$ and $x \in R$

$$F^{-1}(q) \leq x \Leftrightarrow q \leq F(x).$$

Now we can give a short proof of Theorem 7. Obviously, we have to show the only-if part of the assertion. Choose $(x_1, \ldots, x_k) \in R^k$. Then,

$$\begin{aligned}
&P\{(X_{n-i+1:n} - b_n)/a_n \leq x_i, i = 1, \ldots, k\} \\
&= P\{F^{-1}(U_{n-i+1:n}) \leq a_n x_i + b_n, i = 1, \ldots, k\} \\
&= P\{U_{n-i+1:n} \leq F(a_n x_i + b_n), i = 1, \ldots, k\} \\
&= P\{n(U_{n-i+1:n} - 1) \leq n(F(a_n x_i + b_n) - 1), i = 1, \ldots, k\}.
\end{aligned}$$

Now, $n(F(a_n x_i + b_n) - 1) \to_{n\to\infty} \log(G(x_i))$, $i = 1, \ldots, k$, by (3). Moreover, it is easy to see that $(n(U_{n-i+1:n} - 1))_{i=1}^k$ converges in distribution to $G_{2,1}^{(k)}$ with k-dimensional Lebesgue-density $g_{2,1}^{(k)} = \exp(x_k)$ if $0 > x_1 > \ldots > x_k$ and zero elsewhere. Hence,

$$P\{(X_{n-i+1:n} - b_n)/a_n \le x_i, i = 1, \ldots, k\} \to_{n\to\infty} G_{2,1}^{(k)}((\log\{G(x_i)\})_{i=1}^k)$$

from which the result follows. ∎

The next basic result is due to Reiss (1981, Theorems 2.6 and 3.2). It is formulated only for the uniform distribution on (-1,0) but suitable transformations immediately entail the assertion for any gPd.

THEOREM 8. *Suppose that* $F = W_{(n)}$. *Then, for any* $k, n \in N$

$$\sup_{B\in\mathcal{B}^k} |P\{(X_{n-i+1:n})_{i=1}^k \in B\} - G^{(k)}(B)| \le Ck/n$$

where C *is a universal positive constant.*

Notice that the assumption $F = W_{(n)}$ in the preceding result can be formulated as follows: Suppose that F is itself a gPd, i.e. $F = W$ and put

$$(9)\qquad \begin{array}{lll} a_n := n^{1/\alpha}, & b_n := 0 & \text{if } W = W_{1,\alpha}, \\ a_n := n^{-1/\alpha}, & b_n := 0 & \text{if } W = W_{2,\alpha}, \\ a_n := 1, & b_n := \log(n) & \text{if } W = W_3. \end{array}$$

The joint distribution of the k largest order statistics under $W_{(n)}$ is then given by the one of $((X_{n-i+1:n} - b_n)/a_n)_{i=1}^k$ under W.

Notice that the preceding result implies in particular that the rate of convergence of the largest order statistic is of order $O(1/n)$ in case of the exponential distribution as well as in case of the uniform distribution on $(0, 1)$.

Theorem 8 raises in particular the question, whether there exists a distribution F such that the distance $\sup_{B\in\mathcal{B}^k} |P\{((X_{n-i+1:n} - b_n)/a_n)_{i=1}^k \in B\} - G^{(k)}(B)|$ is strictly less than in the case of a gPd, i.e. $o(k/n)$ in place of $O(k/n)$.

The following result denies this question. It shows that there are no distributions which yield a better rate of convergence than $O(k/n)$ and this rate is attained by the class of gPds only. Notice that this result is achieved under the assumption that the k largest order statistics are equally standardized; different normalizing constants depending on k may lead to better convergence rates (cf. Kohne and Reiss (1983)).

Denote by $G_{(k)}$ the k-th one-dimensional marginal distribution of $G^{(k)}$, i.e. $G_{(k)}(x) = G(x) \cdot \sum_{i=0}^{k-1}(-\log(G(x)))^i/i!$, $x \in G^{-1}(0, 1)$. Then our main result is the following one.

THEOREM 10. *Let* F *be continuous and strictly increasing in a left neighborhood of* $\omega(F) := \sup\{x \in R : F(x) < 1\}$. *Suppose that there exist* $C > 0$ *and normalizing constants* $a_n > 0$, $b_n \in R$ *such that for any* $k \in \{1, \ldots, n\}$, $n \in N$,

$$\sup_{t\in R} |P\{(X_{n-k+1:n} - b_n)/a_n \le t\} - G_{(k)}(t)| \;\le\; Ck/n.$$

Then there exist $c > 0$, $d \in R$ *such that* $F(x) = W(cx + d)$ *for* x *near* $\omega(F)$, *where* W *is the gPd pertaining to* G.

It is easy to see that the bound $O(k/n)$ in Theorem 8 also holds if F is only ultimately a gPd and thus, we have established a characterization of the class of gPds in terms of the best attainable rate of joint convergence of extremes.

Moreover, recall that the significant property of an extreme value distribution G is its max-stability. This stability does no longer hold if in place of the maximum the k-th largest order statistic is considered, $k \geq 2$. The preceding result shows in particular that under G the distance between the distribution of $((X_{n-i+1:n} - b_n)/a_n)_{i=1}^k$ and $G^{(k)}$ is strictly larger than $O(k/n)$. On the other hand, this distance is minimized under W; this is now a property of the class of gPds which is comparable to the max-stablility of extreme value distributions in case of the largest observation.

It is therefore clear that gPds play the central role in extreme value theory if the joint distribution of extremes is considered. In particular, the preceding considerations suggest the following concept if we are dealing with a statistical functional based on the upper k observations in a sample of size n.

To approximate the given nonparametric model underlying the observations by a parametric one, one can try to approximate the distribution of the k upper extremes by their limiting distribution $G^{(k)}$ which is a k-dimensional problem. On the other hand, one might replace at first the underlying df F by a shifted gPd $W_{(n)}$ and consider the n observations as coming from $W_{(n)}$. Turning to the joint distribution of the k upper extremes, it is then intuitively clear that the error of this parametric approximation is now determined by the distance of F and $W_{(n)}$ which is only a one-dimensional problem. This *semiparametric* model was already observed by Pickands (1975) and extensively studied in recent papers by Smith (1987) and Hosking and Wallis (1987), among others.

Our next considerations are in this spirit. Suppose that (6) holds, i.e. suppose that F has a density f such that $a_n f(a_n x + b_n) = w_{(n)}(x)(1 + h_n(x))$. Under suitable conditions one can show (Falk (1986, Theorem 2.35), Reiss (1988, Theorem 5.5.4)) that

$$\sup_{B \in \mathcal{B}^k} |P\{((X_{n-i+1:n} - b_n)/a_n)_{i=1}^k \in B\} - G^{(k)}(B)|$$

$$= O\left(k/n + (\int h^2(x)(\sum_{i=1}^k G_{(i)})(dx))^{1/2}\right). \tag{11}$$

This result reveals that the variational distance between the distribution of the k largest order statistics and $G^{(k)}$ is determined by the distance of $F(a_n x + b_n)$ and $W_{(n)}$ in terms of the remainder function h_n.

The following consequence of the preceding formula is a modification of Corollary 2.48 in Falk (1986). A detailed proof is given in Reiss (1988, Corollary 5.5.5).

PROPOSITION 12. *Suppose that there exists $x_0 < \omega(F)$ such that the density f of F has the representation*

$$f(x) = w(x)e^{h(x)}, \qquad x_0 < x < \omega(F)$$

and equal to zero if $x > \omega(F)$. If h satisfies the condition

$$|h(x)| \leq \begin{cases} Lx^{-\alpha\delta}, & W = W_{1,\alpha}, \\ L(-x)^{\alpha\delta}, & W = W_{2,\alpha}, \\ Le^{-\delta x}, & W = W_3, \end{cases}$$

where L and δ are positive constants, then

$$\sup_{B \in \mathcal{B}^k} |P\{((X_{n-i+1:n} - b_n)/a_n)_{i=1}^k \in B\} - G^{(k)}| = O\left((k/n)^\delta k^{1/2} + k/n\right),$$

where a_n, b_n are the constants as given in (9).

As an application of the preceding result one can for instance easily establish a sharp bound for the rate of uniform convergence of Hill's (1975) estimator of α in the case $F \in \mathcal{D}(G_{1,1/\alpha})$ (see Falk (1985, section 5), Reiss (1988, section 9.5)).

2. Proof of Theorem 10

Although some key steps of the proofs are rather similar we will have to prove the assertion explicitely for any of the three cases $G \in \{G_1, G_2, G_3\}$ since several details are quite different.

Let in the following $X^{(i)}_{n-k+1:n}$, $i = 1, 2, \ldots$, be independent replica of $X_{n-k+1:n}$ and let $Y_1, Y_2, \ldots$ be independent standard exponential rvs.

(A) Case $G = G_{1,\alpha}$. Since $G_{1,\alpha,(k)}$ is the distribution of $(\sum_{j=1}^k Y_j)^{-1/\alpha}$, we obtain from our assumptions and Fubini's theorem that for $M \in N$

$$\sup_{t\in R} |P\{\sum_{i=1}^M |(X^{(i)}_{n-k+1:n} - b_n)/a_n|^{-\alpha} \le t\} - P\{\sum_{j=1}^{Mk} Y_j \le t\}|$$

$$\le M \sup_{t\in R} |P\{|(X_{n-k+1:n} - b_n)/a_n|^{-\alpha} \le t\} - P\{\sum_{j=1}^{k} Y_j \le t\}|$$

$$\le 2CMk/n.$$

Since on the other hand also

$$\sup_{t\in R} |P\{|(X_{n-Mk+1:n} - b_n)/a_n|^{-\alpha} \le t\} - P\{\sum_{j=1}^{Mk} Y_j \le t\}| \le 2CMk/n$$

if $n - Mk + 1 \in \{1, \ldots, n\}$, we obtain

$$\sup_{t\in R} |P\{\sum_{i=1}^M |X_{n-k+1:n} - b_n|^{-\alpha} \le t\} - P\{|X_{n-Mk+1:n} - b_n|^{-\alpha} \le t\}| \tag{13}$$

$$\le 4CMk/n.$$

Choose now $\varepsilon > 0$ and put $k = k(n) = [\varepsilon n/M]$, were $[x]$ denotes the integer part of $x \in R$. Then, $(n-k+1)/n \to_{n\to\infty} 1 - \varepsilon/M$, $(n - Mk + 1)/n \to_{n\to\infty} 1 - \varepsilon$ and thus, if ε is small enough, say $\varepsilon \le \varepsilon_0 \le 1/(8C)$, we obtain

$$\begin{aligned} X^{(i)}_{n-k+1:n} &\to_{n\to\infty} F^{-1}(1 - \varepsilon/M), \qquad i = 1, \ldots, M, \\ X_{n-Mk+1:n} &\to_{n\to\infty} F^{-1}(1 - \varepsilon) \qquad \text{in probability.} \end{aligned} \tag{14}$$

Next we show by a contradiction that there exists a subsequence b_m of b_n which converges to a constant $b \in R$. Suppose that $|b_n| \to_{n\to\infty} \infty$. From (13) and (14) we obtain

$$1/2 \ge \sup_{t\in R} |P\{\sum_{i=1}^M |X^{(i)}_{n-k+1:n}/b_n - 1|^{-\alpha} \le t\} - P\{|X_{n-Mk+1:n}/b_n - 1|^{-\alpha} \le t\}|$$

$$\to_{n\to\infty} \quad 1$$

which is a contradiction. Consequently, we have in probability

$$\sum_{i=1}^M |X^{(i)}_{m-k+1:m} - b_m|^{-\alpha} \to_{m\to\infty} M|F^{-1}(1 - \varepsilon/M) - b|^{-\alpha},$$

$$|X_{m-Mk+1:m} - b_m|^{-\alpha} \to_{m\to\infty} |F^{-1}(1 - \varepsilon) - b|^{-\alpha}.$$

(13) implies

$$1/2 \geq \sup_{t\in R} |P\{\sum_{i=1}^{M} |X^{(i)}_{m-k+1:m} - b_m|^{-\alpha} \leq t\} - P\{|X_{m-Mk+1:m} - b_m|^{-\alpha} \leq t\}|$$

and thus, we conclude that for any $\varepsilon \in (0, \varepsilon_0)$ and $M \in N$

$$M|F^{-1}(1-\varepsilon/M) - b|^{-\alpha} = |F^{-1}(1-\varepsilon) - b|^{-\alpha}. \tag{15}$$

This implies for $0 < \varepsilon, \varepsilon' < \varepsilon_0$

$$\frac{\varepsilon}{\varepsilon'} = \frac{|F^{-1}(1-\varepsilon) - b|^{-\alpha}}{|F^{-1}(1-\varepsilon') - b|^{-\alpha}} \tag{16}$$

as can be seen as follows. Let M_n, M'_n, $n \in N$, be sequences in N such that $M_n/n \to_{n\to\infty} \varepsilon$, $M'_n/n \to_{n\to\infty} \varepsilon'$. Then, by (15) if n is large

$$\begin{aligned} M_n|F^{-1}(1-1/n) - b|^{-\alpha} &= |F^{-1}(1-M_n/n) - b_n|^{-\alpha} \to_{n\to\infty} |F^{-1}(1-\varepsilon) - b|^{-\alpha}, \\ M'_n|F^{-1}(1-1/n) - b|^{-\alpha} &= |F^{-1}(1-M'_n/n) - b|^{-\alpha} \to_{n\to\infty} |F^{-1}(1-\varepsilon') - b|^{-\alpha}. \end{aligned}$$

Since $M_n/M'_n \to_{n\to\infty} \varepsilon/\varepsilon'$, this implies (16).

Recall that $F \in \mathcal{D}(G_1)$ implies $F^{-1}(1-\varepsilon) \to_{\varepsilon\to 0} \infty$ (see Theorem 2.4.3 in the book by Galambos (1987)). Hence, putting $\varepsilon = 1 - F(x)$, we obtain from (16) that for $x, x' > x_0 \in R$

$$\frac{1-F(x)}{1-F(x')} = \frac{(x'-b)^\alpha}{(x-b)^\alpha}$$

which yields $F(x) = 1 - (1-F(x'))(x'-b)^\alpha(x-b)^{-\alpha} = W_{1,\alpha}(cx+d)$, $x \geq x'$.

(B) Case $G = G_{2,\alpha}$. $G_{2,\alpha,(k)}$ is the distribution of $(-\sum_{j=1}^{k} Y_j)^{1/\alpha}$ and hence, we obtain like in the case (A) that for $M \in N$

$$\sup_{t\in R} |P\{\sum_{i=1}^{M} |X^{(i)}_{n-k+1:n} - b_n|^{\alpha} \leq t\} - P\{|X_{n-Mk+1:n} - b_n|^{\alpha} \leq t\}| \leq 4CMk/n.$$

From classical extreme value theory (see Theorems 2.1.2, 2.4.3 and Lemma 2.2.3 of Galambos (1987)) we conclude that $\omega(F) < \infty$ and

$$\frac{a_n}{\omega(F) - F^{-1}(1-n^{-1})} \to_{n\to\infty} 1, \quad \frac{\omega(F) - b_n}{a_n} \to_{n\to\infty} 0.$$

This implies in particular that $b_n \to_{n\to\infty} \omega(F)$. Consequently, by repeating the arguments in the proof of (A), we derive that for $\varepsilon, \varepsilon'$ small enough

$$\frac{\varepsilon}{\varepsilon'} = \frac{(\omega(F) - F^{-1}(1-\varepsilon))^\alpha}{(\omega(F) - F^{-1}(1-\varepsilon'))^\alpha}.$$

By putting $\varepsilon = 1 - F(x)$ we obtain the assertion.

(C) Case $G = G_3$. In this case, the distribution of $-\log(\sum_{j=1}^k Y_j)$ equals $G_{3,(k)}$. In complete analogy to the preceding proofs we deduce from the assumption

$$\sup_{t\in R} |P\{\sum_{i=1}^{M} \exp(-X_{n-k+1:n}^{(i)}/a_n) \le t\} - P\{\exp(-X_{n-Mk+1:n}/a_n) \le t\}|$$
$$\le 4CMk/n$$

for $M \in N$ if $n - Mk + 1 \in \{1, \ldots, n\}$. Put $k = k(n) = [\varepsilon n/M]$. As in part (A) one shows by a contradiction that there exists a subsequence a_m of a_n such that $a_m \to_{m\to\infty} a \in (0, \infty)$. Consequently, we conclude that for $\varepsilon, \varepsilon'$ small enough

$$\frac{\varepsilon}{\varepsilon'} = \frac{\exp(-F^{-1}(1-\varepsilon)/a)}{\exp(-F^{-1}(1-\varepsilon')/a)}$$

which implies the assertion. ∎

Acknowledgment. I am grateful to R.-D. Reiss for drawing my attention to the problem investigated in Theorem 10 of the present article; its solution was conjectured in an earlier version of his forthcoming monograph.

References

Cohen, J.P. (1982), *Convergence rates for the ultimate and penultimate approximation in extreme-value theory*, Adv. in Appl. Probab. **14**, 833–854.

Dwass, M. (1966), *Extremal processes II*, Illinois J. Math. **10**, 381–391.

Falk, M. (1985), "Uniform Convergence of Extreme Order Statistics," Habilitationsschrift, Siegen.

Falk, M. (1986), *Rates of uniform convergence of extreme order statistics*, Ann. Inst. Statist. Math. A **38**, 245–262.

Fisher, R.A. and Tippett, L.H.C. (1928), *Limiting forms of the frequence distribution of the largest or smallest member of a sample*, Proc. Camb. Phil. Soc. **24**, 180–190.

Galambos, J. (1987), "The Asymptotic Theory of Extreme Order Statistics," 2nd. edition, Robert E. Krieger, Melbourne, Florida.

Gnedenko, B. (1943), *Sur la distribution limite du terme maximum d'une série aléatoire*, Ann. Math. **44**, 423–453.

Hill, B.M. (1975), *A simple approach to inference about the tail of a distribution*, Ann. Statist. **3**, 1163–1174.

Hosking, J.R.M. and Wallis, J.R. (1987), *Parameter and quantile estimation for the generalized Pareto distribution*, Techometrics **29**, 339–349.

Khintchine, A. (1938), *Théorèmes limites pour les sommes des variables aléatoires indépendentes*, Moscow (in Russian).

Kohne, W. and Reiss, R.-D. (1983), *A note on uniform approximation to distributions of extreme order statistics*, Ann. Inst. Statist. Math. A **35**, 343–345.

Pickands, J. III (1975), *Statistical inference using extreme order statistics*, Ann. Statist **3**, 119–131.

Pickands, J. III (1986), *The continuous and differentiable domains of attraction of the extreme value distributions*, Ann. Probab. **14**, 996–1004.

Reiss, R.-D. (1974), *On the accuracy of the normal approximation for quantiles*, Ann. Probab. **2**, 741–744.

Reiss, R.-D. (1981), *Uniform approximation to distributions of extreme order statistics*, Adv. in Appl. Probab. **13**, 533–547.

Reiss, R.-D. (1988), "Approximate Distributions of Order Statistics (with Applications to Non-parametric Statistics)," Springer Series in Statistics (to appear).

Smith, R.L. (1982), *Uniform rates of convergence in extreme-value theory*, Adv. in Appl. Probab. **14**, 600–622.

Smith, R.L. (1987), *Estimating tails of probability distributions*, Ann. Statist. **15**, 1174–1207.

Sweeting, T.J. (1987), *On domains of uniform local attraction in extreme value theory*, Ann. Probab. **13**, 196–205.

Weissman, I. (1975), *Multivariate extremal processes generated by independent nonidentically distributed random variables*, J. Appl. Probab. **12**, 477–487.

RECENT RESULTS ON ASYMPTOTIC EXPANSIONS IN EXTREME VALUE THEORY

Trevor J. Sweeting

Department of Mathematics, University of Surrey, Guildford.

Abstract. We present an overview of some current work on asymptotic expansions in extreme value theory. It is shown that in the case of local convergence the problem of obtaining an asymptotic expansion for the quantile function amounts to that of finding an expansion for the integral of a regularly varying function. The class of regularly varying functions of *differentiable order* k is introduced, and it is shown how one can develop expansions for such functions. Asymptotic expansions for distribution functions can in principle be obtained by inversion; in particular, we give conditions for the validity of the formal expansion of Uzgören (1954) for the case $F \in \mathcal{D}(\Lambda)$.

1. Introduction

Let F be an arbitrary distribution function, and suppose that there exist sequences (a_n) and (b_n) with $b_n > 0$, such that

$$F^n(a_n + b_n x) \to G(x) \tag{1.1}$$

for some nondegenerate distribution G. Classical extreme value theory (see for example Galambos (1987)) states that G must be one of the three types $\Lambda(x) = \exp(-e^{-x})$, $\Phi_\alpha(x) = \exp(-x^{-\alpha})$, $x>0$, or $\Psi_\alpha(x) = \exp(-(-x)^\alpha)$, $x<0$ where α is a positive parameter. Alternatively, these three types may be combined into the single form

$$G_\gamma(x) = \exp(-(1+\gamma x)_+^{-1/\gamma}),$$

the generalised extreme value (GEV) distribution. When F satisfies the limit relation (1.1), we write $F \in \mathcal{D}(G)$, the domain of (max) attraction of G.

There is a very large literature concerning improved approximations for distributions of maxima, dating back to Fisher and Tippett (1928) who demonstrated the superiority of a *penultimate* approximation in the case of the normal distribution. The recent work related to asymptotic expansions includes

---

Key words. Extreme value theory, asymptotic expansions, slow variation with remainder.

AMS 1980 subject classifications. Primary 60F05, Secondary 41A60.

Anderson (1971), who obtained the first term in an expansion, and later Gomes (1978) who considered penultimate approximations. See also Cohen (1982) who obtained global uniform error rates. Smith (1982) considered the Φ_α and Ψ_α domains, and obtained more general results of this type assuming a "slow variation with remainder" condition. An early paper on asymptotic expansions in the case $F \in \mathcal{D}(\Lambda)$ was Uzgören (1954), where a formal expansion was developed. No indication was given of the conditions required for its validity however.

In the present paper we develop an approach to asymptotic expansions based on the idea of slow variation with remainder; see for example Goldie and Smith (1987) for general theory. We work with the GEV distribution, in an attempt to unify the theory. We shall only consider here "regular" cases where F possesses a certain number of derivatives. In particular we assume that we are in the domain of *local* attraction of G_γ; we show in Section 2 that in this case the constants a_n and b_n for convergence to the GEV limit G_γ may be chosen in a particularly simple way. In addition, the (normalized) quantile function can be expressed simply in terms of the integral of a certain regularly varying function. Thus unified improved estimates of the quantile function can be obtained from estimates for integrals of regularly varying functions. In Section 3 we consider asymptotic expansions for such integrals; these are obtained using the idea of slow variation with remainder. The class of regularly varying functions of *differentiable order* k is introduced, and it is shown how to develop an asymptotic expansion for integrals of such functions.

The general ideas in Section 3 are applied to the normalized quantile function in Section 4. An asymptotic expansion for the distribution function may also be obtained by inversion. In general, the expressions are quite complex, but on specializing to $F \in \mathcal{D}(\Lambda)$ it is shown how the simple Uzgören (1954) expansion arises as a valid asymptotic expansion under certain restrictive conditions.

2 LOCAL CONVERGENCE AND APPROXIMATION OF THE QUANTILE FUNCTION.

We assume that F possesses a positive density f. Define the functions $a(t) = F^{-1}(1-t^{-1})$, $b(t) = [tf(a(t))]^{-1}$. We say that F is in the *domain of local attraction* of G, written $F \in \mathcal{L}(G)$, if the density of $F^n(a_n+b_n x)$ converges to the density of G locally uniformly (l.u.) as $n\to\infty$. Necessary and sufficient conditions in terms of the functions a and b for $F \in \mathcal{L}(G)$ for the three types of extreme value distribution were given by Sweeting (1985). (Note the reciprocal definition of b used in this paper.) In the case $\mathcal{L}(\Lambda)$, the condition is that b is a *slowly varying* function. Furthermore, the conditions given for the other types can be recast in terms of the regular variation of b; an equivalent condition also appears in

Pickands (1986). That is $F \in \mathcal{L}(G_\gamma)$ if and only if $b \in R_\gamma$, the class of regularly varying functions with index γ.

When $F \in \mathcal{L}(G_\gamma)$ one can always choose the norming constants to be $a_n = a(n)$ and $b_n = b(n)$. For define

$$u_t(x) = -\log t[1-F(a(t) + x\, b(t))].$$

Standard manipulations give $F^n(a_n+b_n) \to G_\gamma(x)$ if and only if $u_t(x) \to \gamma^{-1}\log(1+\gamma x)$ l.u in the set $1 + \gamma x > 0$, which in turn is true if and only if $u_t^{-1}(x) \to \gamma^{-1}(e^{\gamma x}-1)$ l.u in $\mathbb{R}$. But we have

$$u_t^{-1}(x) = [a(te^x)-a(t)]/b(t) = \int_0^x [b(te^u)/b(t)]du \tag{2.1}$$

on using $a'(t) = b(t)/t$. The link between the regular variation of b and convergence of $F^n(a_n+b_nx)$ to $G_\gamma(x)$ is clearly exhibited in (2.1).

The normalized quantile function $q_n(p)$ satisfies the relation $F^n(a_n+b_nq_n(p)) = p$, and it is readily verified that

$$q_n(p) = [a(ne^{x_n}) - a(n)]/b(n) \tag{2.2}$$

where $x_n = -\log(-\log p) + O(n^{-1})$, uniformly in compacts of (0,1). It follows from (2.1) and (2.2) that the problem of obtaining an asymptotic expansion for $q_n(p)$ is equivalent to the problem of deriving an asymptotic expansion for the integral of the regularly varying function b, provided the substitution of the $O(n^{-1})$ approximation $x = -\log(-\log p)$ for x_n contributes a negligible error. In principle, an expansion for $u_t^{-1}(x)$ can be inverted to obtain an expansion for $u_t(x)$, and ultimately $F^n(a_n+b_nx)$. We return to these questions in Section 3 after investigating the general problem of expansions for regularly varying functions.

3. EXPANSIONS FOR REGULARLY VARYING FUNCTIONS

A real measurable function g on $(0,\infty)$ will be said to be *regularly varying with index* ρ, written $g \in R_\rho$, if it is either eventually positive or eventually negative and, for all $x>0$, $g(tx)/g(t) \to x^\rho$ as $t \to \infty$. This trite extension of the usual class of regularly varying functions will prove to be convenient in the present context. We say that $g \in R_\rho$ is *regularly varying with remainder* if there are functions g_1 and k such that

$$x^{-\rho} g(tx)/g(t) - 1 = k(x)g_1(t) + o(g_1(t)) \tag{3.1}$$

as $t \to \infty$, where we will always assume that $g_1(t) \to 0$; see for example Goldie and Smith (1987) for details. It is usually the case that $g_1 \in R_{\rho_1}$; in particular if, for some $x, k(x) \neq 0$ and $k(xy) \neq k(y)$ for all y, then necessarily $g_1 \in R_{\rho_1}$ for some $\rho_1 \leq 0$ (Seneta (1976)). In this case one necessarily has $k = ch_{\rho_1}$ where $h_\rho(x) = \int_1^x u^{\rho-1}du$ and c is constant. Furthermore, when $g \in R_{\rho_1}$ the asymptotic relation (3.1) holds uniformly in compacts of $(0,\infty)$.

The simplest way of exploring further terms in the asymptotic development of a regularly varying function is to specialize to "regular" cases where g has a given number of derivatives . Writing $u(t) = \log\{g(e^t)\}$, $T = \log t$, $X = \log x$, the regular variation of g is equivalent to $u(T+X)-u(T) \to \rho X$, which is suggestive of the relation $u'(T) \to \rho$, or $g_1(t) \to 0$ where

$$g_1(t) = u'(\log t)-\rho = tg'(t)/g(t)-\rho. \tag{3.2}$$

We note that this is precisely the condition that g is *normalized* regularly varying. In fact, if the function g_1 defined in (3.2) is regularly varying, then it is not hard to show that (3.1) holds with this definition of g_1; we have the following result.

LEMMA 3.1 *Suppose* $g \in R_\rho$ *and* $g_1 \in R_{\rho_1}$ *where* g_1 *is given by* (3.2). *Then* (3.1) *holds with this definition of* g_1, *and* $k = h_{\rho_1}$.

REMARK. The result implies that $g_1(t) \to 0$, and $\rho_1 \leq 0$ necessarily.

PROOF. Write $\delta_t = u(\log(tx)) - u(\log t) - \rho\log x$. We have

$$\delta_t = \int_0^{\log x} [u'(\log t + v)-\rho]dv = \int_1^x u^{-1}g_1(tu)du$$

$$= g_1(t)\{h_{\rho_1}(x)+ \int_1^x [g_1(tu)/g_1(t) - u^{\rho_1}]du\}$$

$$= g_1(t)\ \{h_{\rho_1}(x) + o(1)\}\ .$$

Note that the regular variation of both g and g_1 imply that $g_1(t) \to 0$, and $\rho_1 \leq 0$. The result follows since $x^{-\rho}g(tx)/g(t) = e^{\delta_t}$.

We shall find it convenient to work with e^x in (3.1) rather than x. Define

$$H_\rho(x) = \int_0^x e^{\rho u}\ du$$

and for $k \geq 1$ recursively define

$$H_{\rho_0\rho_1\ldots\rho_k}(x) = \int_0^x H_{\rho_1\ldots\rho_k}(u)\, e^{\rho_0 u}\, du.$$

Suppose that $g \in R_{\rho_0}$ and $g_1 \in R_{\rho_1}$ whhere g_1 is given by (3.2). Then it follows from Lemma 3.1 that

$$\int_0^x \{g(te^u)/g(t)\}du = A_o(x,t) + A_1(x,t) + \eta_1(x,t) \qquad (3.3)$$

where $A_o(x,t) = H_{\rho_0}(x)$, $A_1(x,t),= H_{\rho_0\rho_1}(x)g_1(t)$ and $\eta_1(x,t) = o(g_1(t))$, uniformly in compacts of $(0,\infty)$. These considerations motivate the following definition. Let k be a nonnegative integer and $\rho = (\rho_0,\ldots,\rho_k)$. Assume $g^{(k)}$ exists and define $g_0 = g$, $g_i(x) = xg'_{i-1}(x)/g_{i-1}(x) - \rho_{i-1}$, $i=1,\ldots,k$. We say that g is *regularly varying of differentiable order* k *index* ρ, and write $g \in R^k_\rho$, if $g_i \in R_{\rho_i}$, $i=0,1,\ldots,k$. Note that from Lemma 3.1 it follows that $g_i(t) \to 0$ and $\rho_i \leqslant 0$ for $i=1,\ldots,k$. We refer to g_i as the ith *remainder* of g. Rather than give a completely general formulation at this stage, it will be more illuminating to show how the next term in the expansion (3.3) may be generated. We write $H_{01} = H_{\rho_0\rho_1}(x)$, $g_i=g_i(t)$, $A_i = A_i(t,x)$ etc for brevity.

LEMMA 3.2. *Let* $\rho = (\rho_0,\rho_1,\rho_2)$ *and suppose* $g \in R_\rho^{\,2}$. *Then*

$$\int_0^x \{g(te^u)/g(t)\}du = A_o + A_1 + A_2 + \eta_2$$

where $A_0 = H_0$, $A_1 = H_{01}g_1$, $A_2 = H_{011}g_1^2 + H_{012}g_1g_2$ *and* $\eta_2 = o(\max(g_1^2, g_1g_2))$, *uniformly in compacts of* $(0,\infty)$.

PROOF. Let u and δ_t be as in the proof of Lemma 3.1; then we have

$$\delta_t = g_1 \int_0^x \{g_1(te^u)/g_1(t)\}du.$$

But now $g_1 \in R^1_{\rho_1\rho_2}$, and so from (3.3) it follows that

$$\delta_t = H_1g_1 + H_{12}g_1g_2 + o(g_1g_2).$$

Now

$$\int_0^x \{g(te^u)/g(t)\}du = \int_0^x e^{\delta_t}\, e^{\rho_0 u}\, du$$

$$= H_0 + H_{01}g_1 + H_{01}g_1g_2 + \tfrac{1}{2}\, g_1^2 \int_0^x H_1^2\, e^{\rho_0 u}du + o(g_1g_2) + o(g_1^2)$$

Finally note that $\frac{1}{2}H_1^2 = H_{11}$ to give the result.

The method contained in the proof of Lemma 3.2 can be used to generate an expansion to any number of terms. In general, we obtain an expansion of the form

$$\int_0^x \{g(te^u)/g(t)\}du = \sum_{i=0}^{k} A_i + \eta_k \tag{3.4}$$

where each A_i is composed of a sum of terms, each of which is a product of a remainder of g and an H_ρ term. Furthermore, each term in A_i is of smaller order than each term in A_{i-1}, and η_k is of smaller order than each term in A_k, uniformly in compacts of $(0,\infty)$. We shall refer to such an asymptotic expansion as a *block expansion*. The next term in the expansion, when $g_0 \in R_\rho^3$, is given by

$$A_3 = H_{0111}\, g_1^3 + (2H_{0112} + H_{0121})g_1^2 g_2 + H_{0122}g_1 g_2 + H_{0123}g_1 g_3$$

In general, the terms are complex; in particular it is necessary to use identities for the H functions. In the calculation of A_3 above we need $H_1^3 = 6H_{111}$ and $H_1H_{12} = 2H_{112} + H_{121}$. In order to apply the expansion in any given case, it is necessary to regroup terms in decreasing order of magnitude. Note also that (3.4) may *not* produce a strictly valid asymptotic expansion to the kth order if any *cancellation* occurs amongst the terms in A_i, since then η_k may not be $o(A_i)$. Thus it may be that a complete term A_i must be included in the error term; a valid expansion can be obtained to some lesser order by suitable contraction.

We turn next to the special case $g \in R_\rho^k$, $\rho=0$; that is, g and all its remainders are slowly varying. In this case we write $g \in S^k$. Write

$G_t(x) = \int_0^x \{g(te^u)/g(t)\}du$. A formal series expansion of G_t in *powers of* x is of the form

$$G_t(x) = \sum_{i=0}^{\infty} \frac{x^{i+1}}{(i+1)!} G_t^{(i+1)}(0) = \sum_{i=0}^{\infty} \frac{x^{i+1}}{(i+1)!} \{g(t)\}^{-1}[(d^i/dz^i)g(te^z)]_{z=0}.$$

We show that when $g \in S^k$, the asymptotic expansion of G_t coincides with its power series expansion. To see this, we use an inductive argument. Let $g \in S^k$ and denote by $A_i(g)$ the ith term in the block expansion of g, $i \leq k$, generated as described above. The inductive hypothesis is that, for any $g \in S^k$, $A_i(g) = [(i+1)!]^{-1} x^{i+1} G_t^{(i+1)}(0)$, $i=0,\dots,k$. The hypothesis is true for k=1 from Lemma 3.1; assume it is true for some $k \geq 1$ and suppose $g \in S^{k+1}$. We have

$$G_t(x) = \int_0^x \exp[\int_0^u g_1(te^v)dv]du = \int_0^x \exp[g_1G_{1t}(u)]du$$

say. Now since $g_1 \in S^k$, the expansion of $G_{1t}(x)$ is $\sum_{i=1}^{k} A_i(g_1)$ where $A_i(g_1) =$

$[(i+1)!]^{-1}\, x^{i+1}\, G_{1t}{}^{(i+1)}(0)$ by assumption. Introduce the dummy variable z. Then as in the proof of Lemma 3.2, we see that $A_i(g)$ is the coefficient of z^i in the formal expansion of

$$\int_0^x \exp[g_1 \sum_{i=0}^{k} A_i(g_1) z^{i+1}]du$$

in powers of z. But this is the same as the coefficient of z^i in the expansion of

$$\int_0^x \exp[g_1 G_{1t}(zu)]du = z^{-1} \int_0^{zx} \exp[g_1 G_{1t}(v)]dv = z^{-1} G_t(zx)$$

and hence $A_i(g) = [(i+1)!]^{-1}\, x^{i+1} G_t{}^{(i+1)}(0)$, $i=1,\ldots,k+1$, as asserted.

Although the ith error in the above expansion has the simple form $\{g(t)\}^{-1}[d^i/dz^i)g(te^z)]_{z=0}$, we stress again that this is not <u>necessarily</u> an asymptotic sequence, since there might be cancellation occuring amongst the basic terms of which this error is composed. We give an example in Section 4 where this occurs.

Finally, in this section we consider the inversion of the expansions discussed above. Again, we give only an outline of the main development. Consider an asymptotic expansion of the form

$$G_t(x) = \sum_{i=0}^{\infty} a_i(x)\, \epsilon_i(t)$$

where $(\epsilon_i(t))$ is an asymptotic sequence, $\epsilon_o(t) = 1$, $G_t(x)$ is increasing in x for each t, and $a_o(x)$ is increasing. Without loss of generality we may assume $a_0(x) = x$ (for if not, consider $G_t(a_0^{-1}(x))$). The successive terms in the asymptotic expansion of $G_t(x)$ are the coefficients of z^i in the formal power series expansion

$$G_t(x,z) = \sum_{i=0}^{\infty} a_i(x)\, \epsilon_i(t) z^i.$$

Consider a formal expansion $\sum_{i=0}^{\infty} D_i(x,t)z^i$, where $D_0(x,t) = x$. In order for the coefficients of this expansion to be those of an asymptotic (block) expansion of G_t^{-1}, we require

$$x = G_t(G_t^{-1}(x,z),z) = \sum_{i=0}^{\infty} a_i(G_t^{-1}(x,z))\epsilon_i(t)z^i$$

$$= \sum_{i=0}^{\infty} \sum_{k=0}^{\infty} (k!)^{-1}\, a_i{}^{(k)}(x)\epsilon_i(t)[\sum_{j=1}^{\infty} D_j(x,t)z^j]^k z^i \qquad (3.5)$$

assuming $a_i(x)$ admits a Taylor expansion. Comparing coefficients of z^i yields $D_i(x,t)$ in terms of $a_j(x)$ and their derivatives, and $\epsilon_j(t)$ for $j=1,\ldots,i$. For example the first two terms are

$$D_1 = -a_1\epsilon_1,\ D_2 = -a_2\epsilon_2 + a_1a_1'\ \epsilon_1^2 .$$

Each term in D_i is of smaller order than each term in D_{i-1}. The above discussion is informal and it must of course be shown that the resulting expansion is a valid block expansion; full details of this will appear elsewhere. In particular, the error $G_t^{-1}(x) - \sum_{i=0}^{k} D_i(x,t)$ is of smaller order than the *maximum* error term in $D_k(x,t)$.

We shall only discuss here the special case $g \in S^k$. In this case we have an expansion of the form

$$G_t(x) = \int_0^x \{g(te^u)/g(t)\}du = \sum_{i=0}^{\infty} \frac{x^{i+1}}{(i+1)!}\epsilon_i(t)$$

where $\epsilon_i(t) = \{g(t)\}^{-1}[(d^i/dz^i)g(te^z)]_{z=0}$, which will be a valid asymptotic expansion under the conditions described earlier. This is identical to the formal power series expansion of $G_t(x)$, and an obvious question is whether the inverse asymptotic expansion is equivalent to the formal power series expansion.

$$\sum_{i=0}^{\infty} \frac{x^{i+1}}{(i+1)!} \left[G_t^{-1}\right]^{(i+1)}(0)$$

of G_t^{-1}. Given the results outlined above, it is only necessary to check that the choice $D_i(x,t)=[(i+1)!]^{-1}x^{i+1}(G_t^{-1})^{(i+1)}(0)$ satisfies the power series identity (3.5) generated by the identity $x = G_t(G_t^{-1}(x,z),z)$. But with this choice we have $G_t(x,z) = z^{-1}G_t(zx)$ and $G_t^{-1}(x,z) = z^{-1}G_t^{-1}(zx)$, and so the identity is satisfied. Again, we repeat that despite the simple form of error sequence obtained in this case, the "component" sequences must be investigated in order to assert that we have a valid asymptotic expansion.

4. APPLICATION TO EXTREME VALUE THEORY.

We return now to the discussion in Section 2 on the approximation of the quantile function $q_n(p)$. In view of the results of the previous section, from (2.1) we can develop a block expansion for $u_t^{-1}(x)$ whenever $b \in R_\rho^k$. Let $g_k \in R_{\rho_k}$ be the kth remainder of b. Then from (2.2) it follows that this expansion will also be a block expansion for $q_n(p)$ as $n\to\infty$, provided that $ng_k(n) \to \infty$. In the special case $b \in S^k$, the kth order expansion becomes

$$u_t^{-1}(x) = \sum_{i=0}^{k} \frac{x^{i+1}}{(i+1)!} \{b(t)\}^{-1}[(d^i/dz^i)b(te^z)]_{z=0} + \eta_k.$$

Next consider the problem of obtaining an expansion for the distribution function. Suppose that, after possible rearrangement, we have an expansion of the form

$$u_n^{-1}(x) = H_\gamma(x) + \sum_{i=1}^{\infty} a_i(x)\, \epsilon_i(n).$$

Then, writing $v_n = u_n^{-1}H_\gamma^{-1}$, we have

$$v_n(x) = x + \sum_{j=1}^{\infty} c_i(x)\, \epsilon_i(n)$$

where $c_i = a_i H_\gamma^{-1}$. Since $v_n^{-1} = H_\gamma\, u_n$, inversion yields a block expansion for $\gamma^{-1}[(n\{1-F(a_n + b_n x)\})^\gamma - 1]$, as described in the previous section. Under the further condition $ng_k(n) \to \infty$, this will also be an expansion for $\gamma^{-1}[\{-\log F^n(a_n + b_n x)\}^\gamma - 1]$. In the special case $b \in S^k$ the kth order expansion becomes

$$\log[-\log F^n(a_n + b_n x)] = \sum_{i=0}^{k} \frac{x^{i+1}}{(i+1)!} \delta_i(n) + \eta_k(x,n) \tag{4.1}$$

where $\delta_i(n) = u_n^{(i+1)}(0)$. But $u_n'(x) = b_n h(a_n + b_n x)$ where $h(x) = f(x)/[1-F(x)]$ is the hazard function and so

$$\delta_i(n) = b_n^{(i+1)} [(d^i/dz^i)h(z)]_{z=a_n}.$$

The expansion (4.1) was formally stated as long ago as 1954 by Uzgören, although no conditions for validity were given, apart from a condition equivalent to $ng_1(n) \to \infty$. The present derivation shows that (4.1) is a valid asymptotic expansion in the case $F \in \mathcal{L}(\Lambda)$, but only under the conditions (i) $ng_k(n) \to \infty$, (ii) $b \in S^k$, and (iii) no cancellation occurs in the "component" terms of $\delta_i(n)$, as described in Section 3. This final condition is the most cumbersome to check in particular applications. Note that even when $\delta_{k+1}(n) = o(\delta_k(n))$, it is still not necessarily the case that (4.1) is a valid kth order asymptotic expansion. It might be conjectured that when $\delta_{i+1}(n) = o(\delta_i(n))$ for *all* i, then (4.1) is a valid expansion for all k, but this possibility has not been investigated.

When $\gamma > 0$ an alternative strategy would be to apply the results of Section 3 directly to $1-F(x)$, which is in R_γ. The connection between the remainder terms generated in this way, and those obtained by inversion has not yet been explored. One disadvantage of proceeding in this way is that we do not produce a unified expansion of $F \in \mathcal{D}(G_\gamma)$ with a common definition of the normalizing sequences.

We also remark that the constants a_n and b_n may be chosen in alternative ways; for example there is a parallel development using the definitions $a(t) = F^{-1}(e^{-t^{-1}})$ and $b(t) = ta'(t)$. If these definitions are used, then a slightly different asymptotic sequence results, and in particular there is no need for the rather artificial condition $ng_k(n) \to \infty$. On the other hand, one purpose of the present paper was to explore the validity of the Uzgören expansion, which arises under the present choice of constants. In practice, the choice of constants is largely a matter of convenience.

We end by giving two examples to demonstrate the need for conditions (ii) and (iii) above. First note that any positive function $b(t)$ on $[1,\infty)$ defines a a probability distribution, whose inverse is given by $F^{-1}(p) = \int_1^{(1-p)^{-1}} u^{-1}b(u)du$.

EXAMPLE 4.1 Take $b(t) = \exp(-2t^{-\frac{1}{2}})$, which is slowly varying and so $F \in \mathcal{L}(\Lambda)$. Here Uzgören's expansion (4.1) gives $\delta_1(n) = n^{-\frac{1}{2}}$ and $\delta_2(n) = 2n^{-1} + \frac{1}{2}n^{-\frac{1}{2}}$.

The expansion is valid only to the first term in this case; note that $\eta_2(x,n)$ must also be $O(n^{-\frac{1}{2}})$. The reason for the failure is that $g_1(t) = t^{-\frac{1}{2}}$, so that $g_1 \in R_{-\frac{1}{2}}$.

EXAMPLE 4.2 For each t define $g_1(t)$ to be the unique positive root α of the equation

$$\log(1+\alpha^{-1}) = \alpha^{-1} + 2\log t.$$

It is easy to see that $g_1(t) \downarrow 0$, and $g_1(t) \sim -(2\log t)^{-1}$. Then $b(t) = \exp\{\int_1^t s^{-1}g(s)ds\}$ is slowly varying. Furthermore, $g_2 = tg_1'/g_1 = 2g_1(1+g_1)$, $g_3 = tg_2'/g_2 = 2g_1(1+2g_1)$, and so $b \in S^3$. However, the first three error terms in the expansion (4.1) are

$$\delta_1 = g_1, \ \delta_2 = -2g_1^3, \ \delta_3 = -14g_1^3 + O(g_1^4).$$

The two-term expansion in this case is not a strictly valid asymptotic expansion; the reason here is that cancellation has occurred amongst the component terms of δ_2.

REFERENCES

Anderson, C.W. (1971). Contributions to the Asymptotic Theory of Extreme Values. Ph.D. thesis, University of London, England.

Cohen, J.P. (1982). Convergence rates for the ultimate and penultimate approximations in extreme value theory. Adv. Appl. Prob. 14, 833–854.

Fisher R.A. and Tippett, L.H.C. (1928). Limiting forms of the frequency distributions of the largest or smallest member of a sample. Proc. Camb. Phil. Soc. 24, 180–190.

Galambos, J. (1987). The Asymptotic Theory of Extreme Order Statistics. Wiley, New York.

Goldie, C.M. and Smith, R.L. (1987). Slow variation with remainder: Theory and Application. Quart. J. Math. Oxford 38, 45–71.

Gomes, M.I. (1978). Some Probabilistic and Statistical Problems in Extreme Value Theory. Ph.D. Thesis, University of Sheffield, England.

Pickands, J. (1986). The continuous and differentiable domains of attraction of the extreme-value distributions. Ann. Probab. 14, 996–1004.

Seneta, E. (1976). Regularly Varying Functions. Lecture Notes in Math. 508, Springer-Verlag, Berlin.

Smith, R.L. (1982). Uniform rates of convergence in extreme value theory. Adv. Appl. Prob. 14, 600–622.

Sweeting, T.J. (1985). On domains of uniform local attraction in extreme value theory. Ann. Probab 13, 196–205.

Uzgören, N.T. (1954). The asymptotic development of the distribution of the extreme values of a sample. Studies in Mathematics and Mechanics Presented to R. Von Mises, Academic Press, New York.

STRONG LAWS FOR THE k-TH ORDER STATISTIC WHEN $k \leq c \log_2 n$ (II)

Paul Deheuvels

Université Paris VI

Abstract. We give several bounds and characterizations for the lower almost sure classes of $U_{k,n}$, where $k=k_n$ denotes a nondecreasing integer sequence such that $1 \leq k_n = O(\log_2 n)$ as $n \to \infty$, and where $U_{1,n} \leq \dots \leq U_{n,n}$ denote the order statistics of the first n observations from a sequence of independent and uniformly distributed random variables on (0,1).

1. INTRODUCTION AND RESULTS

Let $U_1, U_2, \dots$ be a sequence of independent and uniformly distributed on (0,1) random variables. For $n \geq 1$ denote by $U_{1,n} \leq \dots \leq U_{n,n}$ the order statistics of the first n of these random variables. In this paper we are concerned with the limiting almost sure behavior of $U_{k,n}$ as $n \to \infty$, where $1 \leq k=k_n \leq n$ is a nondecreasing integer sequence satisfying additional assumptions specified below.

We will concentrate our interest in the case where $k_n = O(\log_2 n)$ as $n \to \infty$, where $\log_j$ denotes the j-th iterated logarithm. The reason for this restriction is to be found in the following Theorem A due to Kiefer (1972), which shows that $U_{k,n}$ exhibits a different behavior in each of the ranges $(k_n/\log_2 n \to \infty)$, $(k_n/\log_2 n \to v \in (0,\infty))$ and $(k_n/\log_2 n \to 0)$.

Theorem A. Assume that there exists a sequence $p_n \sim n^{-1}k_n$ such that $p_n \downarrow 0$, $np_n \uparrow$ and $np_n/\log_2 n$ is monotone. Then

(i) If $k_n/\log_2 n \to \infty$ as $n \to \infty$, we have

$$\limsup_{n\to\infty} \pm(2k\log_2 n)^{-1/2}(nU_{k,n}-k) = 1 \text{ a.s..} \tag{1.1}$$

(ii) If $k_n/\log_2 n \to v \in (0,\infty)$ as $n \to \infty$, and if $-1<\delta_v^-<0<\delta_v^+$ denote the two roots of the equation in δ: $v^{-1}=\delta-\log(1+\delta)$, we have

$$\limsup_{n\to\infty} \pm k^{-1}nU_{k,n} = \pm(1+\delta_v^{\pm}) \quad \text{a.s..} \tag{1.2}$$

(iii) If $k_n/\log_2 n \to 0$ as $n \to \infty$, we have

$$\limsup_{n\to\infty} (\log_2 n)^{-1}nU_{k,n} = 1 \text{ a.s.,} \tag{1.3}$$

and

$$\lim_{n\to\infty} \{k^{-1}nU_{k,n} \exp(\tfrac{1+\varepsilon}{k}\log_2 n)\} = \begin{cases} \infty & \text{for } \varepsilon>0 \\ 0 & \text{for } \varepsilon<0 \end{cases} \quad \text{a.s..} \square \tag{1.4}$$

Key words. Strong laws, order statistics. AMS 1980 subject classification. 60F05

These results have been precised for a fixed $k\geq 1$ by Barndorff-Nielsen (1961), Robbins and Siegmund (1972), Shorack and Wellner (1978) and Klass (1984, 1985). Further characterizations of strong limiting upper and lower bounds for $U_{k,n}$ are given in Deheuvels (1986) and Deheuvels and Mason (1988), the latter two of whom proved that (1.4) holds for any sequence $1\leq k=k_n=o(\log_2 n)$ without any regularity condition such as imposed in Theorem A. Some related results are given in Brito (1986).

In this paper, we will obtain refinements of (1.2)-(1.4) under appropriate conditions imposed on k_n. In the first place, we consider the case where k_n remains close of $\kappa(n)$, where κ is an auxiliary function. Namely, we assume that:

(K1) $\kappa(t)$ is a positive nondecreasing concave function of $t\geq t_0$, having right derivative $\kappa'(t)$ such that, for all $p \in \mathbf{R}$

$$t\kappa'(t)\kappa^p(t) \to 0 \text{ and } \kappa(t)\to\infty \text{ as } t\to\infty. \tag{1.5}$$

The sequence k_n will be assumed to satisfy:

(K2) (i) $1\leq k=k_n\leq n$ is nondecreasing and such that $k_n/\log_2 n \to 0$ as $n\to\infty$;

(ii) $k_n\to\infty$ as $n\to\infty$;

(K3) $$\lim_{n\to\infty} \frac{\log(n\kappa'(n))}{k_n} = -\infty \quad \text{and} \quad \lim_{n\to\infty} \frac{\log \kappa(n)}{k_n} = 0.$$

Remark 1.1. The concavity of κ is equivalent to the existence of a nonincreasing right derivative $\kappa'(t)=\lim_{h\downarrow 0} h^{-1}(\kappa(t+h)-\kappa(t))$ of κ for all $t\geq t_0$. Taking $p=-1$ in (1.5), the Karamata-type representation

$$\kappa(t) = \kappa(t_0) \exp\left(\int_{t_0}^{t} \frac{s\kappa'(s)}{\kappa(s)} \frac{ds}{s}\right) \quad \text{for} \quad t\geq t_0, \tag{1.6}$$

shows that $\kappa(t)$ is slowly varying as $t\to\infty$. Moreover (1.5) also implies that

$$\lim_{t\to\infty} \frac{\log(t\kappa'(t))}{\log \kappa(t)} = -\infty. \tag{1.7}$$

Remark 1.2. Simple examples of functions κ which satisfy (1.5) are given, for suitable constants $\lambda>0$ and θ_j, $j=2,\ldots,q$, by

$$\kappa(t) = \lambda \prod_{j=2}^{q} (\log_j t)^{\theta_j}.$$

Having κ, in view of (1.7), one obtains by (K2-3) all sequences k_n related to κ through direct inequalities (see Remark 1.5 in the sequel).

Remark 1.3. Any nondecreasing positive function κ such that (as in Theorem A) $\kappa(t)/\log_2 t\downarrow$ is slowly varying at infinity. In addition, if $\kappa'(t)$ exists, then

$$t\kappa'(t)/\kappa(t) \leq 1/(\log t)(\log_2 t) = o(\kappa^{-p-1}(t)) \text{ as } t\to\infty,$$

for an arbitrary $p \in \mathbf{R}$. Hence for such a κ, (1.5) holds.□

Our first result is stated in Theorem 1 below.

Theorem 1. Under (K1-2-3),

$$\liminf_{n\to\infty} \{k^{-1} nU_{k,n} \exp(1 - \frac{1}{k} \log(n\kappa'(n)))\} = 1 \text{ a.s.}. \square \tag{1.8}$$

Remark 1.4. Let $Y_{n-k+1,n}=-\log U_{k,n}$ denote the k-th upper order statistic generated by the first n observations from an i.i.d. sequence of exponentially distributed random variables with mean one. A result equivalent to (1.8) is that:

$$\limsup_{n\to\infty} \{Y_{n-k+1,n} - \log(n/k) - 1 + \frac{1}{k}\log(n\kappa'(n))\} = 0 \text{ a.s..} \tag{1.9}$$

A statement close to (1.9) has been obtained by Griffin (1986) who showed, under (K2), that for any nondecreasing sequence $u_n>0$ such that $1/\log^2 n \le u_n \le 1/\log n$ and

$$\sum_n \frac{1}{n} u_n \exp(\varepsilon k_n) < \infty \text{ or } =\infty \text{ according as } \varepsilon<0 \text{ or } \varepsilon>0, \tag{1.10}$$

it holds that

$$\limsup_{n\to\infty} \{Y_{n-k+1,n} - \log(n/k) - 1 + \frac{1}{k}\log u_n\} = 0 \text{ a.s..} \tag{1.11}$$

Griffin (1986) also proves the existence of a sequence u_n satisfying (1.10) and (1.11) for an arbitrary sequence k_n satisfying (K2). In view of his results, the main novelty of Theorem 1 is to provide a simple form for u_n in (1.11) under weak additional regularity conditions.

Notice that (1.11) is invalid for $k=k_n$ constant, for in this case Kiefer (1972) has shown that $P(nU_{k,n}\le c_n)=0$ or 1 according as $\sum_n \frac{1}{n} c_n^k <\infty$ or $=\infty$. Since there does not exist a constant $0<\varepsilon<1$ such that simultaneously $\sum_n \frac{1}{n} c_n^k(1+\varepsilon)^k=\infty$ and $\sum_n \frac{1}{n} c_n^k(1-\varepsilon)^k<\infty$, we see that in this case (1.11) cannot possibly hold.

Remark 1.5. Consider the example where $\kappa(t)=\log_q t$ in Theorem 1 for some $q\ge 2$. Since then $\log(t\kappa'(t)) \sim -\log_2 t$ and $\log \kappa(t)=\log_{q+1} t$, an easy consequence of this theorem is that, for any sequence k_n which satisfies (K2)(i) jointly with

$$k_n/\log_{q+1} n \to \infty \text{ as } n\to\infty, \tag{1.12}$$

we have

$$\limsup_{n\to\infty} \{Y_{n-k+1,n} - \log(n/k) - 1 - \frac{1}{k}\sum_{j=2}^{q} \log_j n\} = 0 \text{ a.s..} \tag{1.13}$$

Even though (1.13) holds under the sole conditions, in addition to (K2), that $k_n/\log_{q+1} n \to \infty$, it is worthwhile to obtain similar bounds as given in (1.13) without this additional assumption. This is achieved in Theorem 2 below.

Theorem 2. Under (K2)(i), for any fixed $p\ge 2$, we have

$$\liminf_{n\to\infty} \{k^{-1} nU_{k,n} \exp(1 + \frac{1}{k}(\sum_{j=2}^{p-1} \log_j n + (1+\varepsilon)\log_p n))\} \ge 1 \text{ or } \le 1 \text{ a.s.,} \tag{1.14}$$

according as $\varepsilon>0$ or $\varepsilon<0$ (Here, we use the notation $\sum_\emptyset := 0$).

Remark 1.6. 1°) With the notation introduced in Remark 1.4, (1.14) is equivalent to

$$\limsup_{n\to\infty} \{Y_{n-k+1,n} - \log(n/k) - 1 - \frac{1}{k}(\sum_{j=2}^{p-1} \log_j n + (1+\varepsilon)\log_p n)\} \le 0 \text{ or } \ge 0 \text{ a.s.,} \tag{1.15}$$

according as $\varepsilon>0$ or $\varepsilon<0$.

2°) Take successfully in (1.15) p=q+1 and p=q+2. It is obvious from (1.15) that (1.13) holds if and only if (1.12) holds in addition to (K2)(i). Thus we see that the statement in Theorem 1 is sharp.

3°) It is noteworthy that Theorem 2 does not make use of the assumption that $k_n \to \infty$ (i.e. of (K2)(ii)). In the case of a fixed $k=k_n$ the statement of this theorem is a simple consequence of Kiefer's characterizations (see e.g. Remark 1.4).

4°) Another consequence of Theorem 2 taken with p=2 is that, under (K2)(i),

$$\limsup_{n\to\infty} \{-k\log(k^{-1}nU_{k,n})/\log_2 n\} = 1 \text{ a.s.,} \tag{1.16}$$

which is the statement of Corollary 4 in Deheuvels and Mason (1988).□

In spite of the fact that the bounds given in Theorems 1 and 2 are sufficient for all practical purposes, they do not achieve a complete characterization of the lower a.s. class of $U_{k,n}$. In Deheuvels (1986), the following partial solution of this problem has been given, along with similar results for the upper a.s. class of $U_{k,n}$:

Theorem B. Under (K2)(i), for any sequence $c_n>0$ such that

$$n^{-1}c_n \downarrow \text{ on any interval } N_1 \le n \le N_2 \text{ where } k_n \text{ is constant,} \tag{1.17}$$

$$c_n/k_n \to 0 \text{ as } n\to\infty, \tag{1.18}$$

$$\sum_n \frac{1}{n} k^{1/2} \left(\frac{e}{k} c_n\right)^k \exp(-c_n) < \infty, \tag{1.19}$$

we have $P(nU_{k,n} \le c_n \text{ i.o.}) = 0$.□

Our next theorem gives a converse of Theorem B, namely by proving that the divergence of the series in (1.19) implies, under suitable regularity conditions imposed upon k_n and c_n, that $P(nU_{k,n} \le c_n \text{ i.o.})=1$.

Theorem 3. Assume that:

(K4)
(i) $1 \le k=k_n \le n$ is nondecreasing and such that $n^{-1}k_n \to 0$ as $n\to\infty$;
(ii) $k_n/\log k_{2n} \to \infty$ as $n\to\infty$.

Let also $c_n>0$ be a sequence such that

(C1)
(i) $c_n/k_n \to 0$ as $n\to\infty$;
(ii) $\limsup_{n\to\infty} \left(nk_n\left(\frac{c_{n+1}-c_n}{c_n}\right)1_{\{k_{n+1}=k_n\}}\right) < 1.$

Then the condition

$$\sum_n \frac{1}{n} k^{1/2} \left(\frac{e}{k} c_n\right)^k \exp(-c_n) = \infty, \tag{1.20}$$

implies that $P(nU_{k,n} \le c_n \text{ i.o.}) = 1$.□

Remark 1.7. 1°) By (1.1), a simple argument shows that if $\liminf_{n\to\infty}(k_n/\log_2 n)>0$, then $\liminf_{n\to\infty}(nU_{k,n}/\log_2 n)>0$ a.s.. Thus, if $k_n/\log_2 n \to v \in (0,\infty)$, for any sequence $c_n>0$ such that (C1)(i) holds, i.e. $c_n/k_n \to 0$ as $n\to\infty$, we have $P(nU_{k,n} \le c_n \text{ i.o.})=0$. This shows that the main range of interest of Theorem 3 corresponds to the case where $k_n/\log_2 n \to 0$ as $n\to\infty$.

2°) The condition (C1)(ii) concerns the variation of c_n on the intervals $N_1 \le n \le N_2$ where $k=k_n$ is constant. In such an interval, $U_{k,n}\downarrow$ and therefore the assumption (1.17) that $n^{-1}c_n\downarrow$ in such an interval is natural. This assumption is equivalent to $n(c_{n+1}-c_n)c_n^{-1} 1_{\{k_{n+1}=k_n\}} \le 1$, so that we see that (C1)(ii) is a slightly stronger condition. A close look at the proofs in the sequel shows that (C1)(ii) which corresponds to the fact that $\frac{1}{n} k^{1/2}(\frac{e}{k} c_n)^k \exp(-c_n)\downarrow$ on the intervals $N_1 \le n \le N_2$ can be partly relaxed. However, we will limit ourselves for sake of concision to this case which covers all situations of interest. Let for instance $c_n = k \exp(-\frac{a}{k}\log_2 n)$, where $a=a_n$ is a sequence bounded from zero and infinity. We have for $k=k_n=k_{n+1}$

$$nk_n\left(\frac{c_{n+1}-c_n}{c_n}\right)1_{\{k_{n+1}=k_n\}} = O\left(\frac{1}{\log n}\right) - (1+o(1))n(a_{n+1}-a_n)\log_2 n \text{ as } n\to\infty.$$

In all examples discussed below (see (1.26) and (1.27)), this last expression tends to zero as n tends to infinity, so that (C1)(ii) holds.

3°) As will be seen later on (see Lemma 2.1 in the sequel) the condition (K4)(i) can be weakened to $k_n/\log k_{[\alpha n]} \to \infty$ for some fixed $\alpha>1$, or for some suitable sequences $\alpha=\alpha_n\downarrow 1$ as $n\to\infty$. This condition imposes a minimal regularity on the growth of k_n and is always satisfied for instance under the assumptions of Theorem A.$\square$

We will now present a corollary of Theorem 3 and Theorem B, using the function $\Delta(x)=-\log(1+\delta^-_{1/(x-1)})$ where δ^-_v is as in Theorem A. An alternative definition of $\Delta=\Delta(x)$ is obtained via the equation in Δ

$$\Delta + e^{-\Delta} = x \quad \text{for } \Delta\ge 0 \text{ and } x\ge 1. \tag{1.21}$$

Obviously, (1.21) defines $\Delta(x)$, which also satisfies the asymptotic expansion $\Delta(x) = x - e^{-x} - e^{-2x} - \frac{3}{2}e^{-3x} + O(e^{-4x})$ as $x\to\infty$. Such an expansion in powers of e^{-x} may be achieved up to an arbitrary order.

The relevance of $\Delta(x)$ with respect to our study follows from the fact that, if $c_n=k_n e^{-\Delta}$, then

$$\frac{1}{n} k^{1/2} (\frac{e}{k} c_n)^k \exp(-c_n) = \frac{1}{n}\exp(-\log k + k - k(\Delta + e^{-\Delta})). \tag{1.22}$$

Moreover, if we equate the right-hand-side of (1.22) to $1/\{n(\log n)\ldots(\log_p^{1+\varepsilon} n)\}$, the resulting c_n satisfies the conditions of Theorem 3 for all sequences k_n such that (K4) holds together with (K2)(i)(i.e. $k_n/\log_2 n \to 0$). Since then the series in (1.20) diverges if and only if $\varepsilon\le 0$, we have by Theorem 3 and Theorem B:

<u>Corollary</u> 1. Assume that (K2)(i) and (K4) hold, and let $\beta_{n,\varepsilon}>0$ be defined via

$$\beta_{n,\varepsilon} + e^{-\beta_{n,\varepsilon}} = \frac{1}{k}\{k + \tfrac{1}{2}\log k + \sum_{j=2}^{p-1}\log_j n + (1+\varepsilon)\log_p n\}, \tag{1.23}$$

where $p\ge 2$ and $\varepsilon \in \mathbf{R}$ are arbitrary constants. Then

$$P(nU_{k,n} \le k\exp(-\beta_{n,\varepsilon}) \text{ i.o.}) = 0 \text{ or } 1, \tag{1.24}$$

according as $\varepsilon>0$ or $\varepsilon\le 0$.

<u>Remark</u> 1.8. By (1.21) and (1.23), $\beta_{n,\varepsilon}=\Delta(1+\frac{1}{k}\{\frac{1}{2}\log k + \sum_{j=2}^{p-1}\log_j n + (1+\varepsilon)\log_p n\})$.

Hence, in (1.24) we may replace $P(nU_{k,n} \leq k \exp(-\beta_{n,\varepsilon})$ i.o.) by $P(nU_{k,n} \leq k(1+\omega_n^{\varepsilon})$ i.o.), where $\omega_n^{\varepsilon} = \delta_{v_{n,\varepsilon}}^{-}$ and

$$(1.25) \qquad v_{n,\varepsilon}^{-1} = \frac{1}{k}\{\tfrac{1}{2}\log k + \sum_{j=2}^{p-1} \log_j n + (1+\varepsilon)\log_p n\}.$$

Refined asymptotic expansions for sequences c_n^{ε} such that $P(nU_{k,n} \leq c_n^{\varepsilon}$ i.o.)=0 or 1, according as $\varepsilon>0$ or $\varepsilon<0$ may be obtained from Corollary 1, (1.23) or (1.25) after routine computations. Let $x_{n,\varepsilon} = 1 + \frac{1}{k}\{\frac{1}{2}\log k + \sum_{j=2}^{p-1}\log_j n + (1+\varepsilon)\log_p n\}$. It is noteworthy that, whenever $k_n/\log_2 n \to 0$ and $p \geq 3$, $x_{n,\varepsilon} \sim \frac{1}{k}\log_2 n \to \infty$. Since $\beta_{n,\varepsilon} = \Delta(x_{n,\varepsilon})$ and $\Delta(x) = x - e^{-x}(1+O(e^{-x}))$ as $x\to\infty$, we need discuss the relative magnitude of the components of $x_{n,\varepsilon}$ with respect to $\exp(-x_{n,\varepsilon})$. Because of the large number of possible cases, we will only present a few examples listed below.

<u>Case</u> 1. Assume that $\limsup_{n\to\infty} \{(k_n \log_3 n)/\log_2 n\} < 1$. Then, we may take for $p \geq 4$

$$(1.26) \qquad c_n^{\varepsilon} = k\exp(-\frac{1}{k}\{\log(e^k\sqrt{k}\log n) + \sum_{j=3}^{p-1}\log_j n + (1+\varepsilon)\log_p n\}).$$

<u>Case</u> 2. Assume that $\liminf_{n\to\infty} \{(k_n \log_2 n)/\log_2 n\} > 1$ and $\lim_{n\to\infty} \{k_n/\log_2 n\} = 0$. Then we may take

$$(1.27) \qquad c_n^{\varepsilon} = k\exp(-1 - \frac{1}{k}\log_2 n + (1-\varepsilon)\exp(-\frac{1}{k}\log_2 n)).$$

Interestingly, (1.26) and (1.27) show that the bounding sequences c_n^{ε} for the lower class of $U_{k,n}$ exhibit a different behavior according as k_n is above or below the "critical" order of $(\log_2 n)/\log_3 n$. Sharper evaluations can be obtained for Case 2 and the "intermediate case" between Case 1 and Case 2. We omit details.

The characterization given in Theorem 3 completes the description of the four upper-upper, upper-lower, lower-upper and lower-lower classes of $U_{k,n}$ presented in Deheuvels (1986). This description is now complete for $k=k_n=o(\log_2 n)$ as $n\to\infty$. Our methods may also be extended to the other cases. This will be published elsewhere.

The remarkable feature of these characterizations is the role played in each of the four classes by the convergence or the divergence of the series in (1.20). An open problem worth of interest is to see how far the technical assumptions in our theorems (such as (C1)(ii)) may be relaxed.

In the remainder of our paper, we prove Theorems 1, 2 and 3.

2. <u>PROOFS OF THE THEOREMS</u>

2.1. <u>Proof of Theorem</u> 1. First we show, under (K1-2-3), that for any $\varepsilon>0$ and $p\in\mathbf{R}$,

$$(2.1) \qquad P(nU_{k,n} \leq k\exp(-1-\varepsilon+\frac{1}{k}\log(n\kappa'(n)/\kappa^p(n)))\text{ i.o.}) = 0.$$

Let $c_n := d_{n,-\varepsilon} = k\exp(-1-\varepsilon+\frac{1}{k}\log(n\kappa'(n)/\kappa^p(n)))$. By (K3),

$$\frac{1}{k}\log(n\kappa'(n)/\kappa^p(n)) = \frac{1}{k}\log(n\kappa'(n)) + o(1) \text{ as } n\to\infty.$$

Hence, using again (K3), we see that $c_n/k_n \to 0$ as $n\to\infty$, so that condition (1.18) in Theorem B is satisfied. Moreover, if (2.1) holds for some $p \in \mathbf{R}$, then it holds for all $p \in \mathbf{R}$. In the sequel, we shall assume without loss of generality that $p \geq 0$.

Next we see that on any interval $N_1 \leq n \leq N_2$ where $k_n = k$ is constant, c_n is of the form $C(n\kappa'(n)/\kappa^p(n))^{1/k}$, where $C = ke^{-1-\varepsilon}$ is constant. Since (K1) implies that $\kappa'(t)\downarrow$ and $\kappa(t)\uparrow$, if follows that $n^{-1}c_n = n^{-1}d_{n,-\varepsilon}\downarrow$ in this interval, so that (1.17) holds.

Finally we see that

$$\frac{1}{n} k^{1/2} \left(\frac{e}{k} c_n\right)^k \exp(-c_n) = \kappa'(n)\exp\left(-k\left\{\varepsilon - \frac{1}{2}\frac{\log k}{k} + p\frac{\log \kappa(n)}{k} + \frac{1}{k}c_n\right\}\right),$$

which, since $k^{-1}\log k$, $k^{-1}\log \kappa(n)$ and $k^{-1}c_n$ all tend to zero, is by (K3) for n sufficiently large less than or equal to

$$\kappa'(n)\exp(-\tfrac{1}{2}\varepsilon k_n) \leq \kappa'(n)/\kappa^2(n) \leq \int_{n-1}^{n} \frac{\kappa'(t)}{\kappa^2(t)}\,dt = \kappa^{-1}(n-1) - \kappa^{-1}(n). \tag{2.2}$$

Since (2.2) gives the general term of a summable series, (1.19) holds. Hence the assumptions of Theorem B are satisfied and therefore (2.1) holds.□

Our next step is to show, under (K1-2-3), that for any $\varepsilon > 0$ and $p \in \mathbf{R}$,

$$P\left(nU_{k,n} \leq k\exp\left(-1+\varepsilon+\frac{1}{k}\log(n\kappa'(n)/\kappa^p(n))\right) \text{ i.o.}\right) = 1. \tag{2.3}$$

For some arbitrary but fixed $c > 1$, let $n_j = 1 + [c^j]$ for $j \geq 0$, let $d_{n,\varepsilon}$ be as in the proof of (2.1), and set $\ell = \ell_j = k_{n_j}$ and $\delta_j^\varepsilon = d_{n_j,\varepsilon}$ for $j = 0,1,\ldots$. Denote by V_j the ℓ_j-th lower order statistic from the sample $U_{n_{j-1}+1},\ldots,U_{n_j}$. Since $\ell_j/(n_j - n_{j-1}) \sim (\frac{c}{c-1})\ell_j/n_j \to 0$ as $n\to\infty$, V_j is defined for all $j \geq j_0$ sufficiently large and is such that $U_{\ell_j,n_j} \leq V_j$. Hence (2.3) is a consequence of the statement

$$P(n_jV_j \leq \delta_j^\varepsilon \text{ i.o.}) = 1. \tag{2.4}$$

In view of the independence of the events $\{n_jV_j \leq \delta_j^\varepsilon\}$, the Borel-Cantelli lemma implies that (2.4) holds if and only if

$$\sum_j P(n_jV_j \leq \delta_j^\varepsilon) = \infty. \tag{2.5}$$

Let $N = n_j - n_{j-1}$ and $\gamma = N\delta_j^\varepsilon/n_j$. Observe that $\ell/N \sim (\frac{c}{c-1})k_{n_j}^2/n_j \to 0$ as $j\to\infty$, and that $\gamma/\ell \sim (\frac{c-1}{c})d_{n_j,\varepsilon}/k_{n_j} \to 0$ as $j\to\infty$. By Lemma 4 in Deheuvels (1986) and Stirling's formula it follows that, as $j\to\infty$,

$$P(n_jV_j \leq \delta_j^\varepsilon) = P(NU_{\ell,N} \leq \gamma) \geq \binom{N}{\ell}\left(\frac{\gamma}{N}\right)^\ell\left(1-\frac{\gamma}{N}\right)^{N-\ell} \sim \frac{\ell^{-1/2}}{\sqrt{2\pi}}\left(\frac{e\gamma}{\ell}\right)^\ell e^{-\gamma} := P_j/\sqrt{2\pi}\,. \tag{2.6}$$

Next, we evaluate

$$P_j = \ell^{-1/2}\left\{\frac{e}{\ell}\left(\frac{n_j - n_{j-1}}{n_j}\right)\delta_j^\varepsilon\right\}^\ell \exp\left(-\left(\frac{n_j-n_{j-1}}{n_j}\right)\delta_j^\varepsilon\right).$$

Up to now, the choice of $c > 1$ has remained open. We choose c in such a way that $\frac{c-1}{c} = e^{-\frac{1}{4}\varepsilon}$, so that for all j sufficiently large

$$\delta^{\varepsilon/2} \leq \left(\frac{n_j - n_{j-1}}{n_j}\right)\delta_j^\varepsilon \leq \delta_j^\varepsilon.$$

This, jointly with (K3), implies that for all j sufficiently large,

$$\begin{aligned} P_j &\geq \ell^{-1/2}\left(\frac{e}{\ell}\delta_j^{\varepsilon/2}\right)^\ell \exp(-\delta_j^\varepsilon) = n_j\kappa'(n_j)\exp\left(\ell\left\{\tfrac{1}{2}\varepsilon - \frac{\log \ell}{2\ell} - \frac{p}{\ell}\log \kappa(n_j) - \tfrac{1}{2}\delta_j^\varepsilon\right\}\right) \\ &\geq n_j\kappa'(n_j)\exp(\tfrac{1}{4}\varepsilon k_{n_j}) \geq n_j\kappa'(n_j)\kappa^\alpha(n_j), \end{aligned}$$

where $\alpha>0$ is an arbitrary constant.

By Remark 1.1, (K1) implies that κ is slowly varying. This, in turn, implies that for all j sufficiently large $\kappa^{\alpha}(n_j) \geq \frac{1}{2}\kappa^{\alpha}(n_{j+1})$. Since κ is nondecreasing, it follows that

$$(2.7)\qquad P_j \geq \frac{1}{2}\frac{n_j}{n_{j+1}-n_j}\int_{n_j}^{n_{j+1}} \kappa^{\alpha}(t)\kappa'(t)dt = \frac{1+o(1)}{2(c-1)(\alpha+1)}\{\kappa^{\alpha+1}(n_{j+1})-\kappa^{\alpha+1}(n_j)\}.$$

It is straightforward that the right-hand-side of (2.7) is not summable in j, so that, in view of (2.6), we see that (2.5) holds. This completes the proof of (2.3).□

The proof of Theorem 1 follows from a joint application of (2.1) and (2.3) for an arbitrary choice of $\varepsilon>0$.□

2.2. Proof of Theorem 2. We follow the same steps as in the proof of Theorem 1. Let, for $p\geq 3$, $\theta\in\mathbb{R}$ and $\varepsilon\in\mathbb{R}$

$$c_n = k\exp(-1 + \frac{1}{k}(\sum_{j=2}^{p-1}\log_j n + (1+\theta)\log_p n) - \varepsilon).$$

Since $c_n/k_n = \exp(-\frac{1+o(1)}{k_n}\log_2 n) \to 0$ as $n\to\infty$, we see that (1.18) holds. (1.17) is straightforward and

$$(2.8)\qquad \frac{1}{n}k^{1/2}(\frac{e}{k}c_n)^k\exp(-c_n) = \frac{1}{n}\{\prod_{j=1}^{p-2}(\log_j n)^{-1}\}(\log_{p-1}n)^{-1-\theta}\exp(-k\varepsilon(1+o(1))$$

as $n\to\infty$. Obviously (2.8) is the general term of a convergent (resp. divergent) series when both θ and ε are >0 (resp. <0). By Theorem B, this suffices to show that

$$\liminf_{n\to\infty} k^{-1}nU_{k,n}\exp(-1 + \frac{1}{k}(\sum_{j=2}^{p-1}\log_j n + (1+\theta)\log_p n)) \geq 1 \quad \text{a.s.}$$

for all $\theta>0$, thus proving the first half of (1.14).

The proof of the second half of (1.14) follows along the same lines of the proof of Theorem 1 with the formal replacement of $d_{n,\varepsilon}$ by c_n as given above. We omit the routine details.□

2.3. Proof of Theorem 3. Let $A>0$ be an arbitrary but fixed constant, and denote by n_0 the smallest integer $n\geq 1$ such that for all $m\geq n$, $Am/k_m\geq 1$. The finiteness of n_0 is a consequence of the assumption that $n^{-1}k_n\to 0$ as $n\to\infty$. Define recursively the integer sequence $n_0<n_1<\dots$ by

$$(2.9)\qquad n_{j+1} = [n_j(1 + A/k_{n_j})],\ j=0,1,\dots,$$

with $[u]$ denoting the integer part of u.

Let $\mathcal{H} = \{j\geq 0:\ k_{n_j} < k_{n_{j+1}}\}$ and $\mathcal{E} = \{i\geq 0:\ j\leq i\leq j+k^2_{n_{j+1}}$ for some $j\in\mathcal{H}\}$

Lemma 2.1. Assume that $1\leq k=k_n\leq n$, $c_n>0$, $\limsup_{n\to\infty}(c_n/k_n)=\theta<1$, $k_n\uparrow\infty$ and that $k_{n_j}/\log k_{n_{j+1}} \to\infty$ as $n\to\infty$. Then

$$(2.10)\qquad S = \sum_{j\in\mathcal{H}}\ \sum_{n_j\leq n<n_{j+1}} \frac{1}{n}k^{1/2}(\frac{e}{k}c_n)^k\exp(-c_n) < \infty.$$

Proof. Let $\theta_n=c_n/k_n$. For any $j\in\mathcal{H}$ we have $1\leq k_{n_{j+1}}-k_{n_j}$, so that for $n_j\leq n$

$$\frac{1}{n}k_n^{1/2}(\frac{e}{k_n}c_n)^{k_n}\exp(-c_n) \leq (k_{n_{j+1}}-k_{n_j})\frac{1}{n_j}k_n^{1/2}\{\theta_n\exp(1-\theta_n)\}^{k_n}.$$

Observe that the function $\alpha \to \alpha e^{1-\alpha}$ increases from 0 to 1 as α increases from 0 to 1. This ensures the existence of a constant $\rho>0$ such that $\theta e^{1-\theta}<e^{-\rho}<1$. For such a ρ, our assumptions imply that for all $j\in\mathcal{K}$ sufficiently large and $n_j\le n<n_{j+1}$

$$(2.11) \qquad \frac{1}{n} k_n^{1/2} (\frac{e}{k_n} c_n)^{k_n} \exp(-c_n) \le (k_{n_{j+1}}-k_{n_j}) \frac{1}{n_j} k_n^{1/2} \exp(-\rho c_n),$$

which, since the function $\tau \to \tau^{1/2}\exp(-\rho\tau)$ is ultimately decreasing, is for $j\in\mathcal{K}$ sufficiently large less than or equal to

$$(k_{n_{j+1}}-k_{n_j}) \frac{1}{n_j} k_{n_j}^{1/2}\exp(-\rho k_{n_j}) \le \frac{1}{n_j} (k_{n_{j+1}}-k_{n_j})/k_{n_{j+1}}^3 .$$

It follows that, for some suitable constant S_0, we have (by (2.10) and (2.11))

$$S \le S_0 + 2\sum_{j=0}^{\infty} (\frac{n_{j+1}-n_j}{n_j})(k_{n_{j+1}}-k_{n_j})/k_{n_{j+1}} = S_0 + O(\sum_{j=0}^{\infty} (k_{n_j}^{-1}-k_{n_{j+1}}^{-1})) < \infty,$$

where we have used the fact that $(n_{j+1}-n_j)/n_j \sim Ak_{n_j}^{-1}$ as $j\to\infty$. This completes the proof of Lemma 2.1.□

<u>Lemma</u> 2.2. Assume that $1\le k=k_n\le n$ and $c_n>0$ satisfy

$$(2.12) \qquad \limsup_{n\to\infty} (c_n/k_n) < 1 \quad\text{and}\quad \limsup_{n\to\infty}\left\{ nk_n(\frac{c_{n+1}-c_n}{c_n}) 1_{\{k_{n+1}=k_n\}}\right\} < 1.$$

Then, for all $j \notin \mathcal{K}$ sufficiently large, we have

$$(2.13) \qquad \sum_{n_j\le n<n_{j+1}} \frac{1}{n} k_n^{1/2} (\frac{e}{k_n} c_n)^{k_n} \exp(-c_n) \le 2Ak_{n_j}^{-1/2} (\frac{e}{k_{n_j}} c_{n_j})^{k_{n_j}} \exp(-c_{n_j}).$$

<u>Proof</u>. For $j\notin\mathcal{K}$, $k_n=k$ is constant when n varies in the interval $n_j\le n<n_{j+1}$. Set $\psi_n = \frac{1}{n} k^{1/2} (\frac{e}{k}c_n)^k \exp(-c_n)$. We have

$$n\log(\psi_{n+1}/\psi_n) = n\log(\frac{n}{n+1}) + nk\log(c_{n+1}/c_n) - n(c_{n+1}-c_n)$$
$$\le n\log(\frac{n}{n+1}) + n(\frac{k}{c_n}-1)(c_{n+1}-c_n).$$

Since our assumptions imply that $\liminf_{n\to\infty} (k_n/c_n)>1$, this last expression is ultimately negative whenever $c_{n+1}-c_n\le 0$ or when $c_{n+1}-c_n>0$ jointly with $nk_n(c_{n+1}-c_n)/c_n + n\log(\frac{n}{n+1}) < 0$. Since $n\log(\frac{n}{n+1}) \to -1$ as $n\to\infty$, we see that (2.12) implies that ψ_n is decreasing in the interval $n_j\le n<n_{j+1}$ for all j sufficiently large, so that

$$(2.14) \qquad \sum_{n_j\le n<n_{j+1}} \frac{1}{n} k_n^{1/2} (\frac{e}{k_n} c_n)^{k_n} \exp(-c_n) \le (\frac{n_{j+1}-n_j}{n_j}) k_{n_j}^{1/2} (\frac{e}{k_{n_j}} c_{n_j})^{k_{n_j}} \exp(-c_{n_j}).$$

The proof of (2.13) now follows from (2.14) used jointly with the fact that, as $j\to\infty$, $(n_{j+1}-n_j)/n_j \sim Ak_{n_j}^{-1} < 2Ak_{n_j}^{-1}$.□

By all this, we see that, under the assumptions of Lemmas 2.1 and 2.2,

$$(2.15) \qquad \sum_n \frac{1}{n} k^{1/2} (\frac{e}{k} c_n)^k \exp(-c_n) = \infty \Rightarrow \sum_{j\notin\mathcal{K}} k_{n_j}^{-1/2} (\frac{e}{k_{n_j}} c_{n_j})^{k_{n_j}} \exp(-c_{n_j}) = \infty.$$

Our next step will be to show that in (2.15) we can replace $\mathcal{K}$ by $\mathcal{E}$.

<u>Lemma</u> 2.3. Under the assumptions of Lemmas 2.1 and 2.2, we have

$$(2.16) \qquad \sum_{i\in\mathcal{E}} k_{n_i}^{-1/2} (\frac{e}{k_{n_i}} c_{n_i})^{k_{n_i}} \exp(-c_{n_i}) < \infty.$$

Proof. Let $\rho>0$ and $\theta_n=c_n/k_n$ be as in the proof of Lemma 2.1. We have, as $n\to\infty$,

$$(2.17)\qquad k_n^{-1/2}\left(\frac{e}{k_n}c_n\right)^{k_n}\exp(-c_n) = k_n^{-1/2}\{\theta_n\exp(1-\theta_n)\}^{k_n} \le k_n^{-1/2}\exp(-\rho k_n) \le \exp(-\rho k_n).$$

Recall that $\mathcal{E} = \bigcup_{j\in\mathcal{H}} [j, j+k^2_{n_{j+1}}]$. Hence, using again the remark that for $j\in\mathcal{H}$, we have $k_{n_{j+1}}-k_{n_j}\ge 1$, we see from (2.17) that the series in (2.16) is bounded above by a constant plus

$$\sum_{j=0}^{\infty}(k_{n_{j+1}}-k_{n_j})(2k^2_{n_{j+1}}\exp(-\rho k_{n_j})) = O\left(\sum_{j=0}^{\infty}(k_{n_{j+1}}-k_{n_j})k_{n_j}^{-1}k_{n_{j+1}}^{-1}\right) < \infty$$

as desired. Thus we have (2.16).□

An immediate consequence of (2.15) and (2.16) is that, under the assumptions of Lemmas 2.1 and 2.2,

$$(2.18)\qquad \sum_n \frac{1}{n}k^{1/2}\left(\frac{e}{k}c_n\right)^k\exp(-c_n) = \infty \Rightarrow \sum_{i\notin\mathcal{E}} k_{n_i}^{-1/2}\left(\frac{e}{k_{n_i}}c_{n_i}\right)^{k_{n_i}}\exp(-c_{n_i}) = \infty.$$

From now on, we assume that $k_n\uparrow\infty$ (the case where k_n is fixed is covered by the results of Kiefer (1972)). Let $B=A/2$ and set $\ell=\ell_j=k_{n_j}$, $M=M_j=[n_j(1-B/\ell_j)]$ and $\lambda=\lambda_j=(M_j/n_j)c_{n_j}$. Since $\ell_j^2/n_j\to 0$ as $j\to\infty$, there exists a $j_1<\infty$ such that $j\ge j_1$ ensures that $n_{j-1}<M_j<n_j-\ell_j<n_j$. Hence, for all $j\ge j_1$, we may define W_j to be the ℓ_j-th lower order statistic generated by $\{U_i: n_j-M_j<i\le n_j\}$. Moreover, since $W_j\ge U_{\ell_j,n_j}$, we have

$$(2.19)\qquad P(M_jW_j\le\lambda_j \text{ i.o.})=1 \Rightarrow P(nU_{k,n}\le c_n \text{ i.o.})=1.$$

In the sequel, we shall prove that the divergence of the series in (2.18) imply that the left-hand-side of (2.19) holds. We start with the lemma:

Lemma 2.4. Under the assumptions of Lemma 2.1, if $n^{-1/2}k_n\to 0$ and $k_n\to\infty$ as $n\to\infty$, we have

$$(2.20)\qquad P(M_jW_j\le\lambda_j) = (1+o(1))\frac{e^{-B}}{\sqrt{2\pi}}k_{n_j}^{-1/2}\left(\frac{e}{k_{n_j}}c_{n_j}\right)^{k_{n_j}}\exp\left(-c_{n_j}\left(1-\frac{B}{k_{n_j}}\right)\right).$$

Proof. We have $P(M_jW_j\le\lambda_j) = P(MU_{\ell,M}\le\lambda) = \sum_{\ell\le i\le M}\binom{M}{i}\left(\frac{\lambda}{M}\right)^i\left(1-\frac{\lambda}{M}\right)^{M-i}$. Note that, as $j\to\infty$, $M_j/n_j\to 1$ and $\lambda=\lambda_j\sim c_{n_j}$. Hence both $\lambda=\lambda_j$ and $\ell=\ell_j$ are $o(M^{1/2})=o(M_j^{1/2})$ as $j\to\infty$. This, jointly with an application of Stirling's formula, shows that, as $j\to\infty$,

$$P(M_jW_j\le\lambda_j) \sim \frac{\ell^{-1/2}}{\sqrt{2\pi}}\left(\frac{e}{\ell}\lambda\right)^{\ell}\exp(-\lambda) \sim \frac{1}{\sqrt{2\pi}}k_{n_j}^{-1/2}\left(\frac{e}{k_{n_j}}c_{n_j}\right)^{k_{n_j}}e^{-B}\exp\left(-c_{n_j}\left(1-\frac{B}{k_{n_j}}\right)\right),$$

where we have used the assumption that $c_n/k_n=O(1)$ as $n\to\infty$. This proves (2.20).□

For $j\ge j_1$, let $B_j = \{M_jW_j\le\lambda_j\}$ and let for $N\ge R\ge j_1$, $\Upsilon_N^R = \{i\notin\mathcal{E}: R\le i\le N\}$. By (2.18) and (2.20), we see that, for any fixed $R\ge j_1$ (under the assumptions of Lemmas 2.1 and 2.2)

$$(2.21)\qquad \sum_n\frac{1}{n}k^{1/2}\left(\frac{e}{k}c_n\right)^k\exp(-c_n) = \infty \Rightarrow \lim_{N\to\infty}\sum_{j\in\Upsilon_N^R}P(B_j) = \infty.$$

In view of (2.20), all we need is to prove that (2.21) implies that $\lim_{N\to\infty}P(\bigcup_{j\in\Upsilon_N^R}B_j) = 1$. We will make use of the following inequality (Chung and Erdös (1952))

$$(2.22)\qquad P\Big(\sum_{j\in\Upsilon_N^R} B_j\Big) \geq \Big[\sum_{j\in\Upsilon_N^R} P(B_j)\Big]^2 \Big/ \Big[\sum_{j\in\Upsilon_N^R} P(B_j) + \sum_{i\neq j;\, i,j\in\Upsilon_N^R} P(B_i \cap B_j)\Big].$$

By (2.22), our proof boils down to the evaluation of $P(B_i \cap B_j)$.

Lemma 2.5. There exist constants $j_2<\infty$ and $0<C<\infty$ such that, if $K_j=C\ell_j\log \ell_j$, then, for all $N\geq R\geq j_2$ and $i,j \in \Upsilon_N^R$,

$$(2.23)\qquad \ell_i \neq \ell_j \Rightarrow P(B_i \cap B_j) = P(B_i)P(B_j), \text{ and}$$

$$(2.24)\qquad |i-j|\geq K_j \Rightarrow P(B_i \cap B_j) = P(B_i)P(B_j).$$

Proof. Consider $i\in\Upsilon_N^R$ and $j\in\Upsilon_N^R$ such that $i\neq j$. We have the following two cases:

Case 1. Assume that $\ell_i\neq\ell_j$ and (without loss of generality) that $i<j$. Denote by r the largest integer such that $r\in\mathcal{H}$ and $i<r<j$. We must have the inequalities $i<r<r+\ell_{r+1}^2$ $<j$ otherwise i or j would belong to the set $\mathcal{E}$. Moreover, the maximality of r ensures that ℓ_m is constant when m varies in the interval $r+1\leq m\leq j$, so that $\ell_{r+1}=\ell_j$. Take now j_3 so large that for all $m\geq j_3$ we have $n_{m+1}\geq n_m(1+ A/(2\ell_m))$. Then for $j>i\geq R\geq j_3$,

$$n_i < n_r < n_r(1+ A/(2\ell_j))^{j-r} \leq n_j,$$

and hence, since $B=A/2$ and $j-r>\ell_j^2$,

$$n_i < n_j(1 + B/\ell_j)^{-\ell_j^2} \leq n_j\exp(-B\ell_j + \frac{B^2}{2}).$$

This, in turn is for all $j\geq j_4$ sufficiently large less that $n_j-M_j \sim Bn_j/\ell_j$. Hence B_i and B_j are independent as sought.

Case 2. Assume that $\ell_i=\ell_j$ together with $|i-j|\geq K_j$. Denote by $\ell=\ell_i=\ell_j$ the common value of ℓ_i and ℓ_j. We have $K=K_i=K_j=\ell\log \ell$. Hence we may assume without loss of generality that $i<j$. For $j<i\leq j_3$, the same argument as in Case 3 yields

$$n_i < n_i(1+ B/\ell)^K \leq n_j,$$

and hence

$$n_i < n_j(1+ B/\ell)^{-C\ell\log \ell} \leq n_j\exp(-BC\log \ell_j + \frac{B^2C}{2\ell_j} \log \ell_j).$$

The choice of $C=2/B$ ensures that for some suitable j_5, the expression above is less than $n_j-M_j \sim Bn_j/\ell_j$ whenever $j>i\geq j_5$. This implies the independence of B_i and B_j.

The proof of Lemma 2.5 is completed by taking $j_2=\max(j_3,j_4,j_5)$.□

We will now concentrate on the remaining case where $i\in\Upsilon_N^R$, $j\in\Upsilon_N^R$, $i<j$, $\ell=\ell_i=\ell_j$ and $|i-j|\leq K=C\ell\log \ell$. Here, $C=2/B$ as in the proof of Lemma 2.5. Throughout, we assume that the assumptions of Lemmas 2.1 and 2.2 hold. We will prove the lemma:

Lemma 2.6. There exist $R_0<\infty$ and $\Delta<\infty$ such that, for all $N\geq R\geq R_0$,

$$(2.25)\qquad \sum_{j=i+1}^{C\ell\log \ell} P(B_i \cap B_j) \leq \Delta P(B_i),$$

where $\ell=\ell_i=\ell_j$ and i,j are restricted to vary in the set Υ_N^R.

Proof. In the first place, note that $|n_{m+1}-n_m(1+A/\ell)|\leq 1$ for $i\leq m\leq j$ (since $\ell_m=\ell$).

It is therefore obvious that, uniformly over $j>i\geq R$, we have, as $R\to\infty$,

$$\frac{\ell}{j-i}\log(n_j/n_i)\to A.$$

Hence, there exists an $R_1<\infty$ such that $R\geq R_1$ implies that, for $j>i\geq R$ and $\ell=\ell_i=\ell_j$,

$$\exp(\tfrac{3}{4}A\,\tfrac{j-i}{\ell})\leq n_j/n_i\leq\exp(\tfrac{5}{4}A\,\tfrac{j-i}{\ell}) \tag{2.26}$$

Moreover, (2.12) implies that for $k_n=k_{n+1}$ one has ultimately in n:

$$\frac{nk_n(c_{n+1}-c_n)}{c_n}<1,\text{ or equivalently, }\frac{c_{n+1}}{n+1}<\frac{c_n}{n}(1-\frac{1}{n+1}(1-\frac{1}{k_n}))<\frac{c_n}{n}, \tag{2.27}$$

so that R_1 can be chosen so large that for $j>i\geq R\geq R_1$,

$$c_{n_j}/n_j<(c_{n_i}/n_i)\prod_{n_i<m\leq n_j}(1-\frac{1}{m}(1-\frac{1}{\ell}))<(c_{n_i}/n_i)\exp((1-\frac{1}{\ell})\log(n_i/n_j)) \tag{2.28}$$

$$=(c_{n_i}/n_j)\exp(-\frac{1}{\ell}\log(n_i/n_j))\leq(c_{n_i}/n_j)\exp(\tfrac{5}{4}A\,\frac{j-i}{\ell^2}).$$

In order to evaluate $P(B_i\cap B_j)$, we introduce the following notation. Let

$$\eta=\#\{U_m\leq c_{n_i}/n_i,\ n_i-M_i<m\leq n_j-M_j\},\ \xi=\#\{U_m\leq c_{n_j}/n_j,\ n_j-M_j<m\leq n_i\},$$

$$\xi'=\#\{U_m\leq c_{n_i}/n_i,\ n_j-M_j<m\leq n_i\}\qquad,\ \zeta=\#\{U_m\leq c_{n_j}/n_j,\ n_i<m\leq n_j\}.$$

By (2.27), $c_{n_j}/n_j<c_{n_i}/n_i$ and hence $\xi\leq\xi'$. It follows that we have, for an arbitrary q (note that ζ and (η,ξ') are independent)

$$P(B_i\cap B_j)=P(\eta+\xi'\geq\ell,\xi+\zeta\geq\ell)\leq P(\eta+\xi'\geq\ell,\xi+\zeta\geq\ell,\xi<\ell-q)+P(\xi\geq\ell-q) \tag{2.29}$$

$$\leq P(\eta+\xi'\geq\ell)P(\zeta>q)+P(\xi\geq\ell-q)=P(B_i)P(\zeta>q)+P(\xi\geq\ell-q).$$

Next, we need obtain suitable bounds for $P(\zeta>q)$ and $P(\xi\geq\ell-q)$. Observe that ζ and ξ follow binomial distributions. It turns out that the following inequality (see Bennett (1962), Shorack and Wellner (1986) p.440) is appropriate for our needs. Here, Z denotes a random variable with a binomial B(N,p) distribution, $0\leq p\leq 1/2$ and $t\geq 0$.

$$P(Z\geq t)\leq\{(\frac{eNp}{t})^t e^{-Np}\}^{1/(1-p)}=\exp(-\frac{Np}{1-p}h(\frac{t}{Np})), \tag{2.30}$$

where $h(u)=1+u(-1+\log u)$.

We will cut the rest of our proof in successive steps. In the sequel, we assume throughout that $c_n/k_n\to 0$ as $n\to\infty$ and that $i,j\in\Upsilon_N^R$.

<u>Step</u> 1. Take R_2 such that for $j\geq R\geq R_2$ we have $c_{n_j}/n_j\leq 1/2$. Let $q=\dfrac{K\ell\log(n_j/n_i)}{-\log(c_{n_j}/\ell)}$, where $K>0$ is a constant which will be precised later on.

By (2.30) taken for $Z=\zeta$, $t=q$, $p=c_{n_j}/n_j$ and $N=n_j-n_i$, using the fact that $1-p\geq 1/2$, we have

$$P^2(\zeta\geq q)\leq(\frac{e}{q}(\frac{n_j-n_i}{n_j})c_{n_j})^q\exp(-(\frac{n_j-n_i}{n_j})c_{n_j}).$$

Assume, without loss of generality, that $R_2\geq R_1$. By (2.26) for $j>i\geq R\geq R_2$, we have

$$\tfrac{3}{4}AK(j-i)\leq q(-\log(c_{n_j}/\ell))\leq\tfrac{5}{4}AK(j-i), \tag{2.31}$$

and

$$\frac{n_j-n_i}{n_j}\leq 1-\exp(-\tfrac{5}{4}A\,\frac{j-i}{\ell})\leq\tfrac{5}{4}A(\frac{j-i}{\ell}).$$

Hence

$$P^2(\zeta\geq q) \leq \left(\frac{e}{q}\left(\frac{n_j-n_i}{n_j}\right)c_{n_j}\right)^q = \left(\frac{e}{K\log(n_j/n_i)}\left(\frac{n_j-n_i}{n_j}\right)(c_{n_j}/\ell)(-\log(c_{n_j}/\ell))\right)^q \leq$$

$$\left((-\log(c_{n_j}/\ell))\frac{e\frac{5}{4}A(\frac{j-i}{\ell})}{K\frac{3}{4}A(\frac{j-i}{\ell})}(c_{n_j}/\ell)\right)^q = \exp(-q(-\log(c_{n_j}/\ell))\{1-\frac{\log(5e/3K)}{-\log(c_{n_j}/\ell)}+o(1)\}).$$

Since $c_n/k_n\to 0 \Rightarrow c_{n_j}/\ell\to 0 \Rightarrow (\log(5e/3K))/(-\log(c_{n_j}/\ell))\to 0$, there exists an $R_3\geq R_2$ such that $j>i\geq R\geq R_3$ implies the inequality

$$(2.32)\qquad P(\zeta\geq q) \leq \exp(\tfrac{1}{3}q(-\log(c_{n_j}/\ell)) \leq \exp(\tfrac{1}{4}AK(j-i)).$$

Step 2. We now proceed to evaluate $P(\xi\geq\ell-q)$ in the case where $i<j\leq i+L\ell$, and $L>0$ is a fixed constant. By (2.31), our choice of q ensures that for $j>i\geq R$, $q=(j-i)o(1)$, where the "o(1)" holds uniformly over j and i as $R\to\infty$. Therefore, in the range we consider, $q/\ell=o(1)$ uniformly over $i+L\ell\geq j>i\geq R$ as $R\to\infty$.

Next, we see that there exists an $R_4\geq R_2$ such that for $j>i>R\geq R_4$, we have, by (2.26),

$$(2.33)\qquad M_i = [n_i(1-B/\ell)] \geq M'_{ij} := n_i-n_j+[n_j(1-B/\ell)] = n_i-n_j+M_j \geq n_i-\frac{B}{\ell}n_j-1$$

$$= n_i(1-\frac{B}{\ell})(1-\frac{B}{\ell}(\frac{n_j-n_i}{n_i})/(1-\frac{B}{\ell}))-1 \geq M_i(1-\frac{B}{\ell}(e^{\frac{5}{4}AL}-1)/(1-\frac{B}{\ell}))-1$$

$$\geq M_i(1-\frac{B}{\ell}(e^{\frac{3}{2}AL}-1)).$$

Recall that $P(\xi\geq\ell-q) = P(U_{]\ell-q[,M'_{ij}} \leq c_{n_j}/n_j) = P(M'_{ij}U_{]\ell-q[,M'_{ij}} \leq M'_{ij}c_{n_j}/n_j)$, where $]u[$ denotes the smallest integer greater than or equal to u. It follows from a similar argument as used in the proof of Lemma 2.4 that

$$P(\xi\geq\ell-q) \sim \frac{\ell^{-1/2}}{\sqrt{2\pi}}\left(\frac{e}{]\ell-q[}M'_{ij}c_{n_j}/n_j\right)^{]\ell-q[}\exp(-M'_{ij}c_{n_j}/n_j)$$

$$= \frac{\ell^{-1/2}}{\sqrt{2\pi}}\left(\frac{e}{\ell}c_{n_j}\right)^{\ell-q}(n_i/n_j)^{\ell-q}\exp(-c_{n_j}(n_i/n_j)+q+O(1))\ell^{-1/2},$$

where we have used (2.33) and the fact that $c_n/k_n\to 0$ as $n\to\infty$. This, jointly with (2.28) yields

$$P(\xi\geq\ell-q) \leq \left(\frac{e}{\ell}c_{n_i}\right)^{\ell-q}(n_i/n_j)^{\ell-q}\exp(-c_{n_i}(n_i/n_j)+q+O(1)).$$

By (2.22) and (2.26), this last expression is less than or equal to

$$P(B_i)\left(\frac{e}{\ell}c_{n_i}\right)^{-q}(n_i/n_j)^{\ell-q}\exp(c_{n_i}(1-n_i/n_j)+q+O(1)).$$

By (2.26), $(n_i/n_j)^{\ell-q} \leq \exp(-\frac{3}{4}A(j-i)(\frac{\ell-q}{\ell}))$. Next, by (2.28), we have

$$(c_{n_i}/\ell)^{-q} = (c_{n_j}/c_{n_i})^q\exp(K\ell\log(n_j/n_i)) \leq \exp(\tfrac{5}{4}Aq(\tfrac{j-i}{\ell^2})+\tfrac{5}{4}KA(j-i)),$$

where $q(j-i)/\ell^2 \leq Lq/\ell = o(1)$. Finally, using again (2.26), we have

$$c_{n_i}(1-n_i/n_j) = (c_{n_i}/\ell)\ell(1-n_i/n_j) \leq (c_{n_i}/\ell)\ell(1-\exp(-\tfrac{3}{4}A\,\tfrac{j-i}{\ell}))$$

$$\leq (c_{n_i}/\ell)\tfrac{3}{4}A(j-i) = (j-i)o(1).$$

Up to now, the choice of $K>0$ was open. We fix $K=1/10$, so that, by the above inequalities, for any $L>0$, there exists an $R_5\geq R_4$ (depending upon L) such that, for all $j>i\geq R\geq R_5$ such that $j-i\leq L\ell$, we have the inequality

$$(2.34)\qquad P(\xi\geq\ell-q)\ \leq\ \Gamma P(B_i)\exp(-\tfrac{1}{4}A(j-i)),$$

where $\Gamma>0$ is a constant (depending also upon L).

A joint use of (2.32) and (2.34) yields the inequality, for $i\geq R\geq R_5$,

$$(2.35)\qquad \sum_{j=i+1}^{[L\ell]} P(B_i\cap B_j)\leq P(B_i)(1+\Gamma)/(1-\exp(-\tfrac{1}{4}A)).$$

Step 3. We consider now the case where $i+L\ell\leq j\leq i+C\ell\log\ell$. For this, we use again (2.29) with a different choice of q given by $q = M\ell/(-\log(c_{n_j}/\ell))$, where $M>0$ is a constant precised below. Proceeding as in Step 1, we have, for this choice of q,

$$P^2(\zeta\geq q)\leq \left(\frac{e}{q}\left(\frac{n_j-n_i}{n_j}\right)c_{n_j}\right)^q \exp\left(-\left(\frac{n_j-n_i}{n_j}\right)c_{n_j}\right)\leq \left(\frac{e}{M}\left(\frac{n_j-n_i}{n_j}\right)(c_{n_j}/\ell)(-\log(c_{n_j}/\ell))\right)^q$$

$$\leq \exp(-q(-\log(c_{n_j}/\ell))(1-\frac{\log((e/M)(1-\exp(-\frac{3}{4}L))}{-\log(c_{n_j}/\ell)}+o(1)))\leq \exp(-\tfrac{1}{2}M\ell),$$

for all $j>i\geq R\geq R_6$ sufficiently large. Hence, we have

$$(2.36)\qquad P(\zeta\geq q)\leq \exp(-\tfrac{1}{4}M\ell).$$

Step 4. Under the assumptions of Step 3, we evaluate $P(\xi\geq\ell-q)$, using (2.30). Let M_{ij}' be as in (2.33). Take now in (2.30) $Z=\xi$, $t=\ell-q$, $p=c_{n_j}/n_j$ and $N=M_{ij}'$. Assuming that $\gamma = 1 - c_{n_j}/n_j\geq 1/2$, we have

$$P^{\gamma}(\xi\geq\ell-q)\leq \left(\frac{e}{\ell-q}M_{ij}'c_{n_j}/n_j\right)^{\ell-q}\exp(-M_{ij}'c_{n_j}/n_j).$$

Since $M_{ij}'\leq n_i$, we have by (2.26)

$$(2.37)\qquad P^{\gamma}(\xi\geq\ell-q)\leq \left(\frac{e}{\ell}c_{n_i}\right)^{\ell-q}(c_{n_j}/c_{n_i})^{\ell-q}(n_i/n_j)^{\ell-q}\left(1-\frac{q}{\ell}\right)^{-\ell}$$

$$\leq \left(\frac{e}{\ell}c_{n_i}\right)^{\ell-q}(c_{n_j}/c_{n_i})^{\ell-q}\exp\left(-\tfrac{3}{4}A(j-i)\left(\frac{\ell-q}{\ell}\right)+q+o(1)\right)$$

$$\leq \left(\frac{e}{\ell}c_{n_i}\right)^{\ell-q}(c_{n_j}/c_{n_i})^{\ell-q}\exp(-\tfrac{\gamma}{2}AL\ell),$$

for $j>i\geq R\geq R_7$ sufficiently large and $j-i\geq L\ell$.

By (2.20), (2.28) and (2.37), it follows that

$$P(\xi\geq\ell-q)/P(B_i) = O\{\ell^{1/2}\exp((\frac{\ell-q}{\gamma}-1)\log(\frac{e}{\ell}c_{n_i}))+(\frac{\ell-q}{\gamma})\log(c_{n_j}/c_{n_i})$$

$$-\tfrac{1}{2}AL\ell+c_{n_i}\} = O\{\exp(M\ell(1+o(1))-\tfrac{1}{2}AL\ell(1+o(1)))\}.$$

We now choose $M=\frac{1}{3}AL$, so that for $j>i\geq R\geq R_8$ sufficiently large and $j-i\geq L\ell$,

$$(2.38)\qquad P(\xi\geq\ell-q)\leq P(B_i)\exp(-\tfrac{1}{4}M\ell).$$

By (2.29), (2.36) and (2.38), we have

$$(2.39) \qquad \sum_{j=[L\ell]+1}^{C\ell\log \ell} P(B_i \cap B_j) \leq (C\ell\log \ell)P(B_i)(2\exp(-\tfrac{1}{4}M\ell)),$$

which, jointly with (2.35), implies (2.25). The proof of Lemma 2.6 is now completed.□

We may now collect the pieces of our puzzle. Take $R_9 = \max_{0\leq m\leq 8} R_i$ and $R \geq R_9$. By Lemma 2.5 and (2.22)-(2.25), we have

$$P(\sum_{j\in\Gamma_N^R} B_j) \geq \left(\sum_{j\in\Gamma_N^R} P(B_j)\right)^2 \Big/ \left((1+2\Delta)\sum_{j\in\Gamma_N^R} P(B_j) + (\sum_{j\in\Gamma_N^R} P(B_j))^2\right),$$

which by (2.21) tends to one as N tends to infinity. This completes the proof of Theorem 3.□

3. Acknowledgements.

I am grateful to David M. Mason for stimulating discussions on the subject of this paper. I thank Rolf Reiss and Jurg Hüsler for giving me the oportunity of presenting the contents of this paper in Oberwolfach.

REFERENCES

Barndorff-Nielsen, O. (1961). On the rate of growth of the partial maxima of a sequence of independent identically distributed random variables. Math. Scand. 9 383-394.

Bennett, G. (1962). Probability inequalitues for the sum of independent random variables. J. Amer. Statist. Assoc. 57 33-45.

Brito, M. (1986). Sur l'encadrement presque sûr dans un échantillon ordonné. C. R. Acad. Sci. Paris 303 821-824.

Chung, K.L. and Erdös, P. (1952). On the application of the Borel-Cantelli lemma. Trans. Amer. Math. Soc. 72 179-186.

Deheuvels, P. (1986). Strong laws for the k-th order statistic when $k \leq c\log_2 n$. Probab. Th. Rel. Fields 72 179-186.

Deheuvels, P. and Mason, D.M. (1988). The asymptotic behavior of sums of exponential extreme values. Bulletin des Sciences Mathématiques (to appear).

Griffin, P.S. (1986). Non classical laws for the iterated logarithm behavior of trimmed sums. Preprint.

Kiefer, J. (1972). Iterated logarithm analogues for sample quantiles when $p_n \downarrow 0$. Proc. Sixth Berkeley Symposium on Math. Statist. and Probab. 1 227-244. University of California Press.

Klass, M.J. (1984). The minimal growth rate of partial maxima. Ann. Probab. 12 380-389.

Klass, M.J. (1985). The Robbins-Siegmund series criterion for partial maxima. Ann. Probab. 13 1369-1370.

Robbins, H. and Siegmund, D. (1972). On the law of the iterated logarithm for maxima and minima. Proc. Sixth Berkeley Symposium on Math. Statist. and Probab. 3 51-70. University of California Press.

Shorack, G. and Wellner, J.A. (1978). Linear bounds on the empirical distribution function. Ann. Probab. 6 349-353.

Shorack, G. and Wellner, J.A. (1986). Empirical Processes with Applications to Statistics. Wiley, New York.

EXTREME VALUES WITH VERY HEAVY TAILS

P. Révész

Technische Universität Wien

Abstract. Let $X_1, X_2, \ldots$ be a sequence of positive i.i.d.r.v.'s with $S_o = 0$, $S_n = X_1+X_2+\ldots+X_n$ $(n=1,2,\ldots)$ and let τ_t be the largest integer for which $S_{\tau_t} \leqq t$. Further let $M_t^{(1)} \geqq M_t^{(2)} \geqq \ldots \geqq M_t^{(\tau_t+1)}$ be the order statistics of the sequence $X_1, X_2, \ldots, X_{\tau_t}, t-S_{\tau_t}$. The main result says that if $\mathbb{P}(X_1 < x) = \exp(-(\log x)^{\gamma})$ $(x \geqq 1,\ 0 < \gamma < 1)$ then with probability one for all t big enough $t^{-1}(M_t^{(1)} + M_t^{(2)} + \ldots + M_t^{(r)}) \geqq 1-\varepsilon_t$ if $(r-2)/(r-1) \leqq \gamma < (r-1)/r$, $\varepsilon_t \leqq \exp(-(\log t)^{\beta}(\log\log t)^{-3})$ and $\beta = (r-1)/r - \gamma$ whenever $(r-2)/(r-1) < \gamma < (r-1)/r$.

1. INTRODUCTION

Let $X_1, X_2, \ldots$ be a sequence of positive i.i.d.r.v.'s with $S_o = 0$, $S_n = X_1 + X_2 + \ldots + X_n$ $(n=1,2\ldots)$ and let τ_t be the largest integer for which $S_{\tau_t} \leqq t$. In [3] we have studied the properties of the process

$$M_t = M(t) = \max\{X_1, X_2, \ldots, X_{\tau_t}, t-S_{\tau_t}\}.$$

We were especially interested in the case $\mathbb{E}X_1 = \infty$ and in the „liminf properties" of M_t. We recall a few results of [3].

THEOREM A. Let

$$\mathbb{P}\{X_1 > x\} = F(x) = \frac{x^{-\alpha}}{L(x)} \qquad (0 < \alpha < 1)$$

where $L(x)$ is a slowly varying function. Then

$$\liminf_{t \to \infty} t^{-1}(\log\log t)\, M(t) = \beta(\alpha) \qquad \text{a.s.}$$

Key words. Extreme values, the influence of the maximum terms.
AMS 1980 subject classifications. Primary 60F15, Secondary 60K05.

where $\beta(\alpha)$ *is the root of the equation*

$$\sum_{k=1}^{\infty} \frac{\beta^k}{k!} \frac{\alpha}{k-\alpha} = 1.$$

In order to formulate our next result we recall the definition of π-varying function.

Definition. The function $L: R^+ \to R$ is π-varying if there exists a slowly varying function $a(x)$ for which

$$\lim_{x\to\infty} \frac{L(xt)-L(x)}{a(x)} = \log t \qquad (t > 0).$$

THEOREM B. *Let*

$$\mathbb{P}(X_1 > x) = F(x) = \frac{1}{L(x)}$$

where $L(x)$ *is a* π*-varying function. We also assume that*

$$\frac{1}{x} \int_0^x u \, d\, F(u) = o\left(\frac{1-F(x)}{\log\log\log x}\right)$$

and

$$x(1-F(x)) \int_0^x u \left(1-\log\frac{x}{u}\right) d\, F(u) = O\left(\left(\int_0^x u \, d\, F(u)\right)^2\right) \quad (x\to\infty).$$

Then

$$\liminf_{t\to\infty} \frac{M(t)}{a_t} = 1 \qquad \textit{a.s.}$$

where

$$a_t^{(1)} \leq a_t \leq a_t^{(2)},$$

$$(t-a_t^{(1)})\, Q(a_t^{(1)}) = t\, Q(a_t^{(2)}) = \log\log t$$

and

$$Q(a) = \frac{1}{a} \log \frac{1-F(a)}{\frac{1}{a}\int_0^a u \, d\, F(u)}.$$

In order to illuminate the meaning of Theorem B in [3] we considered the following example.

Example 1. If

$$F(x) = \exp\left(-\frac{\log x}{\log\log x}\right) \qquad (x \geq e)$$

then

$$\liminf_{t \to \infty} \frac{\log\log t}{t \log\log\log\log t} M(t) = 1 \qquad \text{a.s.}$$

Applying Theorem B we can also handle the case $F(x)=\exp(-(\log x)^{\gamma})$. However recently Pruitt (1987) obtained a much more complete result than ours. In fact our following Example 2 is a trivial consequence of his theorem (cf. also Theorem 2 below).

Example 2. If $F(x)=\exp(-(\log x)^{\gamma})$ $(x\geq 1,\ 0<\gamma<1)$ then

$$\liminf_{t\to\infty} t^{-1} M(t)=r^{-1} \quad \text{if} \quad \frac{r-2}{r-1} \leq \gamma < \frac{r-1}{r} \qquad (r=2,3,\ldots).$$

REMARK 1. For any $F(x)$ we have

$$\liminf_{t\to\infty} t^{-1} M(t) \leq 1/2 \qquad \text{a.s.}$$

(Proof is trivial).

Let $M_t = M_t^{(1)} \geq M_t^{(2)} \geq \ldots \geq M_t^{(\tau_t+1)}$ be the order statistics of the sequence $X_1, X_2, \ldots, X_{\tau_t}, t-S_{\tau_t}$. Further let $1 < m=m(t) \leq \tau_t+1$ be a function for which

$$\lim_{t\to\infty} \frac{M_t^{(1)}+M_t^{(2)}+\ldots+M_t^{(m)}}{t} = 1 \qquad \underline{\text{a.s.}} \tag{1.1}$$

We ask: how can be characterized the functions $m(t)$ which satisfy (1.1). One can prove

THEOREM C. *Let*

$$F(x) = \frac{x^{-\alpha}}{L(x)} \qquad (0<\alpha<1)$$

where $L(x)$ *is a slowly varying function. Then*

(i) *for any* $\varepsilon>0$ *there exists a* $C=C(\varepsilon)$ *such that*

$$\liminf_{t\to\infty} \frac{M_t^{(1)}+M_t^{(2)}+\ldots+M_t^{(\mu)}}{t} \geq 1-\varepsilon \qquad \text{a.s.}$$

where $\mu=[C\log\log t]$,

(ii)

$$\liminf_{t\to\infty} \frac{M_t^{(1)}+M_t^{(2)}+\ldots+M_t^{(m)}}{t} = 1 \qquad \text{a.s.}$$

where

$$\frac{m(t)}{\log\log t} \nearrow \infty,$$

(iii)

$$\liminf_{t\to\infty} \frac{M_t^{(1)}+M_t^{(2)}+\ldots+M_t^{(\nu)}}{t} < 1 \qquad \text{a.s.}$$

where $\nu=|C\log\log t|$ *and* C *is an arbitrary positive constant.*

REMARK 2. Theorem C was proved in |1| in the case when $\alpha=1/2$ and $L(x)$ = const. The method used in |1| can be applied in the general case.

In the present paper we prove

THEOREM 1. *Let*

$$F(x) = \exp(-(\log x)^{\gamma}) \qquad (x\geq 1,\ 0<\gamma<1). \tag{1.2}$$

Then with probability one for all t *big enough we have*

$$(1.3) \qquad \frac{M_t^{(1)}+M_t^{(2)}+\ldots+M_t^{(m)}}{t} \geq 1-\varepsilon_t$$

where

$$m=r \quad \underline{\text{if}} \quad \frac{r-2}{r-1} \leq \gamma < \frac{r-1}{r} \qquad (r=2,3,\ldots),$$

$$\varepsilon_t \leq \exp\left(-\frac{(\log t)^\beta}{(\log\log t)^3}\right)$$

and

$$\beta = \frac{r-1}{r} - \gamma \quad \underline{\text{if}} \quad \frac{r-2}{r-1} < \gamma < \frac{r-1}{r} \qquad (r=2,3,\ldots).$$

REMARK 3. Pruitt (1987) proved that (1.3) holds with $\varepsilon_t \to 0$ $(t\to\infty)$ without giving any concrete rate of convergence.

Darling ([2]) proposed to investigate the properties of the sequence $\{M(S_n)\}$. In [3] we already observed that having a description of $M(t)$ one can get some results on the behaviour of $M(S_n)$. For example we have proved: if $F(x)$ satisfyies the condition of Theorem A then we also have

$$\liminf_{n\to\infty} \frac{M(S_n)}{S_n} \log\log S_n = \beta$$

with the same β as in Theorem A.

Now we formulate some more exact results on the connection between $M(t)$ and $M(S_n)$.

THEOREM 2. *If*

$$\liminf_{t\to\infty} t^{-1} M(t) = \Theta \quad \underline{\text{a.s.}} \quad (0 < \Theta \leq 1/2)$$

then

$$(1.4) \qquad \liminf_{n\to\infty} S_n^{-1} M(S_n) = \psi = \frac{\Theta}{1-\Theta} \qquad \underline{\text{a.s.}}$$

<u>If</u>

$$(1.5) \qquad \liminf_{t\to\infty} t^{-1} f(t) M(t)=1 \qquad \underline{\text{a.s.}}$$

<u>with a function</u> $f(t) \nearrow \infty$ <u>then</u>

$$(1.6) \qquad \liminf_{n\to\infty} S_n^{-1} f(S_n) M(S_n)=1 \qquad \underline{\text{a.s.}}$$

REMARK 4. Observe that if $\Theta=1/2$ then $\psi=1$. Hence as we told in Remark 1 $\liminf_{t\to\infty} t^{-1} M(t)\leq 1/2$ a.s. however if $\liminf_{t\to\infty} t^{-1} M(t)=1/2$ a.s. then $\lim_{n\to\infty} S_n^{-1} M(S_n)=1$ a.s.

2. PROOF OF THEOREM 1.

Let

$$N=N(t) = \left[C \operatorname{loglog} t \exp((\log t)^{\gamma})\right] \qquad (C>1).$$

Then

$$\mathbb{P}\{\max(X_1,X_2,\ldots,X_N)< 2t\} = (1-\exp(-(\log 2t)^{\gamma}))^N \approx (\log t)^{-C}.$$

Now take $t_k=2^k$ then

$$\max(X_1,X_2,\ldots,X_{N(t_k)})\geq 2t_k \qquad \text{a.s.}$$

for all but finitely many k. Further let $t_k \leq t < t_{k+1}$ then $\max(X_1,X_2,\ldots,X_{N(t)}) \geq \max(X_1,X_2,\ldots,X_{N(t_k)}) \geq 2t_k>t$ a.s. if t is big enough. Consequently $\tau_t\leq N(t)$ a.s. if t is big enough.

Evaluate

$$p(t) = \mathbb{P}\{\frac{t}{f(t)} \leq X_1 \leq 2t\} = \exp(-(\log \frac{t}{f(t)})^{\gamma}) - \exp(-(\log 2t)^{\gamma}).$$

Assuming that

$$(2.1) \qquad (\log\log t)\ (\log t)^{\gamma-1} \log f(t) \to 0 \qquad (t\to\infty)$$

we have

$$(\log t - \log f(t))^{\gamma} = (\log t)^{\gamma} - (\gamma+o(1))(\log t)^{\gamma-1}\log f(t)$$

and

$$p(t) = \exp(-(\log t)^{\gamma}+(\gamma+o(1))(\log t)^{\gamma-1} \log f(t)) - \exp(-(\log 2t)^{\gamma}) =$$

$$= \exp(-(\log t)^{\gamma}) [\exp((\gamma+o(1))(\log t)^{\gamma-1}\log f(t)) -$$

$$- \exp(-(\gamma+o(1))(\log t)^{\gamma-1}\log 2)] =$$

$$= \exp(-(\log t)^{\gamma})(\gamma+o(1))(\log t)^{\gamma-1}(\log f(t)+\log 2).$$

Let

$$\nu=\nu(t)=N(t)p(t) \approx C\gamma(\log\log t)(\log t)^{\gamma-1}(\log f(t)+\log 2)\to 0 \qquad (t\to\infty),$$

$$\xi_t=\#\{i:\ i\leq N(t),\ \frac{t}{f(t)} \leq X_i \leq t\}$$

and

$$\xi_t^*=\#\{i:\ i\leq N(2t),\ \frac{t}{f(t)} \leq X_i \leq 2t\}.$$

Then we have

$$\mathbb{P}\{\xi_t^*>k\}=1 - \sum_{j=0}^{k} \binom{N}{j} p^j(1-p)^{N-j} \approx \binom{N}{k+1} p^{k+1} \approx \frac{\nu^{k+1}}{(k+1)!}$$

$$\approx \frac{(C\gamma)^{k+1}}{(k+1)!} [(\log\log t)(\log t)^{\gamma-1} (\log f(t)+\log 2)]^{k+1}.$$

From now on we assume that $\gamma<1/2$. Observe that having this condition the choice

(2.2) $$\log f_1(t) = (\log t)^{1/2-\gamma} (\log\log t)^{-2}$$

satisfies condition (2.1). Hence with $f=f_1$ we have

$$\mathbb{P}\{\xi_t^* > k\} \approx \frac{(C\gamma)^{k+1}}{(k+1)!} (\log\log t)^{-(k+1)} (\log t)^{-(k+1)/2}$$

Let $t_j=2^j$. Then $\xi_{t_j}^* \leq 1$ a.s. for all but finitely many j. Now take $t_j \leq t < t_{j+1}$. Since $N(t) \leq N(2t_j)$ and

$$\frac{t_j}{f_1(t_j)} \leq \frac{t}{f_1(t)} < t < 2t_j$$

we obtain that

$$\xi_t \leq \xi_{t_j}^* \leq 1 \qquad \text{a.s.}$$

if t is big enough. Since $\tau_t \leq N(t)$ we have that $S_{N(t)} \geq t$. Clearly the the sum of those elements of the sequence $X_1, X_2, \ldots, X_{\tau_t}$, $t-S_{\tau_t}$ which are smaller than $t/f_1(t)$ is smaller than

(2.3) $$N \frac{t}{f_1(t)} \leq Ct\log\log t \exp((\log t)^{\gamma}) \exp\left(-\frac{(\log t)^{1/2-\gamma}}{(\log\log t)^2}\right) \leq$$

$$\leq t \exp\left(-\frac{(\log t)^{1/2-\gamma}}{(\log\log t)^3}\right)$$

provided that $\gamma < 1/4$ and t is big enough. Hence Theorem 1 is proved for $\gamma < 1/4$.

In order to study the case $1/4 \leq \gamma < 1/2$ let $\log f_2(t) = (\log t)^{\gamma} (\log\log t)^2$.

Clearly this choice also satisfies (2.1). Consequently we have

$$\mathbb{P}\{\frac{t}{f_2(t)} \leqq X_1 \leqq 2t\} = \exp(-(\log t)^{\gamma})(\log t)^{2\gamma-1}(\log\log t)^2 \;(\gamma+o(1)).$$

Following the method used in case $0 < \gamma < 1/4$ we obtain

$$\xi_t^{(2)} \leqq k \qquad\qquad \text{a.s.}$$

if t is big enough where

$$\xi_t^{(2)} = \#\{i:\ i \leqq N(t),\ \frac{t}{f_2(t)} \leqq X_i \leqq t\}$$

and k is the smallest integer for which $(k+1)(1-2\gamma)>1$. Clearly the sum of those elements of the sequence $X_1, X_2, \ldots, X_{\tau_t}$, $t-S_{\tau_t}$ which are smaller than $t/f_2(t)$ is smaller than

$$N\frac{t}{f_2(t)} \leqq Ct\log\log t\ \exp((\log t)^{\gamma})\ \exp(-(\log t)^{\gamma}(\log\log t)^2) \leqq$$

$$\leqq t\exp(-(\log t)^{\gamma}\log\log t) \leqq t\exp(-(\log t)^{1/2-\gamma}). \tag{2.4}$$

Let

$$f_3(t) = k\exp((\log t)^{1/2-\gamma}(\log\log t)^{-2}).$$

Then (2.1) holds and repeating the above given proof we obtain $\xi_t^{(3)} \leqq 1$ a.s. if t is big enough where

$$\xi_t^{(3)} = \#\{i:\ i \leqq N(t),\ \frac{t}{f_3(t)} \leqq X_i \leqq t\}.$$

Hence we have: the number of the elements of the sample $X_1, X_2, \ldots, X_{\tau_t}$, $t-S_{\tau_t}$ lying

in $[0, t/f_2(t)]$ is at most N,

in $[t/f_2(t), t/f_3(t)]$ is at most k,

in $[t/f_3(t),\ t]$ is at most 1,

the sum of the sample elements

in $[0,\ t/f_2(t)]$ is at most $\frac{Nt}{f_2(t)} \leqq t\ \exp(-(\log t)^{1/2-\gamma})$

in $[t/f_2(t),\ t/f_3(t)]$ is at most $\frac{kt}{f_3(t)} \leqq t\ \exp(-(\log t)^{1/2-\gamma}(\log\log t)^{-2})$.

Hence we have the Theorem for $1/4<\gamma<1/2$.

Now we sketch the proof for $1/2<\gamma<2/3$. Let $f_4(t)=f_2(t)$. Then

$$\mathbb{P}\{\frac{t}{f_4(t)} \leqq X_1 \leqq t\} = \exp(-(\log t)^{\gamma})\exp((\gamma+o(1))(\log\log t)^2(\log t)^{2\gamma-1})$$

and by the law of large numbers

$$\xi_t^{(4)} = \xi_t^{(2)} \approx C\log\log t\ \exp((\gamma+o(1))(\log\log t)^2(\log t)^{2\gamma-1}).$$

Let $\log f_5(t) = (\log t)^{2/3-\gamma}(\log\log t)^{-2}$. This choice satisfies (2.1) and we obtain

$$\xi_t^{(5)} = \#\{i:\ i\leqq N(t),\ \frac{t}{f_5(t)} \leqq X_i \leqq t\} \leqq 2 \qquad \text{a.s.}$$

if t is big enough. Assuming that $1/2<\gamma\leqq 5/9$ we obtain: the number of the elements of the sample $X_1, X_2, \ldots, X_{\tau_t}$, $t-S_{\tau_t}$ lying

in $[0,\ t/f_4(t)]$ is at most N,

in $[t/f_4(t),\ t/f_5(t)]$ is at most $C\log\log t\ \exp((\gamma+o(1))(\log\log t)^2 (\log t)^{2\gamma-1})$,

in $[t/f_5(t),\ t]$ is at most 2,

the sum of the sample elements

in $[0,\ t/f_4(t)]$ is at most $\frac{Nt}{f_4(t)} \leqq t\ \exp(-(\log t)^{\gamma}) \leqq t\ \exp(-(\log t)^{2/3-\gamma})$

in $[t/f_4(t),\ t/f_5(t)]$ is at most $Ct\log\log t\ \exp((\gamma+o(1))(\log\log t)^2 (\log t)^{2\gamma-1})(f_5(t))^{-1} \leqq t\ \exp(-(\log t)^{2/3-\gamma} (\log\log t)^{-3})$.

Hence we have Theorem 1 for $1/2 < \gamma \leqq 5/9$. In order to investigate the case $5/9 < \gamma < 2/3$ let $\log f_6(t) = (\log t)^{2\gamma-1}(\log\log t)^3$. This choice satisfies (2.1) and we obtain

$$\xi_t^{(6)} = \#\{i:\ i \leqq N(t),\ \frac{t}{f_6(t)} \leqq X_i \leqq t\} \leqq k$$

where k is the smallest integer for which $(k+1)(2-3\gamma) > 1$.
Further let $\log f_7(t) = (\log t)^{2/3-\gamma}(\log\log t)^{-2}$. Then we obtain

$$\xi_t^{(7)} = \#\{i:\ i \leqq N(t),\ \frac{t}{f_7(t)} \leqq X_i \leqq t\} \leqq 2.$$

Consequently the number of the elements of the sample
$X_1, X_2, \ldots, X_{\tau_t},\ t-S_{\tau_t}$ lying
in $[0,\ t/f_4(t)]$ is at most N,
in $[t/f_4(t),\ t/f_6(t)]$ is at most $C\log\log t \exp((\gamma+o(1))(\log\log t)^2$

$$(\log t)^{2\gamma-1}),$$

in $[t/f_6(t),\ t/f_7(t)]$ is at most k,
in $[t/f_7(t),\ t]$ is at most 2
the sum of the sample elements
in $[0,\ t/f_4(t)]$ is at most $t\exp(-(\log t)^{2/3-\gamma})$
in $[t/f_4(t),\ t/f_6(t)]$ is at most $Ct\log\log t \exp((\gamma+o(1))(\log\log t)^2$

$$(\log t)^{2\gamma-1})(f_6(t))^{-1} \leqq t\exp(-(\log t)^{2\gamma-1}) \leqq$$

$$\leqq t\exp(-(\log t)^{2/3-\gamma}).$$

Hence we have Theorem 1 for $5/9 < \gamma < 2/3$.

The general case can be treated similarly the details will be omitted. We only mention that the critical values for γ will be $(r-2)/(r-1) + 1/r^2(r-1)$ $(r=2,3,\ldots)$.

3. PROOF OF THEOREM 2.

Let $S_n \leqq t < S_{n+1}$ then

$$t^{-1}M(t) = \begin{cases} t^{-1}M(S_n) & \text{if } S_n \leq t \leq \min\{S_{n+1}, S_n+M(S_n)\} \\ t^{-1}(t-S_n) & \text{if } \min\{S_{n+1}, S_n+M(S_n)\} \leq t \leq S_{n+1}. \end{cases}$$

Consequently

$$(3.1) \qquad t^{-1}M(t) \geq (S_n+M(S_n))^{-1} M(S_n) = \frac{S_n^{-1} M(S_n)}{1+S_n^{-1} M(S_n)} \geq \frac{M(S_n)}{2S_n}.$$

Clearly we also have

$$(3.2) \qquad \liminf_{t\to\infty} t^{-1} M(t) \leq \liminf_{n\to\infty} S_n^{-1} M(S_n).$$

Assume that

$$(3.3) \quad \liminf_{t\to\infty} t^{-1} M(t)=\Theta \text{ and } \liminf_{n\to\infty} S_n^{-1} M(S_n)=\psi \qquad (0<\Theta<1).$$

Then by (3.1) and (3.2) we have

$$\frac{\psi}{1+\psi} \leq \Theta \leq \psi$$

i.e.

$$(3.4) \qquad \Theta \leq \psi \leq \frac{\Theta}{1-\Theta}.$$

Since if $X_{n+1} < M(S_n)$ then

$$S_{n+1}^{-1} M(S_{n+1}) = S_{n+1}^{-1} M(S_n) < S_n^{-1} M(S_n).$$

Consequently for each m for which $S_m^{-1}M(S_m)$ is close to ψ one can wait for the next n for which $X_{n+1} > M(S_m)$ i.e. for any $\varepsilon > 0$ and for almost all $\omega \in \Omega$ there exist infinitely many positive integers n such that

$$|S_n^{-1} M(S_n) - \psi| \leq \varepsilon \text{ and } X_{n+1} > M(S_n).$$

Consider such an n and let $t=S_n+M(S_n)< S_{n+1}$. Then

$$t^{-1}M(t) = \frac{M(S_n+M(S_n))}{S_n+M(S_n)} = \frac{M(S_n)}{S_n+M(S_n)} = \frac{S_n^{-1}\ M(S_n)}{1+S_n^{-1}\ M(S_n)}$$

and

$$\Theta \leq \frac{\psi}{1+\psi} \text{ i.e. } \psi \geq \frac{\Theta}{1-\Theta}.$$

Our last inequality combined with (3.4) implies

$$\psi = \frac{\Theta}{1-\Theta}$$

and we have (1.4).

Assume (1.5) and let

$$\liminf_{n \to \infty} S_n^{-1}\ f(S_n)\ M(S_n) = A \geq 1 \qquad \text{a.s.}$$

Then there exists a random sequence $\{n_k\}$ such that

$$S_{n_k}^{-1}\ f(S_{n_k})\ M(S_{n_k}) \to A \qquad \text{a.s.} \qquad (k \to \infty).$$

Consider a point t in $S_{n_k} \leq t \leq S_{n_k+1}$ then by (3.1) we have

$$t^{-1}M(t)\ f(t) \geq \frac{S_{n_k}^{-1}\ M(S_{n_k})\ f(S_{n_k})}{1+S_{n_k}^{-1}\ M(S_{n_k})} \to A$$

what implies that $A \leq 1$. Hence we have (1.6) and the proof of Theorem 2 is complete.

REFERENCES

[1] E.Csáki, P.Erdös and P.Révész (1985). On the length of longest excursion. Z.Wahrsch. Verw. Geb. 68, 365-382.

[2] D.A.Darling (1952). The influence of the maximum term in addition of independent random variables. Trans. Amer. Soc. 73, 95-107.

[3] P.Révész and E.Willekens (1988). On the maximal distance between two renewal epochs. Stochastic Processes and their Applications.

[4] W.E.Pruitt (1987). The contribution to the sum of the summand of maximum modulus. The Annals of Probability 15, 885-896.

A SURVEY ON STRONG APPROXIMATION TECHNIQUES IN CONNECTION WITH RECORDS

D. Pfeifer and Y.-S. Zhang

University of Oldenburg and Beijing University of Iron and Steel

__Abstract.__ The intention of the paper is to review various techniques for the strong approximation of record times, inter-record times and record values (essentially by Poisson and Wiener processes) which have been considered in the recent years by different authors. Besides the iid case, we also discuss which of the methods described are suited to treat corresponding problems in more general (non-iid) settings.

1. Introduction.

Let $\{X_n; n \geq 1\}$ be an iid sequence of random variables (r.v.'s) on some probability space $(\Omega, \mathcal{A}, P)$, with a joint continuous cumulative distribution function F. *Record times* $\{U_n; n \geq 0\}$, *inter-record times* $\{\Delta_n; n \geq 0\}$ and *record values* $\{X_{U_n}; n \geq 0\}$ for this sequence are recursively (and by our assumptions, a.s. well-)defined by

$$(1) \qquad U_0 = \Delta_0 = 1,\ U_{n+1} = \inf\{k; X_k > X_{U_n}\},\ \Delta_{n+1} = U_{n+1} - U_n \quad (n \geq 0).$$

These sequences, related to the partial extremes of the original sequence, have been of increasing interest since their first exploration by Chandler (1952). Several survey articles have been published since then, e.g. by Glick (1978) or Nevzorov (1988), to mention some. Besides structural properties of these sequences, asymptotic features (in various meanings) have been the subject of research very early, pointing out relationships with the law of large numbers, the central limit theorem and the law of the iterated logarithm for iid sequences of r.v.'s. In particular, it was shown by several authors that

$$(2) \qquad \frac{1}{n} Z_n \to 1 \text{ a.s. } (n \to \infty)\ ,\quad n^{-1/2}(Z_n - n) \xrightarrow{\mathcal{D}} N(0,1) \text{ and } \lim \genfrac{}{}{0pt}{}{\sup}{\inf} \frac{Z_n - n}{\sqrt{2n \log\log n}} = \pm 1 \text{ a.s.}$$

where $\xrightarrow{\mathcal{D}}$ means convergence in distribution, and Z_n may be replaced by any of the r.v.'s $\log U_n$, $\log \Delta_n$ or $-\log(1-F(X_{U_n}))$, $n \geq 1$. (For further details, c.f. e.g. Rényi (1962), Neuts (1967), Strawderman and Holmes (1970), and Shorrock (1972a, 1972b).) Due to the fact that for the proofs of the forementioned results, mostly different techniques were used, tailored to the specific situation, there was a need for a unifying approach explaining the similarity of the asymptotic behaviour of the three different random sequences in (2). It turned out that certain strong approximation techniques in the spirit of Komlós, Major and Tusnády (1976) were suitable tools for such an explanation. Their main result states that under certain regularity conditions (existence of exponential moments) it is possible to approximate a partial sum process $S_n = \sum_{k=1}^{n} \xi_k$ (with an iid sequence $\{\xi_k\}$) by a partial sum process $T_n = \sum_{k=1}^{n} \eta_k$ where $\{\eta_k\}$ is a sequence of iid normally distributed r.v.'s with the same mean and variance as the $\{\xi_k\}$ sequence, on the same probability space (if rich enough to carry sufficiently many iid sequences) such that

$$(3) \qquad |S_n - T_n| = O(\log n) \text{ a.s.} \quad (n \to \infty).$$

In this paper, such techniques for the strong approximation of record times, inter-record times and record values are reviewed. The possibility of extending some of the results to more general than iid cases is also discussed.
However, before doing so, it will be necessary to expose some well-known structural properties of the sequences involved.

Theorem 1. (Rényi (1962)) Under the conditions specified above, the record time sequence $\{U_n\}$ forms a homogeneous Markov chain (MC) with transition probabilities given by

$$(4) \qquad P(U_{n+1} > k \mid U_n = j) = \frac{j}{k}, \quad 1 \leq j \leq k,\ n \geq 0.$$

Theorem 2. (Shorrock (1972a, 1972b)) The inter-record time sequence $\{\Delta_n\}$ is conditionally independent given the sequence of record values, with a (conditional) geometric distribution of the form

$$(5) \qquad P(\Delta_n = k \mid X_{U_{n-1}}) = \{1-F(X_{U_{n-1}})\}F^{k-1}(X_{U_{n-1}}) \quad \text{a.s.} \quad (n,k \geq 1).$$

Furthermore, the sequence $\{-\log(1-F(X_{U_n}))\}$ forms the arrival time sequence of a unit rate Poisson process (i.e., has independent exponentially distributed increments with unit mean).

From Theorem 2 it follows immediately that the record value sequence itself forms the arrival time sequence of such a Poisson process if the underlying distribution is exponential with unit mean.

In what follows we shall discuss in more detail the different strong approximation approaches for record times, inter-record times and record values, both individually and also jointly.

2. The conditional independence approach.

Deheuvels (1982, 1983) proved that it is possible to define an iid sequence $\{Y_n\}$ on the same probability space (Ω, A, P) (if rich enough) such that

$$(6) \quad \Delta_n = \text{int}\{Y_n/-\log(1-\exp(-S_n))\}+1 \quad \text{with } S_n = -\log(1-F(X_{U_{n-1}})),\ n \geq 1,$$

where $\{Y_n\}$ is exponentially distributed with unit mean, and independent of the record value sequence. This reflects precisely the conditional independence property of Theorem 2 (note that a geometrically distributed r.v. can be generated from an exponentially distributed one by appropriate rounding off). By a suitable Taylor expansion, relation (6) leads to relation

$$(7) \quad \log \Delta_n = \log Y_n + S_n + o(1) \quad \text{a.s.} \quad (n \to \infty)$$

which in turn proves relation (2) for inter-record times via (3), jointly with the transformed record value sequence (i.e. with the same strong approximand $\{T_n\}$ from the Komlós-Major-Tusnády construction). This is due to the fact that by a simple Borel-Cantelli argument,

$$(8) \quad \log Y_n = O(\log n) \quad \text{a.s.} \quad (n \to \infty).$$

(A slightly refined expansion of (7) is given in Pfeifer (1985).)

Unfortunately, this approach does not immediately lead to a nice strong approximation of record times from where a direct proof of (2) could be read off easily (cf. Deheuvels (1982)). The following approach provides such an approximation.

3. The Markov chain approach.

This approach is based on Theorem 1, expanding ideas of Williams (1973) and Westcott (1977). The following result (Pfeifer (1987)) is a key to the procedure.

Theorem 3. Let $\{M_n\}$ be a homogeneous MC with conditional cumulative distribution function $F(.\mid.)$. Let $F_-(.\mid.)$ denote the corresponding left-continuous version. Then there exists an iid sequence of uniformly distributed r.v.'s (over (0,1)) on the same probability space (if rich enough), $\{V_n\}$, say, such that

$$(9) \quad V_{n+1} = (1-W_{n+1})F(M_{n+1} \mid M_n) + W_{n+1}F_-(M_{n+1} \mid M_n) \qquad (n \geq 0)$$

where $\{W_n\}$ is also iid uniformly distributed, and independent of the MC $\{M_n\}$.

Using relation (4), this translates in our case into

$$(10) \quad V_{n+1} = (1-W_{n+1})\frac{U_n}{U_{n+1}} + W_{n+1}\frac{U_n}{U_{n+1}-1} \qquad (n \geq 0)$$

using the fact that with V_n also $1-V_n$ is uniformly distributed. Taking logarithms in (10) one obtains the following strong approximation result (Pfeifer (1987)).

Theorem 4. There exists on the same probability space $(\Omega,\mathcal{A},P)$ (if rich enough) an arrival time sequence $\{S_n^*\}$ of a unit-rate Poisson process and a non-negative r.v. Z possessing all positive moments, with mean $E(Z) = 1-C$ ($C = .577216$ denoting Euler's constant), such that

(11) $\quad Z$ and $\{(S_n^*-n)/\sqrt{n}\}$ are asymptotically independent

(12) $\quad \log U_n = Z+S_n^*+o(1)$ a.s., $\log \Delta_n = Z+S_n^*+\log(1-\exp(S_{n-1}^*-S_n^*))+o(1)$ a.s. $(n \to \infty)$

where $\{-\log(1-\exp(S_{n-1}^*-S_n^*))\}$ again is an iid sequence of exponentially distributed r.v.'s with unit mean.

This again proves relation (2) via (3), this time jointly for record times and inter-record times.

It should be pointed out that the Poisson process in Theorem 4 does not coincide with the one in Deheuvels' approach. Actually, we have

$$(13) \quad S_n^* = \sum_{k=1}^{n} -\log V_k$$

with $\{V_k\}$ as in (10).

We should like to mention that with the approaches 2. and 3., also moment estimations of the logarithms of record times and inter-record times are readily obtained (cf. Pfeifer (1984a), and Nevzorov (1988)).

Recently, Deheuvels (1988) has extended the one-dimensional MC approach to the two-dimensional case (record times and record values jointly form a MC, too). With this approach, he was able to extend the strong approximation by a Poisson process to all three sequences in (2) simultaneously.

Instead of expressing everything in terms of Poisson processes, the Komlós-Major-Tusnády construction also allows for a formulation in terms of (standard) Wiener processes. Namely, if $\{W(t);t \geq 0\}$ stands for such a process, the forementioned results show that it is possible to establish on the same probability space (if rich enough) the following strong relationship, which is even closer to (2):

$$(14)\quad \log U_n = n + W(n) + O(\log n) \quad \text{a.s. } (n \to \infty)$$

$$\log \Delta_n = n + W(n) + O(\log n) \quad \text{a.s. } (n \to \infty)$$

$$-\log(1-F(X_{U_n})) = n + W(n) + O(\log n) \quad \text{a.s. } (n \to \infty)$$

(see Deheuvels (1988), and Pfeifer (1986).)

4. The embedding approach.

Going back to Resnick (1973, 1974, 1975) and Resnick and Rubinovitch (1973), the basic idea here is an appropriate embedding of the partial maxima sequence $\{\max(X_1,\ldots,X_n);\ n \geq 1\}$ derived from $\{X_n\}$ into so-called *extremal processes*. Any such process $\{E(t); t>0\}$ is a pure jump Markov process with right continuous paths and finite-dimensional marginal distributions which in our case are given by

$$(15)\quad P\Big(\bigcap_{i=1}^{k}\{E(t_i) \leq x_i\}\Big) = F^{t_1}(\min\{x_1,\ldots,x_k\}) \prod_{i=2}^{k} F^{t_i - t_{i-1}}(\min\{x_i,\ldots,x_k\})$$

for all selections $0 < t_1 < t_2 < \ldots < t_k$ of time points, and values $x_1,\ldots,x_k \in \mathbb{R}$. Such an extremal process 'interpolates' the partial maxima process in that we have

$$(16)\quad \{\max(X_1,\ldots,X_n)\} \stackrel{\mathcal{D}}{=} \{E(n)\},$$

where $\stackrel{\mathcal{D}}{=}$ means equality in distribution. The interesting point is here that the jump time sequence $\{\tau_n\}$ of the extremal process in the interval $(1,\infty)$ forms a non-homogeneous Poisson point process with intensity $1/t$, $t \geq 1$, such that the points are a.s. clustering in the intervals $(U_n - 1, U_n)$, $n \geq 1$. It follows that the surplus number S of extremal jumps over the record times is a.s. finite, with $E(S|\Sigma) = E(Z|\Sigma)$, where Σ denotes the σ-field generated by the record values (Pfeifer (1986)). Here Z is the r.v. from Theorem 4, such that we also have

$$(17)\quad E(S) = 1 - C.$$

An application of the log function now shows that

$$(18)\quad \log U_n = \log \tau_{n+S} + o(1) = \log \tau_n + O(\log n) \quad \text{a.s. } (n \to \infty),$$

where now $\{\log \tau_n\}$ forms a homogeneous Poisson point process on $(0,\infty)$ with unit intensity. This again proves (2) via (3), for $Z_n = \log U_n$.

5. The time change approach.

It is well-known that if the underlying distribution is doubly-exponential, then $\max(X_1,...,X_n) - \log n$, $n \geq 1$, is also doubly-exponentially distributed (see e.g. Leadbetter et al. (1983)). The same holds true if the indices n are replaced by the random times U_n, i.e. $X_{U_n} - \log U_n$, $n \geq 1$, is also doubly-exponentially distributed (Pfeifer (1986)). Since doubly-exponential and exponential distributions are tail-equivalent, and any doubly-exponentially distributed sequence is $O(\log n)$ a.s., it follows that, in general,

(19) $-\log(1-F(X_{U_n})) - \log U_n$ is asymptotically doubly exponentially distributed

(20) $-\log(1-F(X_{U_n})) - \log U_n = O(\log n)$ a.s. $(n \to \infty)$.

Since by (12), we always have $\log U_n - \log \Delta_n = O(\log n)$ a.s. $(n \to \infty)$, relation (20) is an elegant tool for again proving (2) (or (14), resp.), for all three sequences simultaneously.

6. Generalizations and open problems.

Here we shall shortly discuss which of the above approaches are suitable for treating more general situations than the iid case.

6.1. The Markov case. Not much is known in the case where the underlying r.v.'s form a (homogeneous, say) MC; cf. Biondini and Siddiqui (1975). However, it was shown in Pfeifer (1984b) that in this case, the corresponding inter-record times are still conditionally independent given the σ-field of record values, such that the conditional independence approach 2. is potentially applicable. The main problem here lies in the fact that neither the general structure of the record value process is completely clear, nor is the (conditional) waiting time distribution between successive record values geometric in general. Research in this direction is in progress.

6.2. The non-homogeneous record model. Here one considers the case that after the occurence of a new record value, the underlying distribution is allowed to change, keeping however the independence assumption (Pfeifer (1982a, 1982b)). Here, too, the conditional independence of inter-record times given the record values is preserved, the (conditional) waiting time distributions between records being still (but possibly different) geometric distributions. Here the record value process is connected with general pure birth processes, such that strong approximations in the spirit of approach 2. become available. Some possibilities for this procedure are outlined in Pfeifer (1984c).

Unfortunately, the Markov property for record times goes lost in general for this model, such that approach 3. is not applicable.

6.3. Nevzorov's record model. Here one considers the case that the underlying distributions are of the form $F_n = F^{\alpha_n}$, where $\{\alpha_n\}$ is a sequence of positive real numbers with $\sum_{n=1}^{\infty} \alpha_n = \infty$ (see Nevzorov (1988) and further references therein). Letting $A(n) = \sum_{k=1}^{n} \alpha_k$, $n \geq 1$ it can be proved that the corresponding record times again form a homogeneous MC with transition probabilities given by

$$(21) \qquad P(U_{n+1} > k \mid U_n = j) = \frac{A(j)}{A(k)}, \qquad 1 \leq j \leq k,\ n \geq 0.$$

Hence the Markov chain approach 3. is applicable, and shows that an analogue of Theorem 4 is valid for $\log A(U_n)$ if the conditions $\sum_{k=1}^{\infty} (\alpha_k/A(k))^2 < \infty$, $\sum_{k=1}^{\infty} (\alpha_k/A(k)) = \infty$ are met (see Pfeifer (1988) and Zhang (1988)). Zhang also proved that an embedding approach with non-homogeneous extremal processes works under these conditions, and that similarly the surplus number S of extremal jumps over the record times is a.s. finite with $E(S) \leq \sum_{k=1}^{\infty} (\alpha_k/A(k))^2$. (Similar embeddings have been considered earlier by Ballerini and Resnick (1985), however with a different emphasis.)
Hence for Nevzorov's model, relation (2) is valid for the sequence $Z_n = \log A(U_n)$.

Finally, we should like to mention that a time change approach similar to 5. also applies here, however for the specific situation that F is doubly-exponential. By imitation of the proof of Theorem 3 in Pfeifer (1986), it can be seen that here still $X_{U_n} - \log A(U_n)$ is doubly-exponentially distributed, hence again

$$(22) \qquad X_{U_n} = \log A(U_n) + O(\log n) \quad \text{a.s. } (n \to \infty).$$

Note that the latter relation also holds without the above-mentioned regularity conditions. For instance, if $\alpha_n = e^{cn}$, $n \geq 1$ for some $c > 0$, then $\log A(n) = nc - \log(1-e^{-c}) + O(e^{-nc})$ $(n \to \infty)$, hence we have here

$$(23) \qquad X_{U_n} = cU_n + O(\log n) \quad \text{a.s.} \quad (n \to \infty)$$

with

(24) $$U_n = \sum_{k=1}^{n} I_k + O(1) \quad \text{a.s.} \quad (n \to \infty)$$

where the I_k are iid with $P(I_k=0) = 1-P(I_k=1) = e^{-c}$.

This gives immediately rise to results like (2), with the proper normalizations, via (3), for U_n and X_{U_n}.

References

R. Ballerini and S.I. Resnick (1985): Records from improving populations. J. Appl. Prob. 22, 487 - 502.

R.M. Biondini and M.M. Siddiqui (1975): Record values in Markov sequences. In: Statistical Inference and Related Topics, Vol. 2, 291 - 352. Ac.Press.

K.N. Chandler (1952): The distribution and frequency of record values. J. Roy. Statist. Soc. Ser. B, 14, 220 - 228.

P. Deheuvels (1982): Strong approximation in extreme value theory and applications. Coll. Math. Soc. János Bolyai 36. In: Limit Theorems in Probability and Statistics, 369 - 404. North Holland.

________ (1983): The complete characterization of the upper and lower class of the record and inter-record times of an i.i.d. sequence. Z. Wahrscheinlichkeitsth. verw. Geb. 62, 1 - 6.

________ (1988): Strong approximations of k-th records and k-th record times by Wiener processes. Prob. Th. Rel. Fields 77, 195 - 209.

N. Glick (1978): Breaking records and breaking boards. Amer. Math. Monthly 85, 2 - 26.

J. Komlós, P. Major, and G. Tusnády (1976): An approximation of partial sums of independent rv's, and the sample df. II. Z. Wahrscheinlichkeitsth. verw. Geb. 34, 33 - 58.

M.R. Leadbetter, G. Lindgren, and H. Rootzén (1983): Extremes and Related Properties of Random Sequences and Processes. Springer.

M.F. Neuts (1967): Waiting times between record observations. J. Appl. Prob. 4, 206 - 208.

V.B. Nevzorov (1988): Records. Th. Prob. Appl. 32, 201 - 228.

D. Pfeifer (1982a): Characterizations of exponential distributions by independent non-stationary record increments. J. Appl. Prob. 19, 127 - 135. Corr., 906.

________ (1982b): The structure of elementary pure birth processes. J. Appl. Prob. 19, 664 - 667.

________ (1984a): A note on moments of certain record statistics. Z. Wahrscheinlich-keitsth. verw. Geb. 66, 293 - 296.

________ (1984b): A note on random time changes of Markov chains.

Scand. Act. J. , 127 - 129.

D. Pfeifer (1984c): Limit laws for inter-record times from non-homogeneous record values. J. Org. Behavior Stat. 1, 69 - 74.

_______ (1985): On the rate of convergence for some strong approximation theorems in extremal statistics. European Meeting of Statisticians, Marburg 1984. In: Stat. & Decisions Supp. Iss. No. 2, 99 - 103.

_______ (1986): Extremal processes, record times and strong approximation. Pub. Inst. Stat. Univ. Paris XXXI, 47 - 65.

_______ (1987): On a joint strong approximation theorem for record and inter-record times. Th. Prob. Rel. Fields 75, 212 - 221.

_______ (1988): Extremal processes, secretary problems and the 1/e-law. To appear in: J. Appl. Prob.

A. Rényi (1962): Théorie des éléments saillants d'une suite d'observations. Coll. Comb. Meth. in Prob. Th., August 1 - 10, 1962, Math. Inst., Aarhus Univ., Denmark, 104 - 117.

S.I. Resnick (1973): Extremal processes and record value times. J. Appl. Prob. 10, 863 - 868.

_________ (1974): Inverses of extremal processes. Adv. Appl. Prob. 6, 392 - 406.

_________ (1975): Weak convergence to extremal processes. Ann. Prob. 3, 951 - 960.

S.I. Resnick and M. Rubinovitch (1973): The structure of extremal processes. Adv. Appl. Prob. 5, 287 - 307.

R.W. Shorrock (1972a): On record values and record times. J. Appl. Prob. 9, 316 - 326.

_______ (1972b): A limit theorem for inter-record times. J. Appl. Prob. 9, 219 - 223.

W.E. Strawderman and P.T. Holmes (1970): On the law of the iterated logarithm for inter-record times. J. Appl. Prob. 7, 432 - 439.

M. Westcott (1977): A note on record times. J. Appl. Prob. 14, 637 - 639.

D. Williams (1973): On Rényi's record problem and Engel's series. Bull. London Math. Soc. 5, 235 - 237.

Y.S. Zhang (1988): Strong approximations in extremal statistics (in German). Ph.D. Thesis, Technical University Aachen.

Self-Similar Random Measures, Their Carrying Dimension, and Application to Records

U. Zähle
Sektion Mathematik, Friedrich-Schiller-Universität Jena
6900 Jena, GDR

Abstract. What structure possesses the set $\Xi = \{t: X_t = \sup\{X_s : s \le t\}\}$ of instants at which a random process X on $T = [0,\infty)$ takes its record values. A simple answer in terms of the Hausdorff dimension can be given using an axiomatic approach to statistically self-similar random measures. Let $0 < H \le 1$ and X_t be a continuous "record-H-self-similar" process with $X_0 = 0$. Define a "natural" random measure ξ, concentrated on Ξ, by $\xi[0,t) = \sup\{X_s : s \le t\}$. ξ is H-self-similar, meaning that (a) $\xi[0,t) = r^H \xi[0,t/r)$, (b) ξ is distributed according to a Palm distribution. The theory of self-similar random measures tells then that ξ is carried by sets of Hausdorff dimension H.

1. Introduction

Extreme values can be studied from different points of view. The behaviour of the maximum in a long random sequence is the very topic of extreme value theory. Other aspects are the behaviour of the cumulative maxima, or records, and the time at which the records occur. They are investigated, e.g., in [1] and [2] - to give some recent references.

Different questions arise when we pass over to random processes $(X_t)_{t\ge0}$ with continuous time. Here both the set

$$\Xi = \{t \ge 0: X_s < X_t,\ 0 \le s < t\}$$

of time instants where records occur and the part of the graph over Ξ become uncountable. In case of differentiable processes Ξ is piecewise connected. For continuous non-differentiable processes the structure of Ξ is more complicated. Usually, Ξ as well as the graph $\Gamma = \{(t,X_t): t \ge 0\}$ and the zero set $Z = \{t \ge 0: X_t = 0\}$ are fractal sets in this case.

A first (rough) characteristic describing these sets is their fractal Hausdorff-Besicovitch dimension. Remember that the family $\mathcal{H}^\alpha$,

$\alpha \geq 0$, of Hausdorff measures on $\mathbb{R}^d$ relates to any subset $B \subset \mathbb{R}^d$ a number

$$\dim B = \sup\{\alpha: \mathcal{H}^\alpha(B) = +\infty\} = \inf\{\alpha: \mathcal{H}^\alpha(B) = 0\} \in [0,d],$$

the Hausdorff-Besicovitch dimension of B. dim B reflects some aspects of the local behaviour of B. It measures the local "density", the degree of the scattering of B. So if the compact set $K \subset \mathbb{R}^d$ is the union of sets $K_1,\dots,K_N$, which are similar to K in the ratio $r < 1$, and if $K_i \cap K_j$ is empty, or at least "small" - such sets are called self-similar - then $\dim K = \ln N/\ln(1/r)$. This was made precise by Hutchinson in [4].

Self-similar fractal sets form a well behaved subclass within the family of all fractal sets. So it is not surprising to ask whether there is any self-similarity, when tackling a problem with fractal ingredients. In this paper we will show that self-similarity (section 2) applies to records (section 4), but also to graphs and zeros (section 3), of so-called self-similar processes and report the consequences for the Hausdorff dimension (section 5).

Note that the results, being of local nature, can be obtained for processes which are "local self-similar" as well.

2. Statistical Self-Similarity

The very essence of self-similarity (small parts are the "same", up to a factor, as the whole) and the connection between some self-similarity index and Hausdorff dimension - as formulated in the introduction - can be generalized to random sets. We will give a brief introduction to one such generalization.

The following scale invariance ("scaling property") of the Wiener process W_t is well-known: The process $r^{1/2}W_{t/r}$ has the same distribution as W_t, for any $r > 0$. This scaling property carries over to the random closed sets Γ (graph), Z (zero set), and Ξ (record set), e.g. $rZ \overset{d}{=} Z$, for any $r > 0$. It is well-known that Z has Hausdorff dimension 1/2. However, it is easy to construct examples showing that the scaling property $r\Sigma \overset{d}{=} \Sigma$ alone does not imply $\dim \Sigma = 1/2$. To see this, take for instance

$$\Sigma = a^\beta \bigcup_{n=-\infty}^{\infty} [a^n, ba^n]$$

with $a > 1$ and $1 \leq b < e$ fixed, and β distributed uniformly on [0,1]. Obviously, $r\Sigma \overset{d}{=} \Sigma$ for any $r > 0$. But $\dim \Sigma = 1$ or 0 depending on whether $b > 1$ or $b = 1$.

A suitable notion of statistical self-similarity has to reflect rather the fact that statistical scaling must hold with respect to "any" point of a self-similar random set.

In [10] it was pointed out that it is adequat to pass over from random sets to random measures, and that a suitable notion of self-similarity uses Palm distribution to choose "any" (typical) point of the support of the random measure as a center of scaling, for Palm distribution theory makes the notion of a typical point precise.

Given the distribution $\mathbb{Q}$ of a stationary random measure (i.e. a probability measure on the measurable space $[M,\mathcal{M}]$ of Radon measures on $\mathbb{R}$) of finite intensity

$$0 < \lambda_{\mathbb{Q}} = \int \mu[0,1)\ \mathbb{Q}(d\mu) < \infty,$$

the Palm distribution $\mathbb{Q}_0$ of $\mathbb{Q}$ is defined by

$$\mathbb{Q}_0(E) = \lambda_{\mathbb{Q}}^{-1} \iint_0^1 1_E(T_t\mu)\ \mu(dt)\ \mathbb{Q}(d\mu),\ E \in \mathcal{M},$$

where $(T_t\mu)(B) = \mu(B - t)$. This definition remains meaningful if the probability distribution $\mathbb{Q}$ is replaced by a stationary non-finite σ-finite measure $\mathbb{H}$ on $[M,\mathcal{M}]$ with finite intensity. It is well-known that $\mathbb{Q}_0$ has the probabilistic meaning of choosing a (mass) point at random from the realizations μ of $\mathbb{Q}$ and looking at μ from that point as the new origin (cf. [5], [6]).

Now we can adopt the following

2.1. Definition. A random measure η (on $\mathbb{R}$) with distribution $\mathbb{P}$ is called *D-self-similar*, if

(m.i) $\eta(\cdot) \overset{d}{=} r^D \eta(r^{-1}(\cdot))$ for any $r > 0$,

(m.ii) $\mathbb{P}$ is the Palm distribution of some non-finite σ-finite "distribution" $\mathbb{H}$. ∎

3. An Example: Level Sets

The zero set of the Wiener process is the prototyp of a "self-similar" set whatever it does mean. We will show here that the zero sets of a much larger class of processes are covered by our notation 2.1 of self-similarity.

3.1. Definition. A process $(X_t)_{t\in\mathbb{R}}$ is called *H-self-similar*, for some $H \in (0,1)$, if (p.o) $\quad X_t$ is pathwise continuous,

(p.i) $\quad X_{(\cdot)} \overset{d}{=} r^H X_{(\cdot)/r}$ for any $r > 0$,

(p.ii) $\quad X_t$ has stationary increments. ∎

(p.i) implies $X_0 = 0$. The two-sided Wiener process defined as $W_t^{(1)}$ if $t \geq 0$ and as $W_{-t}^{(2)}$ if $t \leq 0$, where $W_t^{(i)}$ are independent Wiener processes, is 1/2-self-similar. The class of self-similar-processes is very large, as can be seen from the discussion in [9].

Obviously, for a self-similar process, the scaling property (p.i) holds with X_t replaced by $Y_t = X_{t-t_0} - X_{t_0}$, for any $t_0 \in \mathbb{R}$, thus holds "with respect to any instant t_0".

For the graph $\Gamma = \{(t,X_t) : t \in \mathbb{R}\}$ of an H-self-similar process scaling holds with respect to any of its points in the following sense: For any $t_0 \in \mathbb{R}$,

$$\Gamma - (t_0, X_{t_0}) = S_{H,r}[\Gamma - (t_0, X_{t_0})],$$

where $B - (t,x) = \{(s-t, y-x) : (s,y) \in B\}$, $S_{H,r}(t,x) = (rt, r^H x)$. (Such random sets are usually called *self-affine* rather than self-similar. Note that this property of self-affinity can be formulated in terms of Γ only, since X_{t_0} is uniquely determined by Γ.)

The level sets of a process X_t, which has a sufficiently high "variation activity", carry so-called local time measures. Suppose X_t is H-self-similar and X_1 has a bounded density. Then X_t possesses a random occupation kernel $\alpha(y,B)$, $y \in \mathbb{R}$, $B \in \mathcal{B}$, meaning that

$$\int f(t,X_t)\, dt = \iint f(t,y)\, \alpha(y,dt)\, dy$$

for any measurable function $f \geq 0$, cf. [7], Theorems 21.12 and 6.4 for a detailed discussion. Especially, $\alpha(y,\{t : X_t \neq y\}) = 0$. To be short denote in the following by I any interval $[t_1,t_2)$. From the "fundamental theorem of the calculus" (e.g. [3], 2.9.8) it follows that, for almost all realizations X_t, the Radon-Nikodym derivative $\alpha(y,I)$ of $\mathcal{L}\{s \in I : X_s \in dy\}$ - $\mathcal{L}$ denotes Lebesgue measure - with respect to dy coincides for almost every y with

$$\lim_{\varepsilon \to 0+} \frac{1}{2\varepsilon} \mathcal{L}\{s \in I : y-\varepsilon < X_s < y+\varepsilon\}.$$

For (right) continuous processes with stationary increments this limit exists for any fixed y, say for $y = 0$, and almost all $X_t(\omega)$. Thus we can and will take this canonical modification in the following.

3.2. The local time measure $\alpha(0,\cdot)$ of an H-self-similar process is (1-H)-self-similar.

Proof. The random measure $\alpha(0,\cdot)$ satisfies (m.i). In fact, using (p.i),

$$\begin{aligned}\alpha(0,rI) &= \lim_{\varepsilon\to 0+} \frac{1}{2\varepsilon} \mathcal{L}\{s \in rI : -\varepsilon < X_s < \varepsilon\} \\ &= \lim_{\varepsilon\to 0+} \frac{1}{2\varepsilon} r\mathcal{L}\{s' \in I : -r^{-H}\varepsilon < X_{s'} < r^{-H}\varepsilon\} \\ &= r\, r^{-H} \lim_{\varepsilon'\to 0+} \frac{1}{2\varepsilon'} \mathcal{L}\{s' \in I : -\varepsilon' < X_{s'} < \varepsilon'\} \\ &= r^{1-H}\, \alpha(0,I).\end{aligned}$$

The distribution of $\alpha(0,\cdot)$ satisfies (m.ii) as well. To see this define $\mathbb{H}$ by

$$\mathbb{H}(E) = \int_{-\infty}^{\infty} \Pr\{\alpha(y,\cdot) \in E\}\, dy,\ E \in \mathcal{M}.$$

In other words, $\mathbb{H}$ is the (σ-finite non-finite) "distribution" of the occupation density of the realizations of the "flow of X_t", $\mathbb{H}_X$, defined by

$$\mathbb{H}_X(F) = \int_{-\infty}^{\infty} \Pr\{(X_t)-y \in F\}\, dy,$$

F an event for random processes. Since $\mathbb{H}_X$ is stationary (cf. [7], 5.2.2, for the Brownian flow and [12] for the general case), $\mathbb{H}$ is so. Finally, $\mathbb{H}$ has finite intensity and $\alpha(0,\cdot)$ has distribution $\mathbb{H}_0$. Since this example is not the actual object of this paper, the reader is referred to [12] for more details, especially for the proof of the last facts. ■

4. Records

The "record set"

$$\Xi_W = \{t > 0 : W_s < W_t,\ 0 < s < t\}$$

of the Wiener process is scale-invariant: $r\Xi_W \overset{d}{=} \Xi_W$ for any $r > 0$. Moreover, thanks to the strong Markov property, the record set has this scaling property starting at "any" of its points (it is - in the same way as the zero set - "regenerative"). So Ξ_W is likely to be self-similar. We will show in this section that record sets of a larger class of processes are covered by the notion 2.1 of self-similarity.

For a (continuous) process $(X_t)_{t\geq 0}$, some $a > 0$ and $s \geq 0$ define

$$\tau_a(X) = \inf \{t : X_t = a\}$$

and

$$(T_{-s}X)_t = X_{t+s} - X_s, \ t \geq 0.$$

4.1 Definition. We call a process $(X_t)_{t\geq 0}$ *record-H-self-similar*, for some $H \in (0,1)$, if it is continuous and H-scale-invariant, i.e. fulfils (p.o) and (p.i) from 3.1 on $[0,\infty)$ instead on $\mathbb{R}$, and if

(p.ii') $\quad 0 < \tau_a(X) < \infty$ a.s. and

$$[T_{-\tau_a(X)}X]_{(\cdot)} \stackrel{d}{=} X_{(\cdot)}, \text{ for some } a > 0.$$ ■

(p.i) and (p.ii') imply (p.ii') to hold for *any* $a > 0$. Obviously,

$$\tau_{a+b}(X) - \tau_a(X) = \tau_b(T_{-\tau_a(X)}X) \stackrel{d}{=} \tau_b(X).$$

4.2. If X_t is record-self-similar then

$$\mathbb{E}\tau_a(X) = +\infty \text{ and } \lim_{a\to+\infty} \tau_a(X) = +\infty \quad \text{a.s.}$$

Indeed, from (p.i) one obtains

$$\mathbb{E}\tau_{2a}(X) = 2^{1/H}\, \mathbb{E}\tau_a(X),$$

which contradicts

$$\mathbb{E}\tau_{2a}(X) = 2\, \mathbb{E}\tau_a(X)$$

if $\mathbb{E}\tau_a(X)$ is finite.

The second statement is standard, since the sequence

$$(\tau_{(n+1)b}(X) - \tau_{nb}(X))_{n=0}^{\infty}$$

is stationary by (p.ii'). ■

Clearly, the Wiener process is record-1/2-self-similar.

To get closer to the applicability of Definition 2.1 we need record-self-similar processes on the whole axis. (p.ii') supplies us with a method for extending the distribution of a record-self-similar process $(X_t)_{t\geq 0}$ to a distribution of a process on $\mathbb{R}$.

4.3. Any record-H-self-similar process $(X_t)_{t\geq 0}$ can be extended to a record-H-self-similar process $(Y_t)_{t\in\mathbb{R}}$ on the whole axis, meaning that $(X_t)_{t\geq 0} \stackrel{d}{=} (Y_t)_{t\geq 0}$ and Y_t fulfils (p.o), (p.i), and (p.ii') on $\mathbb{R}$. This extension is unique, and $\lim_{t\to-\infty} Y_t = -\infty$.

Proof. Put

$$X_t^{(b)} = X_{t+\tau_b(X)} - b \quad \text{if } t \geq -\tau_b(X),\ b > 0,$$

and = 0 otherwise. Then, for $0 < a < b$, one can check

$$(X_t^{(b)})_{t\geq-\tau_a(X)} \stackrel{d}{=} (X_t^{(a)})_{t\geq-\tau_a(X)},$$

hence the distributions of $(X_t^{(b)})_{t\in\mathbb{R}}$ converge weakly whenever $b \longrightarrow +\infty$. The continuous process distributed according to this limit distribution is the desired $(Y_t)_{t\in\mathbb{R}}$. In fact, by construction it satisfies (p.0) and (p.i). The following rough argument is basic for proving (p.ii'): On $-\tau_a(X) \leq t < 0$ one has

$$\begin{aligned} Y_t \stackrel{d}{=} X_{t+\tau_a(X)} - a &\stackrel{d}{=} r^H X_{[t+\tau_a(r^H X_{(\cdot)/r})]/r} - a \\ &= r^H[X_{t/r+\tau_{ar^{-H}}(X)} - ar^{-H}] \stackrel{d}{=} r^H Y_{t/r}. \end{aligned}$$

This extension is unique in view of (p.ii'). The limit behaviour as $t \longrightarrow -\infty$ follows from the construction: $\tau_a(Y) > -\infty$ a.s. for all a. ∎

With any record-self-similar process $(X_t)_{t\in\mathbb{R}}$ we can associate a natural random measure $\xi = \xi_X$ (call it the "record measure") concentrated on the record set

$$\Xi = \{t \in \mathbb{R} : X_s < X_t,\ s < t\}$$

by defining

$$\xi[t_1,t_2) = \sup\{X_s : s < t_2\} - \sup\{X_s : s < t_1\}.$$

4.4. The random record measure ξ of a record-H-self-similar process is H-self-similar, and $\xi \neq 0$ a.s.

Proof. 1. The last statement is a consequence of 4.3.

2. ξ satisfies (m.i). Indeed, from (p.i) we obtain

$$\begin{aligned} r^H \sup\{X_s : s < t/r\} &= \sup\{r^H X_{s/r} : s < t\} \\ &\stackrel{d}{=} \sup\{X_s : s < t\}. \end{aligned}$$

3.-5. To check (m.ii) we will adopt techniques from point process theory.

3. Let $\mathbb{P}$ be the distribution of a random measure on $\mathbb{R}$ with continuous (i.e. atomless) realizations of infinite mass. (ξ is of that kind!) Define, for a > 0 and a continuous measure μ on $\mathbb{R}$,

$$\tau_a(\mu) = \sup\{t : \mu(0,t] < a\},$$

denote $(T_s\mu)(B) = \mu(B - s)$ and put

$$\Gamma_a\mu = T_{-\tau_a(\mu)}\,\mu$$

If $\mathbb{P} = \mathbb{P}\circ\Gamma_a^{-1}$ for some a > 0 then there exists a σ-finite stationary measure $\mathbb{H}$ on $[M,\mathcal{M}]$ such that the Palm distribution $\mathbb{H}_0$ equals $\mathbb{P}$. Especially, $0 < \lambda_{\mathbb{H}} < \infty$. This statement is part of the analogue (for continuous random measures) of the characterization therorem for Palm measures of point processes ([6], 3.5.2/8.6.1). Although the proof of it is not a straightforward modification of that for point processes, we will not dwell on it here (cf. [12]).

4. $\Gamma_a\xi \overset{d}{=} \xi$. This is a consequence of $\tau_a(\xi_X) = \tau_a(X)$ and (p.ii').

5. It remains to prove that $\mathbb{H}$ is non-finite. Here we will use the analogue (again for continuous random measures) of the inversion formula [6], 3.4.14/8.2.7, for point processes,

$$\mathbb{H}(E) = \lambda_{\mathbb{H}}\frac{1}{a}\int\int_0^{\tau_a(m)} 1_E(T_{-t}\mu)\,dt\,\mathbb{H}_0(d\mu), \quad E \in \mathcal{M},\ 0\text{-meas} \notin E$$

(cf. [12]). Hence, using 4.2,

$$\mathbb{H}(M) = \lambda_{\mathbb{H}}\frac{1}{a}\,\mathbb{E}\tau_a(\xi) = +\infty.$$ ∎

5. Hausdorff Dimension

Knowing that a random measure η is D-self-similar in the sense of Definition 2.1., the minimal possible Hausdorff dimension of random sets carrying η equals D.

5.1. Theorem. ([10], 3.1) If a random measure η (on $\mathbb{R}$) is D-self-similar, and (*) $\mathbb{E}\eta[1,1+\varepsilon) < \infty$ for some $\varepsilon > 0$, then the carrying Hausdorff dimension of η equals min{D,d}, meaning that for almost all η

(d.i) there exists an F_σ-set G with $\eta(\mathbb{R}\setminus G) = 0$, $\dim G \le D$,
(d.ii) $\eta(B) > 0$ implies $\dim B \ge \min\{D,d\}$, for any Borel set $B \subset \mathbb{R}$. ∎

The theorem (which holds in $\mathbb{R}^d$ as well) formulates the announced connection between the self-similarity index D and carrying Hausdorff dimension. Condition (*) is technical, and usually fulfilled.

(d.ii) implies dim supp $\eta \geq \min\{D,d\}$, but equality will not hold in general. However, one can take the view that (minimal, in the above sense) carrying sets for the local time measure of section 3 and the record measure of section 4 are more important than the zero set and the the record set, respectively. Moreover, I conjecture that the zero and the record sets are minimal carrying itself.

Using 3.2 we obtain the following

5.2. Corollary. The local time measure (at level zero) $\alpha(0,\cdot)$ of an H-self-similar process (as defined in 3.1) has carrying dimension $1 - H$, if $\mathbb{E}\alpha(0,[1,2)) < \infty$. ∎

For stable processes with *independent* increments this was proved in [8]. The Wiener process is covered with $H = 1/2$.

4.4 together with 5.1 implies the following

5.3. Corollary. The record measure ξ concentrated on the record set Ξ of a record-H-self-similar process (as defined in 4.1) has carrying dimension H, if $\mathbb{E}\xi[1,2) < \infty$. ∎

Especially, the record set of the Wiener process has essentially Hausdorff dimension 1/2. It may be interesting that for processes which are both self-similar and record-self-similar the sum of the Hausdorff dimensions of the zero set and the record set, in the sense of carrying dimensions of natural measures on them, equals $1 - H + H = 1 = \dim \mathbb{R}$.

To calculate the Hausdorff dimension of the graph or, more precisely, the carrying dimension of a natural measure, the so-called occupation measure, concentrated on the graph, one needs an extension of 5.1 to self-affine measures (cf. section 3), which is given in [12]. Based on that result, the dimension of the graph of an H-self-similar process is computed in [12] to be equal to 2-H.

References

1. Deheuvels, P. (1984), Strong approximations of records and record times, Statistical extremes and applications (Vimeiro 1983), 491-496, NATO Adv. Sci. Inst. Ser. C, 131, Reidel, Dordrecht - Boston.

2. Goldie, Ch. M. & Rogers, L. C. G. (1984), The k-record processes are i.i.d., Z. Wahrscheinlichkeitstheorie verw. Gebiete, 67, 197-211.

3. Federer, H. (1969), Geometric measure theory, Berlin - Heidelberg - New York.

4. Hutchinson, J. E. (1981), Fractals and self-similarity, Indiana Univ. Math. J. 30, 713-747.

5. Kallenberg, O. (1983), Random measures, Berlin.

6. Kerstan, J., Mathes, K. & Mecke, J. (1978), Infinitely divisible point processes, Berlin 1974/Chichester - New York - Brisbane - Toronto.

7. Knight, F. (1981), Essentials of Browninan motion and diffusion, AMS, Providence.

8. Taylor, S. J. & Wendel, J. G. (1966), The exact Hausdorff measure of the zero set of a stable process, Z. Wahrscheinlichkeitstheorie verw. Gebiete 6, 170-180.

9. Verwaat, W. (1985), Sample path properties of self-similar processes with stationary increments, Ann. Probab. 13, 1-27.

10. Zähle, U., Self-similar random measures, I - Notion, carrying Hausdorff dimension, and hyperbolic distribution, Probab. Th. Rel. Fields, to appear.

11. -, -. II - A generalization: self-affine measures, Math. Nachrichten, to appear.

12. -, -. III, in preparation.

ON EXCEEDANCE POINT PROCESSES FOR STATIONARY SEQUENCES UNDER MILD OSCILLATION RESTRICTIONS

M.R. Leadbetter and S. Nandagopalan
Department of Statistics, University of North Carolina
Chapel Hill, NC 27599-3260

Abstract. It is known ([1]) that any point process limit for the (time normalized) exceedances of high levels by a stationary sequence is necessarily Compound Poisson, under general dependence restrictions. This results from the clustering of exceedances where the underlying Poisson points represent cluster positions, and the multiplicities correspond to cluster sizes.

Here we investigate a class of stationary sequences satisfying a mild local dependence condition restricting the extent of local "rapid oscillation". For this class, criteria are given for the existence and value of the so-called "extremal index" which plays a key role in determining the intensity of cluster positions. Cluster size distributions are investigated for this class and in particular shown to be asymptotically equivalent to those for lengths of runs of consecutive exceedances above the level. Relations between the point processes of exceedances, cluster centers, and upcrossings are discussed.

1. Introduction and basic results.

Many of the standard results of classical extreme value theory may be cast in terms of exceedances of levels by the sequence of random variables $\xi_1, \xi_2 \dots$ For example if $\{u_n\}$ are constants, the event $\{\max(\xi_1,\dots,\xi_n) \le u_n\}$ is identical with the event $\{S_n = 0\}$, where S_n denotes the number of *exceedances* of u_n by $\xi_1,\dots,\xi_n$ (i.e. the number of i, $1 \le i \le n$, for which $\xi_i > u_n$). Thus the limiting law for the maximum may be obtained from a (Poisson) limit for the binomial random variable S_n. Similarly the event that the rth largest of $\xi_1,\dots,\xi_n$ does not exceed u_n is equivalent to $\{S_n < r\}$, leading to the limiting distributions of extreme order statistics. Yet more general results may be obtained by considering point processes of exceedances and their convergence to a Poisson process in the classical case. This viewpoint, (explored e.g. in [3] and [4]) is especially useful for dependent

---

Key words: exceedances, stationary sequences, point processes.

AMS Classification: Primary 60G10; Secondary 60-02, 60G15, 60G17, 60G55, 60F05

Research supported by the Air Force Office of Scientific Research Contract No. F49620 85C 0144.

sequences where low dependence can still lead to the classical results, but higher dependence involves clustering of exceedances and leads to compound Poisson limits. Nevertheless the appropriate special results of interest (for maxima and order statistics) are still readily obtained.

Specifically, let $\xi_1, \xi_2 \ldots$ be a stationary sequence. Write $M_n = \max(\xi_1, \xi_2, \ldots, \xi_n)$ and for $\tau > 0$ let $u_n(\tau)$ denote levels such that $n(1-F(u_n(\tau))) \to \tau$, where F is the distribution function (d.f.) of each ξ_i. Then it is often the case that $P\{M_n \leq u_n(\tau)\} \to e^{-\theta\tau}$ where θ is a fixed parameter $(0 \leq \theta \leq 1)$, referred to as the "extremal index" of the sequence. It is known that $\theta = 1$ for i.i.d. sequences and many dependent cases, and that $\theta > 0$ for "almost all" cases ofinterest. For such levels $u_n(\tau)$ it may be shown under general conditions that the intensity for the Poisson limiting cluster positions in N_n is simply $\theta\tau$.

These results require a restriction on the long range dependence of the sequence, and two such conditions ($D(u_n)$, $\Delta(u_n)$, defined below) are useful. It is well known that under a further short range dependence condition ($D'(u_n)$ - cf. [4] Section 3.4) it may be shown that $\theta = 1$ and the Compound Poisson limit for N_n becomes Poisson. In this paper we consider a special but much wider class of sequences subject to a weaker condition which restricts rapid oscillations - here called $D''(u_n)$ - than $D'(u_n)$, and for which all values of θ in $(0,1]$ are possible. It will be shown for this class that the joint distribution of ξ_1 and ξ_2 determines whether the extremal index exists, and gives its value. Finally for this class clusters of exceedances may be simply identified asymptotically as runs of consecutive exceedances and the cluster sizes as run lengths.

Section 2 contains the theory surrounding the maximum and the extremal index when the local dependence condition $D''(u_n)$ holds, and in Section 3 asymptotic properties of point processes of exceedances, upcrossings and cluster centers are discussed. Notation used throughout will include $M(E)$ to denote $\max\{\xi_i: i \in E\}$ for any set $E \subset (0,n]$ ($M_n = M[1,n]$). A time scale normalization by 1/n will be used to define various point processes on the unit interval. In particular the exceedance point process N_n is defined with respect to a sequence of "levels" $\{u_n\}$ by

(1.1) $$N_n(B) = \#\{i,\ 1 \leq i \leq n:\ i/n \in B,\ \xi_i > u_n\}$$

for each Borel subset B of $(0,1]$. This involves a slight awkwardness of notation in that $M(E)$ is defined for subsets E of $(0,n]$, whereas $N_n(B)$ is defined for $B \subset (0,1]$ when writing an equivalence $\{N_n(B) = 0\} = \{M(nB) \leq u_n\}$ but a more intricate notation does not seem worthwhile.

The long range dependence condition $D(u_n)$ is defined as follows. Abbreviate $F_{i_1\ldots i_n}(u,u\ldots u)$ to $F_{i_1\ldots i_n}(u)$. Then for a sequence $\{u_n\}$, $D(u_n)$ is said to hold if for each n, $1\leq i_1<i_2\ldots<i_p<j_1\ldots<j_{p'}\leq n$, $j_1-i_p\geq \ell$ we have

$$|F_{i_1\ldots i_p j_1\ldots j_{p'}}(u_n) - F_{i_1\ldots i_p}(u_n)F_{j_1\ldots j_{p'}}(u_n)| \leq \alpha_{n,\ell}$$

where $\alpha_{n,\ell_n} \to 0$ for some $\ell_n=o(n)$. Frequently integers $k_n \to \infty$ will be chosen so that

$$k_n\alpha_{n,\ell_n} \to 0, \qquad k_n\ell_n/n \to 0. \tag{1.2}$$

Note that, by $D(u_n)$, this holds automatically for bounded k_n-sequences but $k_n \to \infty$ can clearly be chosen so that (1.2) is satisfied. Note also that the condition $D(u_n)$ is of similar type to (but much weaker than) strong mixing. In the following basic result and throughout, m will denote Lebesgue measure. The result is a slightly more general form of Lemma 2.3 of [1].

Lemma 1.1 Let $D(u_n)$ hold and $\{k_n\}$ satisfy (1.2). Let J_i $(=J_{i,n})$, $1\leq i\leq k_n$, be disjoint subintervals of $(0,1]$ with $\frac{n}{k_n\ell_n}\sum_1^{k_n} m(J_i) \to \infty$ (which holds, in particular, if $m(\bigcup_1^{k_n} J_i) \to \alpha > 0$). Then

(i) $\gamma_n = P\{M(\bigcup_1^{k_n} nJ_i) \leq u_n\} - \prod_1^{k_n} P\{M(nJ_i) \leq u_n\} \to 0$ as $n \to \infty$

(ii) If J is a fixed subinterval of $(0,1]$ with $m(J) = \alpha$, $\bigcup_1^{k_n} J_i \subset J$ and $m(\bigcup_1^{k_n} J_i) \to \alpha$, then

$$P\{M(nJ) \leq u_n\} - \prod_1^{k_n} P\{M(nJ_i) \leq u_n\} \to 0. \tag{1.3}$$

Proof: The assertion (i) is proved by arguments very close to those used in Lemma 2.2 of [1]. The main difference is the complicating feature in that here we do not assume that $m(J_i) \geq \ell_n/n$ for each i, but clearly the intervals J_i for which $m(J_i) < \ell_n/n$ form a set whose total measure cannot exceed $k_n\ell_n/n \to 0$. The proof of (i) will not be given in detail, though its flavor may be seen from the sketch for (ii) below. It is in fact very simple in the usual situation where exceedances in short intervals are unlikely in the sense that $k_n P\{M_{\ell_n} > u_n\} \to 0$, and is made more lengthy to cover cases when this does not hold by showing that both terms of (i) actually tend to zero.

(ii) (sketch of proof). By stationarity the intervals J_i may be taken to be abutting and $\bigcup_1^{k_n} J_i = I_n$, $J - I_n = I_n^*$ taken to be intervals without affecting either term of (1.3), and $m(I_n^*) \to 0$. By (i), (ii) will follow if

$$\gamma_n' = P\{M(nI_n) \leq u_n\} - P\{M(nJ) \leq u_n\} \to 0$$

and it is sufficient to show that if γ_n' has a limit as $n \to \infty$ through a subsequence S, then that limit is zero. This is immediate if $P\{M(nI_n^*) > u_n\}$ tends to zero, since this probability dominates γ_n'. Otherwise $P\{M(nI_n^*) > u_n\} \to \delta > 0$ as $n \to \infty$ through some subsequence $S' \subset S$. Clearly $q_n (\to \infty)$ copies $I_{n,j}$ of I_n^*, each separated by at least ℓ_n/n, may be placed in I_n, and $P\{M(nI_n) \leq u_n\}$ thus dominated by $P\{\bigcap_1^{q_n} (M(nI_{n,j}) \leq u_n)\}$. By appropriate choice of q_n this probability may be approximated by $P^{q_n}\{M(nI_n^*) \leq u_n\}$ (using $D(u_n)$) which tends to zero as $n \to \infty$ through S'. Hence the first term in γ_n' tends to zero and it dominates the second, which thus also tends to zero. □

2. Extremal theory under $D''(u_n)$.

If $D(u_n)$ holds, and k_n are integers satisfying (1.2), and $k_n(1-F(u_n)) \to 0$, $r_n = [n/k_n]$, define

$$D''(u_n):\ n \sum_{j=2}^{r_n - 1} P\{\xi_1 > u_n,\ \xi_j \leq u_n < \xi_{j+1}\} \to 0.$$

and write $\mu(u) = P\{\xi_1 \leq u < \xi_2\}$. We say that $\{\xi_n\}$ has an upcrossing of u at j if $\xi_{j-1} \leq u < \xi_j$, so that $\mu(u)$ may obviously be interpreted as the mean number of upcrossings of u per unit time. (This notation will be used throughout this and the next section without comment). The condition $D''(u_n)$ involves a weaker restriction than $D'(u_n)$ of [4] which is used to guarantee that $\theta = 1$, whereas under D" all values $0 \leq \theta \leq 1$ are possible. For most of our purposes D" can be slightly weakened by replacing "$\xi_1 > u_n$" by "$\xi_1 \le u_n < \xi_2$" thus restricting the local occurrence of two or more upcrossings, but the present form is convenient for use here.

Proposition 2.1. Suppose $D(u_n)$, $D''(u_n)$ hold for given constants $\{u_n\}$, $\{k_n\}$, $\{r_n\}$ as above and write

$$\upsilon = \liminf n\mu(u_n),\ \upsilon' = \limsup n\mu(u_n).$$

Then

$$\liminf P\{M_n \le u_n\} = e^{-\upsilon'},\ \limsup P\{M_n \le u_n\} = e^{-\upsilon}.$$

In particular $P\{M_n \le u_n\} \to e^{-\upsilon}$ if and only if $n\mu(u_n\} \to \upsilon$.

Proof: Write $A_j = \{\xi_j \le u_n < \xi_{j+1}\}$. Then $\{M_{r_n} > u_n\} = \{\xi_1 > u_n\} \cup \bigcup_{j=1}^{r_n-1} A_j$ so that

$$\sum_{j=1}^{r_n-1} P(A_j) - \sum_{1 \le i < j \le r_n-1} P(A_i \cap A_j) \le P\{M_{r_n} > u_n\} \le 1-F(u_n) + \sum_{j=1}^{r_n-1} P(A_j)$$

Hence, since $P(A_j) = \mu(u_n)$, (and using stationarity),

$$(r_n-1)\mu(u_n) - S_n \le P\{M_{r_n} > u_n\} \le 1-F(u_n) + (r_n-1)\mu(u_n)$$

in which $S_n = r_n \sum_{j=2}^{r_n-1} P\{\xi_1 > u_n, \xi_j \le u_n < \xi_{j+1}\} = o(k_n^{-1})$ by $D''(u_n)$. Multiplication by k_n yields

$$n\mu(u_n)(1 + o(1)) - o(1) \le k_n P\{M_{r_n} > u_n\} \le n\mu(u_n) + o(1).$$

From which it follows that

$$\limsup k_n P\{M_{r_n} > u_n\} = \upsilon',\ \liminf k_n\ P\{M(r_n) > u_n\} = \upsilon.$$

Now by Lemma 1.1,

$$P\{M_n \le u_n\} = \left(1 - \frac{k_n P\{M_{r_n} > u_n\}}{k_n}\right)^{k_n} + o(1).$$

For $\epsilon > 0$, $k_n P\{M_{r_n} > u_n\} \ge \upsilon - \epsilon$ for sufficiently large n, so that $P\{M_n \le u_n\} \le (1 - \frac{\upsilon-\epsilon}{k})^{k_n} + o(1) \to e^{-\upsilon+\epsilon}$ and hence $\limsup P\{M_n \le u_n\} \le e^{-\upsilon}$.

Similarly $P\{M_n \le u_n\} \ge (1 - \frac{\upsilon+\epsilon}{k_n})^{k_n} + o(1)$ for infinitely many values of n so that $\limsup P\{M_n \le u_n\} \ge e^{-\upsilon-\epsilon}$ and hence $\limsup P\{M_n \le u_n\} \ge e^{-\upsilon}$, showing that $\limsup P\{M_n \le u_n\} = e^{-\upsilon}$. Similarly $\liminf P\{M_n \le u_n\} = e^{-\upsilon'}$, as required. □

<u>Corollary 2.2</u> If $I_j = (a_j, b_j]$ are disjoint subintervals of $(0,1]$, $1 \le j \le k$, then under the conditions of Proposition 2.1, if $n\mu(u_n) \to \upsilon$,

$$P\{\bigcap_1^k (M(nI_j) \le u_n)\} \to \exp\{-\upsilon \sum_1^k (b_j - a_j)\}$$

Proof: It follows from Lemma 1.1 that $P\{\bigcap_1^k (M(nI_j) \le u_n)\} - \prod_1^k P\{M(nI_j) \le u_n\} \to 0$ so that it is only necessary to show the result for k=1. Let k_n be as in Proposition 2.1, $r_n = [n/k_n]$. Then it follows readily from Lemma 1.1 and Proposition 2.1 that

$$P^{k_n}\{M_{r_n} \le u_n\} = P\{M_n \le u_n\} + o(1) \to e^{-\upsilon}$$

and hence that for $0 < a < b \le 1$,

$$\begin{aligned} P\{M((na,nb]) \le u_n\} &= (P\{M_{r_n} \le u_n\})^{([nb]-[na])/r_n} + o(1) \\ &= (P\{M_{r_n} \le u_n\})^{k_n(b-a)(1+o(1))} + o(1) \\ &\to e^{-\upsilon(b-a)} \end{aligned}$$

as required to complete the proof. □

We consider now levels $u_n = u_n(\tau)$ defined to satisfy $n(1-F(u_n(\tau))) \to \tau$. Note first the simply proved relation

$$\begin{aligned} \mu(u) &= P\{\xi_1 \le u < \xi_2\} \\ &= P\{\xi_2 \le u | \xi_1 > u\}\,(1-F(u)) \end{aligned} \tag{2.1}$$

Proposition 2.1 may be applied as follows.

<u>Proposition 2.3</u> Assume $D(u_n)$, $D''(u_n)$ hold for $u_n = u_n(\tau)$, some $\tau > 0$. Write $\theta = \liminf P\{\xi_2 \le u_n(\tau) | \xi_1 > u_n(\tau)\}$, $\theta' = \limsup P\{\xi_2 \le u_n(\tau) | \xi_1 > u_n(\tau)\}$. Then $\limsup P\{M_n \le u_n(\tau)\} = e^{-\theta\tau}$, $\liminf P\{M_n \le u_n(\tau)\} = e^{-\theta'\tau}$. In particular, $P\{\xi_2 \le u_n(\tau) \,|\xi_1 > u_n(\tau)\} \to \theta$ if and only if $P\{M_n \le u_n(\tau)\} \to e^{-\theta\tau}$.

Proof: By (2.1),

$$v = \liminf n\mu(u_n(\tau)) = \theta\tau, \; v' = \limsup n\mu(u_n(\tau)) = \theta'\tau$$

and the results follow at once from Proposition 2.1. □

If $P\{M_n \le u_n(\tau)\} \to e^{-\theta\tau}$ for all $\tau > 0$ the parameter θ will be referred to as the *extremal index* of the sequence $\{\xi_n\}$. It is known (cf. [4] Theorem 3.7.1) that if $D(u_n(\tau))$ holds for each $\tau > 0$ and $P\{M_n \le u_n(\tau)\}$ converges for some $\tau > 0$, then $P\{M_n \le u_n(\tau)\}$ converges for all $\tau > 0$ and the limit has the form $e^{-\theta\tau}$ for fixed θ, $0 \le \theta \le 1$, i.e. the extremal index then exists. The following result, gives a convenient existence criterion assuming also $D''(u_n)$, and follows immediately from Proposition 2.3 and these observations.

Corollary 2.4 Assume $D(u_n(\tau))$, $D''(u_n(\tau))$ hold for each $\tau > 0$. If $P\{\xi_2 \le u_n(\tau) | \xi_1 > u_n(\tau)\} \to \theta$ for some $\tau > 0$ then convergence to θ occurs for all $\tau > 0$, and $\{\xi_n\}$ has extremal index θ. Conversely if $P\{M_n \le u_n(\tau)\} \to e^{-\theta\tau}$ for some $\tau > 0$, $\{\xi_n\}$ has extremal index θ and $P\{\xi_2 \le u_n(\tau) | \xi_1 > u_n(\tau)\} \to \theta$ for all $\tau > 0$. □

The following lemma, giving alternative expressions for θ involves stationarity but does not require any dependence condition.

Lemma 2.5 If $n\mu(u_n) \to v$ the following are equivalent:

(i) $P\{\xi_2 \le u_n | \xi_1 > u_n\} \to \theta$

(ii) $n(1-F(u_n)) \to v/\theta$ (i.e. $u_n = u_n(v/\theta)$)

(iii) $n(1-F_{1,2}(u_n)) \to v + v/\theta$. ($F_{1,2}(u_n) = P\{\xi_1 \le u_n, \xi_2 \le u_n\}$).

Proof: Equivalence of (i) and (ii) is immediate from (2.1). That of (ii) and (iii) follows since

$$\begin{aligned} n\mu(u_n) &= nP\{\xi_1 \le u_n < \xi_2\} = n(F(u_n) - F_{1,2}(u_n)) \\ &= n((1 - F_{1,2}(u_n)) - (1-F(u_n))) \end{aligned}$$ □

Write now $\tilde{u}_n(v)$ to denote a sequence u_n satisfying $n\mu(u_n) \to v$ and $F(u_n) \to 1$. The next result shows that $n(1-F(\tilde{u}_n(v))) \to v/\theta$ when ξ_n has extremal index θ. This will be denoted by the slightly imprecise, but convenient statement "$\tilde{u}_n(v) = u_n(v/\theta)$".

Proposition 2.6 (i) Suppose $D(\tilde{u}_n(v))$, $D''(\tilde{u}_n(v))$ hold for all $v > 0$, and $\{\xi_n\}$ has extremal index $\theta > 0$. Then $\tilde{u}_n(\nu) = u_n(\nu/\theta))$ (i.e. $n(1-F(\tilde{u}_n(\nu))) \to \nu/\theta$ as $n \to \infty$).

(ii) Conversely suppose $D(u_n(\tau))$, $D''(u_n(\tau))$ hold for all $\tau > 0$. If for some τ,θ $u_n(\tau) = \tilde{u}_n(\theta\tau)$, then $u_n(\tau) = \tilde{u}_n(\theta\tau)$ for all $\tau > 0$ and θ is the extremal index of $\{\xi_n\}$.

Proof: To show (i) note that from Proposition 2.1 $P\{M_n \le \tilde{u}_n(\nu)\} \to e^{-\nu}$ and hence $P\{\tilde{M}_n \le \tilde{u}_n(\nu)\} \to e^{-\nu/\theta}$ ([4], Theorem 3.7.2) where $\tilde{M}_n$ is the maximum of n i.i.d. random variables with the same distribution F as the ξ_i. That is $F^n(\tilde{u}_n(\nu)) \to e^{-\nu/\theta}$ from which it follows at once that $n(1-F(\tilde{u}_n(\nu))) \to \nu/\theta$.

(ii) By Lemma 2.5, $P\{\xi_2 \le u_n(\tau) | \xi_1 > u_n(\tau)\} \to \theta$, hence by Corollary 2.4, this holds for all τ and θ is the extremal index. In particular, $P\{M_n \le u_n(\tau)\} \to e^{-\theta\tau}$ for all τ. By Proposition 2.1, therefore, $u_n(\tau) = \tilde{u}_n(\theta\tau)$ which completes the proof. □

3. Point Processes of Exceedances and Upcrossings

Let N_n denote the exceedance point process for a level u_n as defined by (1.1), viz. $N_n(B) = \#\{i, 1 \le i \le n: i/n \in B, \xi_i > u_n\}$ for $B \subset (0,1]$. Further, write $\tilde{N}_n$ for the "point process of upcrossings", defined on (0,1] as the points $\frac{i}{n}$ such that $\xi_{i-1} \le u_n < \xi_i$ i.e. $\tilde{N}_n(B) = \#\{i, 1 \le i \le n: i/n \in B, \xi_{i-1} \le u_n < \xi_i\}$. It is readily shown that $\tilde{N}_n$ converges in distribution to a Poisson Process under D, D".

Proposition 3.1 Suppose $D(u_n)$, $D''(u_n)$ hold for a sequence $u_n = \tilde{u}_n(\nu)$, i.e. $n\mu(u_n) \to \nu$. Then $\tilde{N}_n \xrightarrow{d} N$ where N is a Poisson Process on (0,1] with intensity ν.

Proof: This follows in a standard way from Kallenberg's Theorem ([2] Theorem 4.7):

(i) If $0 < a < b \le 1$, $\mathcal{E}\tilde{N}([a,b]) \sim n(b-a)\mu(u_n) \to (b-a)\nu = \mathcal{E}N((a,b])$

(ii) $0 \le P\{\tilde{N}_n((a,b]) = 0\} - P\{M((na,nb]) \le u_n\} \le P\{\xi_{[na]+1} > u_n\} = 1 - F(u_n)$

and for disjoint subintervals $(a_i,b_i]$ of (0,1] $1 \le i \le k$,

$$0 \le P\{\tilde{N}_n(\bigcup_1^k (a_i,b_i]) = 0\} - P\{M(\bigcup_1^k (na_i,nb_i]) \le u_n\}$$
$$\le k(1-F(u_n)) \to 0$$

and hence by Lemma 1.1 and Corollary 2.2,

$$P\{\tilde{N}_n(\bigcup_1^k (a_i,b_i]) = 0\} = \prod_1^k P\{M((na_i,nb_i]) \le u_n\} + o(1) \to \exp\{-\nu \sum_1^k (b_i-a_i)\}.$$

But this expression is simply $P\{N(\bigcup_1^k (a_i,b_i]) = 0\}$ thus verifying the conditions of Kallenberg's Theorem. □

Corollary 3.2 If $D(u_n)$, $D''(u_n)$ hold for a sequence $u_n = u_n(\tau)$ (i.e. $n(1-F(u_n)) \to \tau$) and $\{\xi_n\}$ has extremal index $\theta > 0$, then $\tilde{N}_n \xrightarrow{d} N$ where N is Poisson with intensity $\theta\tau$.

Proof: Since $\{\xi_n\}$ has extremal index θ and $u_n = u_n(\tau)$, $P(M_n \le u_n) \to e^{-\theta\tau}$ (see [3], for example). Thus by Proposition 2.3, $P\{\xi_2 \le u_n \mid \xi_1 > u_n\} \to \theta$ and hence, by (2.1), $n\mu(u_n) \to \theta\tau$, i.e. $u_n = \tilde{u}_n(\theta\tau)$.

The result now follows from Proposition 3.1 with $\nu = \theta\tau$. □

The above discussion hinges on the assumption $D''(u_n)$. In that case (as will be seen) each run of consecutive exceedances following an upcrossing may be regarded as a "cluster" of exceedances. If $D''(u_n)$ is not assumed, clusters may consist of groups of "exceedance runs". In general a simple and useful definition of clusters is obtained by choosing k_n to satisfy (1.2) and considering the subintervals $J_i = ((i-1)r_n/n, ir_n/n]$, $1 \le i \le k_n$ of $(0,1]$. Then the exceedances in any interval J_i (i.e. points $\frac{j}{n} \in J_i$ with $\xi_j > u_n$) are regarded as forming a cluster. The "cluster centers" may be defined in an arbitrary way as any point in a J_i containing a cluster - here we use the position of the first event in the cluster. The positions of the cluster centers then form a point process N_n^* for which the following convergence holds (proved similarly to Proposition 3.1).

Proposition 3.3. Suppose $D(u_n)$ holds, where $P\{M_n \le u_n\} \to e^{-\nu}$ for some $\nu > 0$. Then $N_n^* \xrightarrow{d} N$ where N is Poisson with intensity ν. As in Corollary 3.2 if $u_n = u_n(\tau)$ and $\{\xi_n\}$ has extremal index $\theta > 0$ then N has intensity $\theta\tau$. □

In cases where $D''(u_n)$ holds, N_n^* and $\tilde{N}_n$ are asymptotically equivalent as might be expected, in the strong sense of the next result. That is the cluster positions essentially coincide with the upcrossings. It will be seen further (in Proposition 3.5) that cluster sizes then also correspond asymptotically to lengths of exceedance runs, so that clusters and exceedance runs may be identified.

Proposition 3.4 Under the conditions of Proposition 3.1 the total variation of the random signed measure $\tilde{N}_n - N_n^*$ satisfies $\mathcal{E}||\tilde{N}_n - N_n^*|| \to 0$ as $n \to \infty$.

Proof: Define a point process N_n' to consist of all points of $\tilde{N}_n$ together with any points $\frac{j}{n}$ of the form $j=(i-1)r_n + 1$ (i.e. the first point of a subinterval J_i) for which $\xi_j > u_n$. Then $N_n'(B) \geq \tilde{N}_n(B)$ for each $B \subset (0,1]$, and $||N_n' - \tilde{N}_n|| = N_n'((0,1]) - \tilde{N}_n((0,1])$ so that

$$\mathcal{E}||N_n' - \tilde{N}_n|| \leq k_n P\{\xi_1 > u_n\} \to 0 \quad \text{by assumption.} \tag{3.1}$$

Clearly also $N_n'(B) \geq N_n^*(B)$ and $||N_n' - N_n^*|| = N_n'(0,1] - N_n^*(0,1]$. But $\mathcal{E}N_n'((0,1]) \leq \mathcal{E}\tilde{N}_n((0,1]) + k_n(1-F(u_n)) = (n-1)\mu(u_n) + o(1) \to \upsilon$ and

$$\mathcal{E}N_n^*(0,1] = k_n P\{M_{r_n} > u_n] + o(1) \to \upsilon$$

by Lemma 1.1 since $P\{M_n \leq u_n\} - P^{k_n}\{M_{r_n} \leq u_n\} \to 0$ and $P\{M_n \leq u_n\} \to e^{-\upsilon}$. Hence $\mathcal{E}(N_n'(0,1] - N_n^*(0,1]) \to 0$ showing that $E||N_n' - N_n^*|| \to 0$ which combines with (3.1) to give the desired conclusion. □

The discussion of the limiting behavior of the actual exceedance point process N_n requires a dependence restriction of similar type, but somewhat stronger than $D(u_n)$. Such a condition $(\Delta(u_n))$ is used in [1] where it is shown that if $P\{M_n \leq u_n\} \to e^{-\upsilon}$ for some $\upsilon > 0$ then N_n converges in distribution to a Compound Poisson Process provided the cluster size distribution $\pi_n(j)$ converges for each j to $\pi(j)$, a probability distribution on (1,2,3,...). Here the $\pi_n(j)$'s are simply defined to be the distribution of the number of events in a cluster (i.e. in an interval $((i-1)r_n/n, ir_n/n]$) given that there is at least one. The Poisson Process underlying this limit has intensity υ and may be regarded as the limiting point process of cluster centers. The distribution for the multiplicity of each event in the Compound Poisson limit is just $\pi(j)$.

It is natural to ask whether the $\pi_n(j)$ may be replaced by the distribution $\pi_n'(j)$ of the length of an exceedance run defined more precisely by

$$\pi_n'(j) = P\{\xi_2 > u_n, \xi_3 > u_n, \ldots \xi_{j+1} > u_n, \xi_{j+2} \leq u_n | \xi_1 \leq u_n < \xi_2\}$$

That this is the case is shown under D"(u) by the following result

Proposition 3.5 Suppose $D(u_n)$, $D''(u_n)$ hold where $u_n = \tilde{u}_n(\upsilon)$ for some $\upsilon > 0$. Then $\pi_n(j) - \pi_n'(j) \to 0$ as $n \to \infty$ for each $j=1,2,\dots$

Proof: It will be more convenient (and clearly equivalent) to show that $Q_n(j)-Q_n'(j) \to 0$ where $Q_n(j) = \sum_{s=j}^{\infty} \pi_n(s)$, $Q_n'(j) = \sum_{s=j}^{\infty} \pi_n'(s)$. Writing J for the interval $(0,r_n/n]$ we have for $j \geq 1$,

$$\begin{aligned} Q_n(j) &= P\{N_n(J) \geq j \mid N_n(J) > 0\} = P\{N_n(J) \geq j\}/P\{N_n(J) > 0\} \\ &= \frac{k_n}{\upsilon} P\{N_n(J) \geq j\} (1+o(1)) \end{aligned}$$

since $P\{N_n(J) > 0\} = P\{M_{r_n} > u_n\} \sim \upsilon/k_n$ (by Lemma 1.1, since $P\{M_n \leq u_n\} \to e^{-\upsilon}$) so that

$$\begin{aligned} Q_n(j) = \frac{k_n}{\upsilon}\Big[&P\{\xi_1 > u_n, N_n((\tfrac{1}{n},\tfrac{r_n}{n}]) \geq j - 1\} \\ &+ \sum_{i=1}^{r_n-j+1} P\{\xi_1 \leq u_n,\dots\xi_{i-1} \leq u_n < \xi_i, N_n((\tfrac{i+1}{n}, \tfrac{r_n}{n}]) \geq j-1\}\Big](1+o(1)) \end{aligned}$$

Now

$$\frac{k_n}{\upsilon} P\{\xi_1 > u_n, N_n((\tfrac{1}{n}, \tfrac{r_n}{n}]) \geq j-1\} \leq \frac{k_n}{n}(1-F(u_n)) = o(1)$$

and

$$\begin{aligned} 0 &\leq P\{\xi_1 \leq u_n,\dots\xi_{i-1} \leq u_n < \xi_i, N_n(\tfrac{i}{n}, \tfrac{r_n}{n}] \geq j-1\} \\ &\quad - P\{\xi_1 \leq u_n\dots\xi_{i-1} \leq u_n < \xi_i, \xi_{i+1} > u_n,\dots\xi_{i+j-1} > u_n\} \\ &\leq P\{\xi_i > u_n, \bigcup_{j=i+2}^{r_n} (\xi_{j-1} \leq u_n < \xi_j)\} \\ &\leq \sum_{j=3}^{r_n} P\{\xi_1 > u_n, \xi_{j-1} \leq u_n < \xi_j\} = o(1/n) \end{aligned}$$

by $D''(u_n)$, so that

$$Q_n(j) = \frac{k_n}{\upsilon} \Big[\sum_{i=1}^{r_n-j+1} P\{\xi_1 \leq u_n,\dots\xi_{i-1} \leq u_n, \xi_i > u_n\dots\xi_{i+j-1} > u_n\}\Big] (1+o(1)) + o(1).$$

Also

$$0 \leq P\{\xi_{i-1} \leq u_n, \xi_i > u_n,\dots\xi_{i+j-1} > u_n\} - P\{\xi_1 \leq u_n,\dots\xi_{i-1} \leq u_n, \xi_i > u_n,\dots\xi_{i+j-1} > u_n\}$$

$$\leq \sum_{j=3}^{r_n} P\{\xi_1 > u_n, \xi_{j-1} \leq u_n < \xi_j\} = o(1/n)$$

so that

$$\begin{aligned} Q_n(j) &= \frac{k_n}{\nu}\left[\sum_{i=1}^{r_n-j+1} P\{\xi_{i-1} \leq u_n, \xi_i > u_n, \ldots \xi_{i+j-1} > u_n\}\right](1 + o(1)) + o(1) \\ &\sim \frac{k_n}{\nu}(r_n - j + 1)\, P\{\xi_2 > u_n, \ldots \xi_{j+1} > u_n \mid \xi_1 \leq u_n < \xi_2\}\, \frac{\nu}{n}(1 + o(1)) + o(1) \\ &= Q_n'(j)(1 + o(1)) + o(1) = Q_n'(j) + o(1) \end{aligned}$$

as required. □

References

[1] Hsing, T., Hüsler, J., Leadbetter, M.R. "On the exceedance point process for a stationary sequence" Prob. Theor. and Rel. Fields, 78, 97-112 (1988).

[2] Kallenberg, O. "Random Measures" Akademie-Verlag (Berlin) and Academic Press (London), 3rd Ed. 1983.

[3] Leadbetter, M.R. and Rootzén, H., "Extremal theory for stochastic processes", Ann. Probability, 16, 431-478 (1988).

[4] Leadbetter, M.R., Lindgren, G. and Rootzén, H., "Extremes and related properties of random sequences and processes" Springer Statistics Series, 1983.

A CENTRAL LIMIT THEOREM FOR EXTREME SOJOURN TIMES OF STATIONARY GAUSSIAN PROCESSES

Simeon M. Berman[1]
Courant Institute of Mathematical Sciences, New York University
251 Mercer Street, New York 10012

Abstract

Let $X(t)$, $t \geq 0$, be a real measurable stationary Gaussian process with mean 0 and covariance function $r(t)$. For a given measurable function $u(t)$ such that $u(t) \to \infty$ for $t \to \infty$, let L_t be the sojourn time of $X(s)$, $0 \leq s \leq t$, above $u(t)$. Assume that the spectral distribution function in the representation of $r(t)$ is absolutely continuous; then $r(t)$ also has the representation $r(t) = \int b(t+s)b(s)ds$, where $b \in L_2$. The main result is: If $b \in L_1$, and if $u(t)$ increases sufficiently slowly, then $(L_t - EL_t)/(\mathrm{Var}(L_t))^{1/2}$ has a limiting standard normal distribution for $t \to \infty$. The allowable rate of increase of $u(t)$ with t is specified.

1. Introduction and Summary.

Let $X(t), t \geq 0$, be a real measurable stationary Gaussian process with mean 0 and covariance function $r(t) = EX(0)X(t)$. For simplicity we take $r(0) = 1$. For $t > 0$, let $L_t(u)$ be the sojourn time of $X(s)$, $0 \leq s \leq t$, above the level $u : L_t(u) = mes(s : 0 \leq s \leq t, X(s) > u)$. Then for a given measurable function $u(t)$, we define

(1.1) $$L_t = L_t(u(t)) = \int_0^t 1_{[X(s) > u(t)]}\, ds.$$

The main result of this paper is a new limit theorem for the distribution of L_t, for $t \to \infty$, where $u(t)$ increases at a specified rate with t. We assume that the spectral distribution function in the representation of $r(t)$ is absolutely continuous. Then $r(t)$ also has the representation (see, for example, [7], page 532),

(1.2) $$r(t) = \int_{-\infty}^{\infty} b(t+s)b(s)ds,$$

where $b(s)$ is the Fourier transform of the square root of the spectral density, and

(1.3) $$\int_{-\infty}^{\infty} |b(s)|^2 ds < \infty.$$

Since $X(t)$ is real valued, $b(s)$ is also real valued. Our main result is:

---

[1] This paper represents results obtained at the Courant Institute of Mathematical Sciences, New York University, under the sponsorship of the National Science Foundation, Grant DMS 85 01512, and the U. S. Army Research Office, Contract DAAL 03 86 K 0127.

THEOREM 1.1. *Assume in addition to* (1.3) *that*

$$\int_{-\infty}^{\infty} |b(s)|ds < \infty. \tag{1.4}$$

If $u(t)$ *increases to* ∞ *sufficiently slowly with* t, *then* $L_t - EL_t/(\mathrm{Var}\ L_t)^{1/2}$ *has, for* $t \to \infty$, *a limiting standard normal distribution. The maximum rate at which* $u(t)$ *may increase with* t *is described by the following condition: For some* $\theta > 1$ *and some* $\delta > 0$,

$$\lim_{t\to\infty} t\ e^{-u^2(t)\theta/2} = \infty, \tag{1.5}$$

and

$$\lim_{t\to\infty} e^{\delta u(t)} \int_{|s|>\sqrt{t}\exp(-u^2(t)\theta/4)} b^2(s)ds = 0. \tag{1.6}$$

Note that the function under the limit in (1.6), with $u(t) = u$, increases with u for fixed t, and decreases to the limit 0 for $t \to \infty$, for fixed u. Thus it is always possible to find a function $u(t)$ satisfying (1.5) and (1.6). We observe that the hypothesis contains no conditions on the local behavior of $r(t)$ at $t = 0$ other than the implied condition of continuity. Thus the theorem holds not only for classes of Gaussian processes with continuous sample functions, but also for those with sample functions which are unbounded in every interval.

The first results on the limiting distribution of L_t involved a level function $u(t)$ satisfying

$$u(t) \sim (2\ log\ t)^{1/2}, \qquad \text{for } t \to \infty. \tag{1.7}$$

The normalization of L_t was done by multiplication by a positive increasing function $v(t)$ determined by the asymptotic form of $1 - r(t)$ for $t \to 0$. The limiting distribution of $v(t)L_t$ was shown to be a compound Poisson distribution, where the compounding distribution is uniquely defined in terms of a parameter α representing the index of variation of $1 - r(t)$ for $t \to 0$. The hypothesis of such a theorem includes conditions on the behavior of $1 - r(t)$, which are related to the local behavior of $X(t)$, and mixing conditions expressed in terms of the rate of decay of $r(t)$ for $t \to \infty$. The results in this area include those of Volkonskii and Rozanov [10], Cramer and Leadbetter [6], page 279, and Berman [2], [4].

In the type of limit theorem described above, where the normalization is done by multiplication of L_t by the increasing function $v(t)$, the local behavior of $r(t)$ plays a decisive role in the limiting operation; indeed, the function $v(t)$ as well as the form of the compounding distribution in the compound Poisson limit depend on α. This is related to the fact that the level $u(t)$ in (1.7) is, in this context, very high, so that the sojourns above it are brief and rare, and so the normalized sojourn $v(t)L_t$ is very sensitive to the local behavior of $X(t)$ near the level value.

In Theorem 1.1 above, the assumption (1.5) is equivalent to

$$2\ \log t - \theta u^2(t) \to \infty, \quad \text{for some } \theta > 1. \tag{1.8}$$

This implies that $\limsup(u^2(t)/2\ \log t) < 1$. Thus the level function $u(t)$ here is smaller than the one in (1.7). Since the level is lower, the sojourns above it are not as brief, and so the local behavior of the sample function is not significant in the limiting distribution of the sojourn time.

In the place of the normalization $v(t)L_t$ we have $(L_t - EL_t)/(\text{Var } L_t)^{1/2}$. An early result in this direction was obtained by the author in [1], where the level $u(t)$ is actually a constant u. (After the publication of [1], Professor Murray Rosenblatt brought to the author's attention the fact that one of the two main results of [1], namely, Theorem 3.2, was actually a special case of a central limit theorem for functionals of stationary Gaussian processes proved by T. C. Sun [9]). The two theorems of that paper require only the mixing conditions $r \in L_2$ and $r \in L_1$, respectively, but no local conditions.

Other theorems of this type for levels $u(t) \to \infty$ at the rate (1.8) were given by the author in [3] and [5]. No local conditions were required for $r(t)$ near $t = 0$. The mixing condition stated that $r(t) \to 0$ for $t \to \infty$ but at a rate which cannot be too rapid. This was part of a more general study using methods of the theory of long-range dependence of Gaussian processes. In particular, these theorems do not cover the case where the conclusion of Theorem 1.1 is obviously true, namely, when $r(t)$ has compact support. Theorem 1.1 above now includes this as a special case.

Let $\phi(z)$ be the standard normal density function, and let $\phi(x, y; \rho)$ be the standard bivariate normal density with correlation coefficient ρ:

$$\phi(x, y; \rho) = (2\pi)^{-1}(1-\rho^2)^{-1/2} \exp\left\{-\frac{x^2 - 2\,\rho xy + y^2}{2(1-\rho^2)}\right\}. \tag{1.9}$$

The normalizing functions EL_t and $\text{Var}(L_t)$ can be expressed in terms of the functions ϕ. Indeed, by stationarity and Fubini's theorem, we have

$$EL_t = \int_0^t E[1_{[X(s)>u(t)]}]\, ds = t\int_{u(t)}^{\infty} \phi(z)\, dz. \tag{1.10}$$

Furthermore, there is a well known formula for $\text{Var}(L_t)$, [6], page 214:

$$\text{Var}(L_t) = 2\int_0^t (t-s)\int_0^{r(s)} \phi(u, u; y)\, dy\, ds. \tag{1.11}$$

The integral on the right hand side of (1.11) may be written as the sum of a "finite" and "infinite" part: For $\epsilon > 0$, it is the sum of the integrals over the domains $[0, \epsilon]$ and $(\epsilon, t]$, respectively. In our previous normal limit theorems in [1], [3] and [5], the finite part was shown to be of smaller order than the infinite part. However, the present theorem is different because, under the present conditions, the finite part is dominant. Furthermore it is shown that the finite part is independent of the values of $r(t)$ outside $[0, \epsilon)$, for any $\epsilon > 0$.

2. Asymptotic Estimate of the Variance Integral.

In this section we estimate the integral,

$$\int_0^t (t-s)\int_0^{r(s)} \phi(u, u; y)\, dy\, ds \tag{2.1}$$

appearing in (1.11) for large values of u and t.

THEOREM 2.1. *For $\epsilon > 0$, consider the expression*

$$t\int_0^{\epsilon}\int_0^{r(s)} \phi(u, u; y)\, dy\, ds. \tag{2.2}$$

If $r \in L_1$, then, for every $\epsilon > 0$,

(2.3) $$\text{Expression (2.1)} \sim \text{Expression (2.2)}$$

under the double limit $u \to \infty$, $t \to \infty$.

PROOF. For the given $\epsilon > 0$, define η as

(2.4) $$\eta = 1 - \max(|r(s)| : s \geq \epsilon).$$

Then, for arbitrary δ such that

(2.5) $$0 < \delta < \eta,$$

there exists ϵ', $0 < \epsilon' < \epsilon$, such that

(2.6) $$\min(r(s) : 0 \leq s \leq \epsilon') \geq 1 - \delta.$$

In order to verify the statement of the theorem it suffices to show that

$$\frac{\int_\epsilon^t (t-s) \int_0^{|r(s)|} \phi(u,u;y)dyds}{\int_0^\epsilon (t-s) \int_0^{r(s)} \phi(u,u;y)dyds}$$

tends to 0 for $u, t \to \infty$. For this purpose it suffices to prove the stronger result that the ratio converges to 0 after the substitution of ϵ' for ϵ in the limit of integration in the denominator because, by continuity, $r(s) \geq 0$ for $0 \leq s \leq \epsilon$:

(2.7) $$\frac{\int_\epsilon^t (t-s) \int_0^{|r(s)|} \phi(u,u;y)dyds}{\int_0^{\epsilon'} (t-s) \int_0^{r(s)} \phi(u,u;y)dyds}.$$

The numerator in (2.7) is at most equal to

(2.8) $$t\phi(u,u;\ 1-\eta) \int_0^t |r(s)|ds$$

because $\phi(u,u;y)$ is increasing for $0 \leq y < 1$, and η is defined by (2.4). The denominator is at least equal to

$$(t-\epsilon') \int_0^{\epsilon'} \int_{1-\delta}^{r(s)} \phi(u,u;y)dyds,$$

which is at least equal to

(2.9) $$\phi(u,u;1-\delta)(t-\epsilon') \int_0^{\epsilon'} (r(s) - 1 + \delta)ds.$$

By (2.8) and (2.9) the ratio (2.7) is at most equal to

$$\frac{t}{t-\epsilon'} \cdot \frac{\int_0^\infty |r(s)|ds}{\int_0^{\epsilon'} (r(s)-1+\delta)ds} \cdot \frac{\phi(u,u;1-\eta)}{\phi(u,u;1-\delta)} \sim constant \left(\frac{2\delta - \delta^2}{2\eta - \eta^2}\right)^{1/2} exp\left[-u^2 \frac{\eta - \delta}{(2-\delta)(2-\eta)}\right],$$

which, by (2.5), converges to 0 for $u, t \to \infty$.

COROLLARY 2.1. *For arbitrary* $\theta > 1$, *the integral* (2.1) *is asymptotically at least equal to a constant times* $te^{-u^2\theta/2}$.

PROOF. According to Theorem 2.1, the integral (2.1) is asymptotically equal to (2.2). In the proof of the theorem we found the lower bound (2.9) for (2.2) for arbitrary $\delta > 0$ and sufficiently small ϵ'. It follows from the definition (1.9) of ϕ that (2.9) is asymptotically equal to a constant times

$$te^{-u^2/(2-\delta)}.$$

For arbitrary $\theta > 1$, choose $\delta = 2(\theta - 1)/\theta$, and this establishes the statement of the corollary.

Theorem 2.1 implies that the asymptotic value of (2.2) is the same for all $\epsilon > 0$. Hence we define $B(u)$ as any function which is asymptotically equal to the coefficient of t in (2.2), and $B(u)$ is independent of ϵ:

$$B(u) \sim \int_0^{\epsilon} \int_0^{r(s)} \phi(u, u; y)\,dy\,ds, \tag{2.10}$$

for $u \to \infty$. It follows from (1.11), (2.10) and Theorem 2.1 that

$$\mathrm{Var}(L_t) \sim 2tB(u), \tag{2.11}$$

for $u, t \to \infty$.

The next lemma is used to estimate the difference in the sojourn time distribution due to a change in the covariance of the process.

LEMMA 2.1. *Let* $r_1(s)$ *and* $r_2(s)$ *be two functions such that* $|r_i(s)| \leq 1$; *then, for* $0 < \epsilon < 1$,

$$\int_0^{\epsilon} \left| \int_{r_1(s)}^{r_2(s)} \phi(u, u; y)\,dy \right| ds \leq (1/\pi)e^{-u^2/2} \left\{ \int_0^{\epsilon} |r_2(s) - r_1(s)|\,ds \right\}^{1/2}. \tag{2.12}$$

PROOF. It follows from the definition (1.9) that

$$\phi(u, u; y) \leq [2\pi(1-y)^{1/2}]^{-1} e^{-u^2/2}, \text{ for } 0 \leq y < 1,$$

so that the left hand member of (2.12) is at most equal to

$$(2\pi)^{-1} e^{-u^2/2} \int_0^{\epsilon} \left| \int_{r_1(s)}^{r_2(s)} (1-y)^{-1/2}\,dy \right| ds,$$

which, by integration over y, is equal to

$$\pi^{-1} e^{-u^2/2} \int_0^{\epsilon} |(1 - r_2(s))^{1/2} - (1 - r_1(s))^{1/2}|\,ds,$$

which, by the moment inequality and the assumption $\epsilon < 1$, is at most equal to

$$\pi^{-1} e^{-u^2/2} \{ \int_0^{\epsilon} |(1 - r_2(s))^{1/2} - (1 - r_1(s))^{1/2}|^2\,ds \}^{1/2}.$$

This is at most equal to

$$\pi^{-1}e^{-u^2/2}\left\{\int_0^{\epsilon}|r_2(s)-r_1(s)|ds\right\}^{1/2}. \tag{2.13}$$

Indeed, this follows from the elementary inequality $(x-y)^2 \leq |x^2-y^2|$ for nonnegative x and y.

3. Asymptotic Analysis of the Covariance Integral and Related Functions.

Let $b(t)$ be a real measurable function such that $b \in L_2$ and define

$$\|b\| = \left(\int_{-\infty}^{\infty} b^2(s)ds\right)^{1/2}. \tag{3.1}$$

For $v > 0$, define

$$b_v(t) = \begin{cases} b(t), & \text{for } |t| \leq v/2 \\ 0, & \text{for } |t| > v/2 \end{cases} \tag{3.2}$$

and the three functions

$$r(t) = \int_{-\infty}^{\infty} b(t+s)b(s)ds \tag{3.3}$$

$$r_{1,v}(t) = \int_{-\infty}^{\infty} b(t+s)(b_v(s)/\|b_v\|)ds \tag{3.4}$$

$$r_{2,v}(t) = \|b_v\|^{-2}\int_{-\infty}^{\infty} b_v(t+s)b_v(s)ds. \tag{3.5}$$

We assume that

$$r(0) = \|b\|^2 = 1, \tag{3.6}$$

so that

$$\|b_v\| \leq 1. \tag{3.7}$$

LEMMA 3.1. *For all v such that $\|b_v\| > 0$,*

$$\sup_t |r(t) - r_{1,v}(t)| \leq \|b_v\|^{-1}\int_{|s|>v/2} b^2(s)ds + \left(\int_{|s|>v/2} b^2(s)ds\right)^{1/2}, \tag{3.8}$$

and

$$\sup_t |r_{1,v}(t) - r_{2,v}(t)| \leq \|b_v\|^{-2}\int_{|s|>v/2} b^2(s)ds + \left(\int_{|s|>v/2} b^2(s)ds\right)^{1/2}\|b_v\|^{-1}. \tag{3.9}$$

PROOf. (3.3) and (3.4) imply

$$|r(t) - r_{1,v}(t)| = |\int_{-\infty}^{\infty} \left[b(t+s) \left\{ b(s) - \frac{b_v(s)}{\|b_v\|} \right\} \right] ds|. \tag{3.10}$$

The right hand member of (3.10) is at most equal to the sum of the two terms

$$(\|b_v\|^{-1} - 1)|\int_{|s| \le v/2} b(t+s)b(s)ds| \tag{3.11}$$

and

$$|\int_{|s|>v/2} b(t+s)b(s)ds|. \tag{3.12}$$

For arbitrary x, $0 < x \le 1$, we have,

$$x^{-1} - 1 = (1-x)/x \le (1-x^2)/x;$$

hence, by (3.6) and (3.7),

$$\begin{aligned}(\|b_v\|^{-1} - 1) &\le \|b_v\|^{-1}(1 - \|b_v\|^2) = \|b_v\|^{-1}(\|b\|^2 - \|b_v\|^2) \\ &= \|b_v\|^{-1} \int_{|s|>v/2} b^2(s)ds.\end{aligned}$$

From this, and an application of the Cauchy-Schwarz inequality to the integral in (3.11), we see that the term (3.11) is at most equal to

$$\|b_v\|^{-1} \int_{|s|>v/2} b^2(s)ds. \tag{3.13}$$

By a similar argument, (3.12) is dominated by

$$\left(\int_{|s|>v/2} b^2(t+s)ds\right)^{1/2} \left(\int_{|s|>v/2} b^2(s)ds\right)^{1/2},$$

which is at most equal to

$$\left(\int_{|s|>v/2} b^2(s)ds\right)^{1/2}. \tag{3.14}$$

The right hand member of (3.8) is now obtained as a consequence of (3.13) and (3.14).

The proof of (3.9) is obtained from the proof above for (3.8) by writing $r_{1,v}(t) - r_{2,v}(t)$ in terms of the integrals (3.4) and (3.5); changing the variable of integration to obtain

$$|r_{1,v}(t) - r_{2,v}(t)| = |\int_{-\infty}^{\infty} \left\{ \frac{b_v(s-t)}{\|b_v\|} \left[b(s) - \frac{b_v(s)}{\|b_v\|} \right] \right\} ds|;$$

noting that $|b_v(s-t)| \le |b(s-t)|$; and then using the estimates of (3.11) and (3.12) after division by $\|b_v\|$.

4. Additional Asymptotic Estimates

Now we apply the results of Section 3 to the estimates of the integral in Section 2.

LEMMA 4.1. *Let* $r, r_{1,v}$ *and* $r_{2,v}$ *be defined as in* (3.3), (3.4) *and* (3.5), *respectively. If* $b \in L_1$, *then, for every* $\epsilon > 0$,

$$\lim_{u,v,t\to\infty} (tB(u))^{-1} \int_\epsilon^t (t-s) \int_0^{|r_{i,v}(s)|} \phi(u,u;y)dyds = 0, \tag{4.1}$$

for $i = 1, 2$, *and where* $B(u)$ *satisfies* (2.10).

PROOF. Lemma 3.1 implies the uniform convergence of $r_{i,v}(t)$ to $r(t)$ for $v \to \infty$. Since $r(t)$ is bounded away from 1 for $t \geq \epsilon$, for arbitrary $\epsilon > 0$, it follows that $r_{i,v}(t)$ is also bounded away from 1 for all $t \geq \epsilon$, uniformly for all large v. Hence, the expression under the limit sign in (4.1) is at most equal to (see (2.8))

$$(B(u))^{-1}\phi(u,u;\rho) \int_0^t |r_{i,v}(s)|ds,$$

where ρ is the uniform upper bound of $r_{i,v}(t)$ for $t \geq \epsilon$, and where $\rho < 1$. By Corollary 2.1, and the definitions (1.9) and (2.10) of ϕ and B, respectively, the expression displayed above is at most equal to a constant times

$$e^{\frac{1}{2}u^2(\theta - \frac{2}{1+\rho})} \int_0^\infty |r_{i,v}(s)|ds. \tag{4.2}$$

Since $\rho < 1$, and θ may be chosen arbitrarily close to 1, we may take θ to satisfy $1 < \theta < 2/(1+\rho)$, so that the exponential factor in (4.2) converges to 0 for $u \to \infty$. By an application of Fubini's theorem and by (3.6), it follows that

$$\int_0^\infty |r_{i,v}(s)|ds \leq \|b_v\|^{-2} \left(\int_{-\infty}^\infty |b(s)|ds\right)^2 \longrightarrow \left(\int_{-\infty}^\infty |b(s)|ds\right)^2.$$

It follows that the expression (4.2) converges to 0, and this completes the proof of (4.1).

LEMMA 4.2. *Suppose that* $u = u(t)$ *and* $v = v(t)$ *are functions of* t *which tend to* ∞ *with* t, *and there exists* $\delta > 0$ *such that*

$$\lim_{t\to\infty} e^{\delta u^2} \int_{|s|>v/2} b^2(s)ds = 0. \tag{4.3}$$

If $b \in L_1$, *then*

$$\lim_{t\to\infty} \frac{\int_0^t (t-s)|\int_{r_{i,v}(s)}^{r(s)} \phi(u,u;y)dy|ds}{tB(u)} = 0. \tag{4.4}$$

PROOF. The expression in the numerator in (4.4) is at most equal to the sum of the three terms,

$$\int_0^\epsilon (t-s)|\int_{r_{i,v}(s)}^{r(s)} \phi(u,u;y)dy|ds \tag{4.5}$$

$$\int_\epsilon^t (t-s)\int_0^{|r(s)|} \phi(u,u;y)dy\, ds, \tag{4.6}$$

$$\int_\epsilon^t (t-s)\int_0^{|r_{i,v}(s)|} \phi(u,u;y)dy\, ds. \tag{4.7}$$

By Theorem 2.1 and Lemma 4.1, and the definition (2.10) of $B(u)$, the terms (4.6) and (4.7) are both of smaller order than $tB(u)$, so that these terms may be ignored in the verification of (4.4). Thus it suffices to show that the term (4.5) is of smaller order than $tB(u)$.

By Lemma 2.1 and Corollary 2.1, the ratio of (4.5) to $tB(u)$ is at most equal to

$$(1/\pi)e^{(u^2/2)(\theta-1)}\left\{\int_0^\epsilon |r(s)-r_{i,v}(s)|ds\right\}^{1/2},$$

for arbitrary $\theta > 1$. By Lemma 3.1, the latter is at most equal to a constant times

$$e^{(u^2/2)(\theta-1)}\|b_v\|^{-1}\left(\int_{|s|>v/2} b^2(s)ds\right)^{1/4}. \tag{4.8}$$

Since there is a δ for which (4.3) holds, then the latter also holds for all δ', $0<\delta'<\delta$. In particular, since θ in (4.8) may be chosen arbitrarily close to 1, it follows that $\delta' = \frac{1}{2}(\theta-1)$ may be chosen to be less than δ, and so the expression (4.8) converges to 0. This completes the proof of (4.4).

5. The Auxiliary Process.

Let $X(t)$, $t \geq 0$, be a stationary measurable Gaussian process with mean 0 and covariance function $r(t)$ with the representation (3.3) and satisfying $r(0) = 1$. Such a process has the stochastic integral representation

$$X(t) = \int_{-\infty}^{\infty} b(t+s)\xi(ds), \tag{5.1}$$

where $\xi(s)$ is the standard Brownian motion. For $v > 0$, we define another process $X_v(t)$ on the same space by means of the stochastic integral

$$X_v(t) = \|b_v\|^{-1}\int_{-\infty}^{\infty} b_v(t+s)\xi(ds). \tag{5.2}$$

Elementary considerations show that the vector process $(X(t), X_v(t))$ has the covariance matrix function

$$E\begin{pmatrix} X(t) \\ X_v(t)\end{pmatrix}(X(0)X_v(0)) = \begin{bmatrix} r(t) & r_{1,v}(t) \\ r_{1,v}(t) & r_{2,v}(t)\end{bmatrix}. \tag{5.3}$$

Let $B(u)$ be a function satisfying (2.10). Then for any function $u(t)$ satisfying (1.5) and (1.6), define

$$v(t) = [tB(u(t))]^{1/2} \tag{5.4}$$

and

$$w(t) = v(t)(u(t))^{1/2}. \tag{5.5}$$

It follows from Theorem 2.1, Corollary 2.1 and the relation (2.10), that

$$B(u) \geq \textit{constant } e^{-u^2\theta/2}, \tag{5.6}$$

for all sufficiently large u, for arbitrary $\theta > 1$. Therefore, the assumed relation (1.5) implies

$$v(t) \to \infty. \tag{5.7}$$

It is obvious from (5.5) that

$$v(t)/w(t) \to 0. \tag{5.8}$$

Since $B(u) \to 0$ for $u \to \infty$, it follows from the definition (5.5) of w and from (1.5) that

$$\frac{w(t)}{t} = \frac{u(t)^{1/4}}{\sqrt{t}} \sqrt{B(u(t))} \to 0. \tag{5.9}$$

In the course of the proof of Theorem 1.1, to be given in Section 6, we shall need estimates of the variances of the sojourns times of the processes $X(s)$ and $X_v(s)$, where $v = v(t)$, over intervals of lengths $v(t)$ and $w(t)$. These sojourn times are defined as

$$\begin{aligned} L_w &= \int_0^w 1_{[X(s)>u]} ds \\ L_w^v &= \int_0^w 1_{[X_v(s)>u]} ds \\ L_v^v &= \int_0^v 1_{[X_v(s)>u]} ds, \end{aligned} \tag{5.10}$$

where $u = u(t)$, $v = v(t)$ and $w = w(t)$ are the functions of t defined above.

If, in the statement of Theorem 2.1, we take w in the place of t, then it follows that

$$\mathrm{Var}(L_w) \sim 2wB(u). \tag{5.11}$$

By the application of the general formula (1.11) to the process X_v, and by the definition of $r_{2,v}$ in (3.5), it follows that

$$\mathrm{Var}(L_w^v) = 2\int_0^w (w-s) \int_0^{r_{2,v}(s)} \phi(u,u;y)dy\, ds. \tag{5.12}$$

Let us establish the relation

$$\mathrm{Var}(L_w^v) \sim \mathrm{Var}(L_w). \tag{5.13}$$

Since

$$\frac{\mathrm{Var}(L_w^v)}{\mathrm{Var}(L_w)} = \frac{\mathrm{Var}(L_w^v) - \mathrm{Var}(L_w)}{\mathrm{Var}(L_w)} + 1,$$

it suffices on the basis of (5.11) to verify

$$\text{Var}(L_w) - \text{Var}(L_w^v) = o(wB(u)). \tag{5.14}$$

The left hand member of (5.14) is, by application of (1.11), at most equal to

$$2\int_0^w (w-s)\left|\int_{r_{2,v}(s)}^{r(s)} \phi(u,u;y)dy\right|ds.$$

By Lemma 4.2, with w in the place of t in the limiting operation in (4.3) and (4.4), the integral displayed above is of smaller order than $wB(u)$. This proves (5.14) and, as a consequence, (5.13).

By the same arguments leading to (5.11) and (5.12) but with v in the place of w, we obtain

$$\text{Var}(L_v^v) \sim 2vB(u). \tag{5.15}$$

A key step in the proof of Theorem 1.1, to be given in Section 6, is showing that the limiting distribution of the sojourn time is the same for the original process $X(s)$ and for the auxiliary process $X_v(s)$ in (5.2), where $v = v(t)$ is defined by (5.4).

LEMMA 5.1. *Put*

$$L_{t,v(t)} = \int_0^t 1_{[X_{v(t)}(s)>u(t)]}ds; \tag{5.16}$$

then,

$$(L_t - EL_t)/(\text{Var}(L_t))^{1/2} \tag{5.17}$$

has the same limiting distribution, if any, as

$$(L_{t,v(t)} - EL_{t,v(t)})/(\text{Var}(L_{t,v(t)}))^{1/2}. \tag{5.18}$$

PROOF. According to the general formula (1.11), with $r_{2,v}(s)$ in the place of $r(s)$, we have

$$\text{Var}(L_{t,v(t)}) = 2\int_0^t (t-s)\int_0^{r_{2,v(t)}(s)} \phi(u,u;y)dy\ ds.$$

Furthermore, by (2.11),

$$\text{Var}(L_t) \sim 2tB(u(t)); \tag{5.19}$$

hence,

$$\frac{\text{Var}(L_t) - \text{Var}(L_{t,v(t)})}{\text{Var}(L_t)}$$
$$\sim \int_0^t (t-s)\int_{r_{2,v(t)}(s)}^{r(s)} \phi(u,u;y)dy\ ds\ /\ [tB(u(t))],$$

which, by Lemma 4.2, converges to 0. Thus, (5.17) and (5.18) are asymptotically equal to

$$(L_t - EL_t)\ /\ [2tB(u(t))]^{1/2} \tag{5.20}$$

and

$$(L_{t,v(t)} - EL_{t,v(t)}) \; / \; [2tB(u(t))]^{1/2}, \tag{5.21}$$

respectively.

Next we observe that, by Fubini's theorem and stationarity, and the fact that $X(0)$ and $X_v(0)$ have the same unit variance,

$$\begin{aligned} EL_t &= \int_0^t P(X(s) > u(t))ds = tP(X(0) > u(t)) \\ &= tP(X_{v(t)}(0) > u(t)) \quad = \int_0^t P(X_{v(t)}(s) > u(t))ds \\ &\qquad\qquad\qquad\qquad\quad = EL_{t,v(t)}. \end{aligned}$$

Therefore, in order to prove that (5.20) and (5.21) have the same limiting distribution, it suffices to show that

$$\mathrm{Var}(L_t - L_{t,v(t)}) \; / \; 2tB(u(t)) \to 0. \tag{5.22}$$

For the proof of (5.22), we note that, according to the calculation in [1], page 729, the numerator in (5.22) is representable as

$$2\int_0^t (t-s)\left(\int_0^{r(s)} + \int_0^{r_{2,v}(s)} -2\int_0^{r_{1,v}(s)}\right) \phi(u,u;y)dy \; ds. \tag{5.23}$$

Writing

$$\int_0^{r(s)} + \int_0^{r_{2,v}(s)} -2\int_0^{r_{1,v}(s)} = 2\int_{r_{1,v}(s)}^{r(s)} - \int_{r_{2,v}(s)}^{r(s)},$$

we see that (5.23) is dominated by

$$\begin{aligned} &4\int_0^t (t-s)\left|\int_{r_{1,v}(s)}^{r(s)} \phi(u,u;y)dy\right|ds \\ &\qquad + 2\int_0^t (t-s)\left|\int_{r_{2,v}(s)}^{r(s)} \phi(u,u;y)dy\right|ds. \end{aligned}$$

The relation (5.22) is now a consequence of Lemma 4.2.

6. Proof of Theorem 1.1.

The proof is based on the classical "blocking" method used in the proofs of central limit theorems for dependent random variables. The novelty of the proof here is in the application of the assumptions (1.5) and (1.6) to the rates of growth of the constructed blocks. With $v(t)$ and $w(t)$ defined in (5.4) and (5.5), respectively, put

$$n(t) = \textit{Integer part of } \frac{t}{v(t)+w(t)}. \tag{6.1}$$

Let $J_1, J_2, \ldots$ and $K_1, K_2, \ldots$ be families of intervals defined for each t as follows:

$$J_j = [(j-1)(v(t)+w(t)),\ (j-1)v(t)+jw(t)],\ \ j = 1,2,\ldots; \tag{6.2}$$

$$K_j = [(j-1)v(t)+jw(t),\ j(v(t)+w(t))],\ \ j = 1,2,\ldots. \tag{6.3}$$

The intervals J_j and K_j are of lengths $w(t)$ and $v(t)$, respectively, and fall on the positive real axis in the order $J_1, K_1, J_2, K_2, \ldots, J_j, K_j, \ldots$. The interval $[0,t]$ contains $J_1, K_1, \ldots, J_n, K_n$, for $n = n(t)$ defined by (6.1), and is contained in the union of $J_1, K_1, \ldots, J_n{+}1, K_n{+}1$.

We define the partial sojourn times of the process $X_v(s)$, $s \geq 0$, for $v = v(t)$, over the intervals J_j and K_j:

$$L(J_j) = \int_{J_j} 1_{[X_v(s)>u(t)]}ds,$$

$$L(K_j) = \int_{K_j} 1_{[X_v(s)>u(t)]}ds,\ \ j = 1,2,\ldots.$$

Note that $r_{2,v}(s)$, which is the covariance function of $X_v(\cdot)$, has support contained in $[-v,v]$; hence, the random variables $X_v(s)$ and $X_v(s')$ are independent for $|s-s'| \geq v$. Since the intervals (J_j) are mutually separated by distances at least equal to v, it follows that the partial sojourn times $L(J_j)$, $j = 1,2,\ldots$ are mutually independent. Furthermore,by the stationarity of the process $X_v(\cdot)$, the random variables $L(J_j)$ are also identically distributed.

LEMMA 6.1. *Let $n = n(t)$ be defined by (6.1); then*

$$\frac{\sum_{j=1}^{n(t)} L(J_j) - \sum_{j=1}^{n(t)} EL(J_j)}{[2tB(u(t))]^{1/2}} \tag{6.4}$$

has, for $t \to \infty$, a limiting standard normal distribution.

PROOF. By the independence of $L(J_j)$, we have

$$\mathrm{Var}\left(\sum_{j=1}^{n(t)} L(J_j)\right) = n(t)\mathrm{Var}(L(J_1)),$$

which, by (5.11) and (6.1), is asymptotically equal to

$$\frac{2t\ w(t)\ B(u(t))}{w(t)+v(t)},$$

which, by (5.8), is asymptotically equal to $2tB(u(t))$:

$$\mathrm{Var}\left(\sum_{j=1}^{n(t)} L(J_j)\right) \sim n(t)\mathrm{Var}(L(J_1)) \sim 2tB(u(t)). \tag{6.5}$$

Define

$$\xi_j = \frac{L(J_j) - EL(J_j)}{[\mathrm{Var}(L(J_j))]^{1/2}},\ \ j = 1,2,\ldots;$$

these are independent with mean 0 and unit variance. The distribution of (6.4) is, by (6.5), asymptotically equivalent to

$$\frac{\xi_1 + \ldots + \xi_{n(t)}}{\sqrt{n(t)}}. \tag{6.6}$$

Now we apply Esseen's theorem [8]: There is a universal constant $C > 0$ such that for every $n \geq 1$,

$$\sup_x |P\left(\frac{\xi_1 + \ldots + \xi_n}{\sqrt{n}} \leq x\right) - \Phi(x)| \leq \frac{CE|\xi_1|^3}{\sqrt{n}},$$

where Φ is the standard normal distribution function. Thus, in order to complete the proof of the lemma it suffices to show that

$$\lim_{t\to\infty} \frac{E|L(J_1) - EL(J_1)|^3}{\{n(t)[\mathrm{Var}(L(J_1))]^3\}^{1/2}} = 0. \tag{6.7}$$

Since the interval J_1 is of length w, we have $|L(J_1) - EL(J_1)|^3 \leq 2w|L(J_1) - EL(J_1)|^2$, and so the ratio in (6.7) is at most equal to $2w/(n(t)\ \mathrm{Var}(L(J_1)))^{1/2}$, which, by (5.8), (5.11), (5.13) and (6.1), is asymptotically equal to

$$\frac{2w}{[(t/w)2wB(u)]^{1/2}},$$

which, by (5.4) and (5.5), is equal to $\sqrt{2}(u(t))^{-1/4}$, which tends to 0 for $t \to \infty$. This completes the proof of (6.7) and of the lemma.

LEMMA 6.2. *Let $n = n(t)$ be defined by* (6.1); *then*

$$\frac{\sum_{j=1}^{n(t)} L(K_j) - \sum_{j=1}^{n(t)} EL(K_j)}{[2tB(u(t))]^{1/2}} \tag{6.8}$$

converges in probability to 0 *for $t \to \infty$.*

PROOF. Since the intervals (K_j) are mutually separated by intervals of lengths at least equal to w, which is greater than v, the random variables $L(K_j)$, like $L(J_j)$, are mutually independent, and are also identically distributed. Then the variance of the numerator in (6.8) is equal to $n(t)\mathrm{Var}(L(K_1))$. Since K_1 is of length v, the relation (5.15) implies $n(t)\mathrm{Var}(L(K_1)) \sim 2n(t)vB(u)$, which, by (6.1), is asymptotically equal to $2tvB(u)/w$, which, by (5.8), is $o(tB(u))$. Thus the variance of the numerator in (6.8) is of smaller order than the square of the denominator. This implies the statement of the lemma.

The proof of the theorem now follows from Lemmas 6.1 and 6.2. Indeed, the normed sojourn time $[L_t - EL_t]/(\mathrm{Var}(L_t))^{1/2}$ is asymptotically equal to

$$(L_t - EL_t)/[2tB(u(t))]^{1/2}.$$

Since

$$\bigcup_{j=1}^{n(t)} (J_j \cup K_j) \subset [0, t] \subset \bigcup_{j=1}^{n(t)+1} (J_j \cup K_j),$$

it follows that

$$
\begin{aligned}
\sum_{j=1}^{n(t)} L(J_j) - \sum_{j=1}^{n(t)+1} EL(J_j) + \sum_{j=1}^{n(t)} L(K_j) - \sum_{j=1}^{n(t)+1} EL(K_j) \\
\leq L_t - EL_t \\
\leq \sum_{j=1}^{n(t)+1} L(J_j) - \sum_{j=1}^{n(t)} EL(J_j) + \sum_{j=1}^{n(t)+1} L(K_j) - \sum_{j=1}^{n(t)} EL(K_j).
\end{aligned}
$$

By Lemma 6.2 the sum of the terms above which involve K_j are, in probability, of smaller order than $[tB(u(t))]^{1/2}$, so that they may be neglected in the estimate of $L_t - EL_t$ above:

$$
\begin{aligned}
&\sum_{j=1}^{n(t)} [L(J_j) - EL(J_j)] - EL(J_{n}+1) \leq L_t - EL_t \\
&\qquad \leq L(J_{n(t)+1}) + \sum_{j=1}^{n(t)} [L(J_j) - EL(J_j)].
\end{aligned} \tag{6.9}
$$

The relation (6.9) implies

$$
\begin{aligned}
&\left|\frac{L_t - EL_t}{[2tB(u(t))]^{1/2}} - \frac{\sum_{j=1}^{n(t)} L(J_j) - EL(J_j)}{[2tB(u(t))]^{1/2}}\right| \\
&\leq \frac{|L(J_{n(t)+1}) - EL(J_{n(t)+1})|}{[2tB(u(t))]^{1/2}}.
\end{aligned} \tag{6.10}
$$

The expected square of the right hand member of (6.10) is, by (5.11) and (5.13), asymptotically equal to $2wB(u)/2tB(u)$, which, by (5.9), converges to 0 for $t \to \infty$. Thus the left hand member of (6.10) also converges to 0 in probability. The asymptotic normality of $(L_t - EL_t)/[2tB(u(t))]^{1/2}$ now follows from Lemma 6.1.

7. A Sufficient Condition on $b(s)$ for the Validity of (1.6).

In the determination of whether a given function $u(t)$ satisfies (1.5) and (1.6), the former is clearly easy to check; however, the latter requires specific information about the tail of $|b(s)|$ for $s \to \infty$. In this section we find a simple sufficient condition on b which, together with (1.5), also implies (1.6). We will prove:

If, for some $q > 0$,

$$
\int_{|s|>t} b^2(s)ds = O(t^{-q}), \quad \text{for } t \to \infty, \tag{7.1}
$$

then any function $u(t)$ which satisfies (1.5) for some $\theta > 1$ also satisfies (1.6) for any $\delta > 0$.

Let $\theta_0 > 1$ be the value of θ for which (1.5) holds. Put

$$
\theta_1 = 1 + \frac{1}{2}(\theta_0 - 1);
$$

it is obviously smaller than θ_0. Then (1.5) also holds for $\theta = \theta_1$. By (7.1) the expression under the limit sign in (1.6) is of the order

$$
t^{-q/2} e^{qu^2(t)\theta_1/4 + \delta u(t)},
$$

which is equal to

$$\{te^{-\frac{1}{2}u^2(t)\theta_1[1+\frac{4\delta}{\theta_1 qu(t)}]}\}^{-q/2}.$$

This converges to 0 under the hypothesis (1.5) for $\theta = \theta_0$ because

$$\theta_1\left(1+\frac{4\delta}{\theta_1 qu(t)}\right) < \theta_0$$

for all sufficiently large t, and so (1.6) is also satisfied.

In general, even without the hypothesis (7.1), $u(t)$ may be selected in the following way. For some $\theta > 1$, let $u(t) = u_1(t)$ satisfy (1.5). For any $d > 0$, put

$$u_2(t) = -\frac{1}{d}\log\left(\int_{|s|>\sqrt{t}\exp(-u_1^2(t)\theta/4)} b^2(s)ds\right)$$

if the integral above is positive, and put $u_2(t) = u_1(t)$ if the integral is equal to 0. If δ satisfies $0 < \delta < d$, then

$$e^{\delta u_2(t)}\int_{|s|>\sqrt{t}\exp(-u_1^2(t)\theta/4)} b^2(s)ds \to 0.$$

Put $u(t) = \min(u_1(t), u_2(t))$; then, $u(t)$ satisfies (1.5) and (1.6) for the given $\delta > 0$ and $\theta > 1$.

8. Asymptotic Properties of the Function $B(u)$

Define $B(u)$ as the function appearing on the right hand side of (2.10); then, by Corollary 2.1, a crude lower asymptotic bound for $B(u)$ is

$$B(u) \geq constant\ e^{-\frac{1}{2}u^2\theta}, \tag{8.1}$$

for every $\theta > 1$. We will now obtain a finer estimate of $B(u)$ which is of independent interest, and then show how it relates to our previous work on sojourns in the compound Poisson limit case.

THEOREM 8.1. *Define*

$$\Psi(x) = \int_x^\infty \phi(z)\,dz, \tag{8.2}$$

where ϕ is the standard normal density; then $B(u)$ has, for arbitrary $\delta > 0$, the upper asymptotic bound

$$2(\frac{2}{2-\delta})^{1/2}\Psi(u)\int_0^\epsilon \Psi(u[\frac{1}{2}(1-r(s))]^{1/2})\,ds, \quad \text{for } u\to\infty, \tag{8.3}$$

and the lower asymptotic bound

$$[2(2-\delta)]^{1/2}\Psi(u)\int_0^\epsilon \Psi(u[\frac{1-r(s)}{2-\delta}]^{1/2})\,ds. \tag{8.4}$$

PROOF. Since $B(u)$ in (2.10) has the same asymptotic value for all $\epsilon > 0$, we may for arbitrary $\delta > 0$, choose ϵ so that

$$1 - r(s) < \frac{1}{2}\delta, \quad \text{for } 0 \leq s < \epsilon. \tag{8.5}$$

Then $B(u)$ is asymptotically equal to

$$\int_0^{\epsilon}\int_{1-\delta}^{r(s)} \phi(u,u;y)\,dy\,ds. \tag{8.6}$$

Indeed, on the one hand, the portion of the integral complementary to (8.6), namely,

$$\int_0^{\epsilon}\int_0^{1-\delta} \phi(u,u;y)\,dy\,ds, \quad \text{or equivalently,} \quad \epsilon\int_0^{1-\delta} \phi(u,u;y)\,dy$$

is at most equal to

$$\epsilon(1-\delta)\phi(u,u;1-\delta) = \frac{\epsilon(1-\delta)}{2\pi[1-(1-\delta)^2]^{1/2}}\exp[-\frac{u^2}{2-\delta}].$$

On the other hand, by (8.1), $B(u)$ is at least of the order $\exp(-\frac{1}{2}u^2\theta)$ for arbitrary $\theta > 1$. Choosing $1 < \theta < 2/(2-\delta)$, we see that $\phi(u,u;1-\delta)/B(u) \to 0$, and this establishes the asymptotic value (8.6).

By the formula (1.9) for the bivariate normal density the integral (8.6) is equal to

$$\phi(u)\int_0^{\epsilon}\int_{1-\delta}^{r(s)} (1-y^2)^{-1/2}\phi(u[\frac{1-y}{1+y}]^{1/2})dy\,ds,$$

where $\phi(z)$ is the standard normal density. By the substitution $z = u^2(1-y)$, the expression above is transformed to

$$\frac{\phi(u)}{u}\int_0^{\epsilon}\int_{u^2(1-r(s))}^{u^2\delta} [z(2-z/u^2)]^{-1/2}\phi\left(\left[\frac{z}{2-z/u^2}\right]^{1/2}\right)dz\,ds. \tag{8.7}$$

An upper bound for (8.7) is

$$(2-\delta)^{-1/2}\frac{\phi(u)}{u}\int_0^{\epsilon}\int_{u^2(1-r(s))}^{\infty} \phi(\sqrt{z/2})\frac{dz}{\sqrt{z}}ds, \tag{8.8}$$

and a lower bound is

$$\phi(u)/u\int_0^{\epsilon}\int_{u^2(1-r(s))}^{u^2\delta} \phi([\frac{z}{2-\delta}]^{1/2})\frac{dz}{\sqrt{2z}}ds. \tag{8.9}$$

Change the variable of integration in (8.8) by the substitution $x = \sqrt{z/2}$; then (8.8) becomes

$$(2-\delta)^{-1/2}\frac{\phi(u)}{u}2\sqrt{2}\int_0^{\epsilon}\int_{u(\frac{1-r(s)}{2})^{1/2}}^{\infty} \phi(x)dx\,ds$$

which is equal to

$$2(\frac{2}{2-\delta})^{1/2}\frac{\phi(u)}{u}\int_0^{\epsilon} \Psi(u(\frac{1-r(s)}{2})^{1/2})ds.$$

By the well known relation $\Psi(u) \sim \phi(u)/u$, for $u \to \infty$, the second expression above is asymptotically equal to the expression (8.3). By the change of variable $x = (z/(2-\delta))^{1/2}$, the expression (8.9) becomes

$$(2(2-\delta))^{1/2}\frac{\phi(u)}{u}\int_0^{\epsilon}\int_{u(\frac{1-r(s)}{2-\delta})^{1/2}}^{u(\delta/(2-\delta))^{1/2}} \phi(x)dx\,ds,$$

which, by the definition of Ψ, is equal to

$$(2(2-\delta))^{1/2}\frac{\phi(u)}{u}\int_0^\epsilon \left[\Psi\left(u(\frac{1-r(s)}{2-\delta})^{1/2}\right)-\Psi\left(u(\frac{\delta}{2-\delta})^{1/2}\right)\right]ds. \tag{8.10}$$

The condition (8.5) implies

$$\lim_{u\to\infty}\sup_{0\le s\le\epsilon}\Psi(u(\frac{\delta}{2-\delta})^{1/2})\ /\ \Psi(u(\frac{1-r(s)}{2-\delta})^{1/2})=0.$$

Hence, the right hand member of (8.10) is asymptotically equal to

$$(2(2-\delta))^{1/2}\frac{\phi(u)}{u}\int_0^\epsilon \Psi(u(\frac{1-r(s)}{2-\delta})^{1/2})ds,$$

which, by the well known relation $\Psi(u)\sim\phi(u)/u$, is asymptotically equal to the expression (8.4).

In the particular case where the covariance function $r(t)$ has the property that $1-r(t)$ is of regular variation for $t\to 0$, we obtain an exact asymptotic value for $B(u)$.

THEOREM 8.2. *Suppose that* $1-r(t)$ *is of regular variation of index* α, $0<\alpha\le 2$, *for* $t\to 0$. *Let* $v=v(u)$, $u>0$, *be a function implicitly defined by the relation*

$$1-r(\frac{1}{v(u)})\sim u^{-2},\quad \text{for } u\to\infty. \tag{8.11}$$

Then

$$B(u)\sim\frac{\Psi(u)}{v(u)}\frac{2^{\alpha/2}}{\sqrt{\pi}}\Gamma(\frac{1}{2}+\frac{1}{\alpha}),\ \text{for } u\to\infty. \tag{8.12}$$

PROOF. Change the variable of integration in (8.3) by means of the substitution $s=t/v$:

$$2(\frac{2}{2-\delta})^{1/2}\frac{\Psi(u)}{v}\int_0^{\epsilon v}\Psi(u(\frac{1-r(t/v)}{2})^{1/2})dt. \tag{8.13}$$

By the argument in the earlier work [4], it follows that $u^2(1-r(t/v))\to t^\alpha$, and, furthermore, the limit may be taken under the integral sign so that (8.13) is asymptotically equal to

$$2(\frac{2}{2-\delta})^{1/2}\frac{\Psi(u)}{v}\int_0^\infty \Psi(s^{\alpha/2}/\sqrt{2})ds. \tag{8.14}$$

Similarly we find that (8.4) is asymptotically equal to

$$(2(2-\delta))^{1/2}\frac{\Psi(u)}{v}\int_0^\infty \Psi(s^{\alpha/2}/(2-\delta)^{1/2})ds. \tag{8.15}$$

Since the upper and lower asymptotic bounds (8.14) and (8.15), respectively, are equal except for factors whose ratio tends to 1 uniformly in u for $\delta\to 0$, and since δ is arbitrary, it follows that the exact asymptotic value exists and is given by the common value of (8.14) and (8.15) for $\delta=0$:

$$\frac{2}{v(u)}\Psi(u)\int_0^\infty \Psi(s^{\alpha/2}/\sqrt{2})ds.$$

The right hand member of (8.12) is obtained from this by evaluating the integral above. The steps are: Transform the integral by means of $t = s^{\alpha/2}/\sqrt{2}$; integrate by parts; and then substitute $y = \frac{1}{2}t^2$ to obtain a Gamma function integral.

We recall that the function $v = v(u(t))$ is the normalizing function for L_t in the case of the compound Poisson limit when $1 - r(t)$ is regularly varying (see [4]). The relation (8.12) does not depend on the rate of increase of u with t so that it may be applied also to the Poisson limit case where $u(t)$ increases as in (1.7). In particular, the level $u(t)$ was chosen in the latter case so that $tv(u(t))\Psi(u(t)) \to 1$, for $t \to \infty$. Thus, it follows by a simple computation that $E[v(u(t))L_t] \to 1$. If we now assume the condition in the hypothesis of Theorem 2.1 above, namely $r \in L_1$, then we may apply Theorem 8.2 to the formula (2.11) for the variance of L_t to obtain the limit of the variance of vL_t:

$$\mathrm{Var}(vL_t) \sim 2v^2 tB(u) \to 2^{\alpha/2+1}\pi^{-1/2}\Gamma(\frac{1}{2} + \frac{1}{\alpha}).$$

REFERENCES

1. Berman, S. M., Occupation times of stationary Gaussian processes, *J. Applied Probability* 7 (1970) 721-733.

2. Berman, S. M., Maxima and high level excursions of stationary Gaussian processes, *Trans. Amer. Math. Soc.* 160 (1971) 65-85.

3. Berman, S. M., High level sojourns for strongly dependent Gaussian processes, *Z. Wahrscheinlichkeitstheorie verw. Gebiete* 50 (1979) 223-236.

4. Berman, S. M., A compound Poisson limit for stationary sums, and sojourns of Gaussian processes, *Ann. Probability* 8 (1980) 511-538.

5. Berman, S. M., Sojourns of vector Gaussian processes inside and outside spheres, *Z. Wahrcheinlichkeitstheorie verw. Gebeite.* 66 (1984) 529-542.

6. Cramer, H. & Leadbetter, M. R., *Stationary and Related Stochastic Processes,* John Wiley, New York, 1967.

7. Doob, J. L., *Stochastic Processes,* John Wiley, New York, 1954.

8. Esseen, C. G., Fourier analysis of distribution functions, *Acta Math.* 77 (1944) 1-125.

9. Sun, T. C., Some further results on central limit theorems for nonlinear functionals of a normal stationary process, *J. Math. Mech.* 14 (1965) 71-85.

10. Volkonskii, V. A. & Rozanov, Yu. A., Some limit theorems for random functions II. *Theory Probability Appl.* 6 (1961) 186-198.

ON THE DISTRIBUTION OF RANDOM WAVES AND CYCLES

Igor Rychlik
Dept. of Mathematical Statistics
University of Lund, Box 118,
S–22100 Lund, Sweden

Abstract: Many technically important problems involving random processes depend on the extremes of the studied functions, which can be regarded as a sequence of "waves" or "cycles". This paper presents some recently proposed approximations for wavelength and amplitude distributions for three commonly used definitions of waves, when the studied function is a sample path of an ergodic, stationary, twice continuously differentiable process.

1. INTRODUCTION

The theory of level crossings of random processes has found applications in many areas of physical sciences, such as fatigue analysis, hydrology, seismology, meteorology, and many others. Many technically important problems depend mostly on the extremes of the studied functions, which can be regarded as a sequence of "waves" or "cycles", where each wave is characterized by its wavelength and amplitude. The observed data can then be presented in the form of the empirical distribution of wavelength and amplitude.

In the case when functions describing for example, water elevation at a fixed point, wind speed, or load history are irregular, there is no obvious definition of a "wave", and the most commonly used definitions are the following.

DEFINITION 1: (crest–to–trough) The wavelength T and amplitude H are the differences in time and height between a crest (local maximum) and the subsequent trough (local minimum); see Fig. 1 (a). □

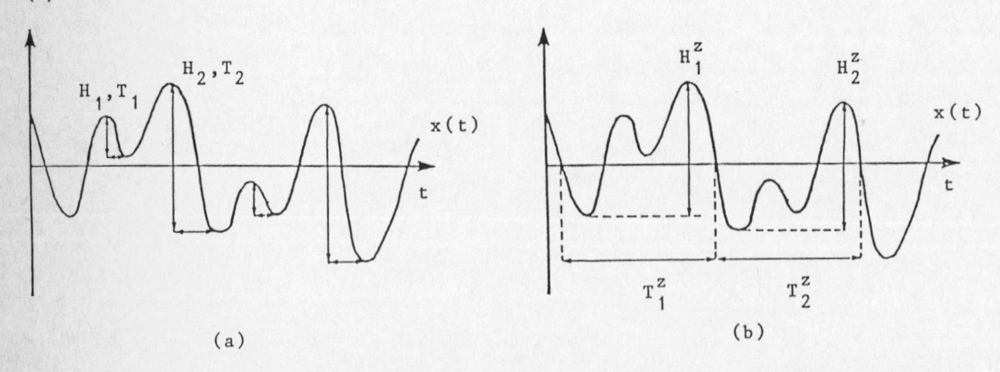

Fig. 1 Definition of crest–to–trough wavelength T and amplitude H (a), and the definition of zerocrossing wavelength T^z and amplitude H^z (b).

Keywords: Crossings, fatigue, periods analysis, Slepian model process, wave–height, wave–length.

DEFINITION 2: (zerocrossing) The zerocrossing wavelength T^z and amplitude H^z are the time interval between consecutive zero–downcrossings and the difference between the highest and lowest value in this interval, respectively; see Fig. 1 (b). □

The crest–to–trough and the zerocrossing waves are often used to describe the wave behaviour of the sea surface. They are also used, in fatigue analysis, in prediction of the number of waves a construction sustains before fatigue failure. However, the fatigue studies disclose some shortcomings of these definitions. For example, consider a function x(t) which consists of two superimposed sine–waves,

$$x(t) = A_0 \sin(\omega_0 t + \phi_0) + A_1 \sin(\omega_1 t + \phi_1),$$

such that $A_0 >> A_1$ and $\omega_0 << \omega_1$. The crest–to–trough wave does not reflect the slow variations of x(t), since it is determined by the fast component, even if A_1 is near zero, leading to very small amplitudes, $H \simeq 2A_1$. The shortcoming of the zerocrossing wave definition is that it takes into account only the "primary" waves, while neglecting the "secondary" ones. In the case when x(t) is the sum of two sines, the zerocrossing waves are mainly determined by the slow component $A_0 \sin(\omega_0 t + \phi_0)$, giving the amplitudes $H^z \simeq 2(A_0 + A_1)$.

Since these are serious disadvantages, when fatigue life is studied, we have developed a more complicated form of wave, the so called Rain–Flow–Cycle. The following analytical definition was given in [14].

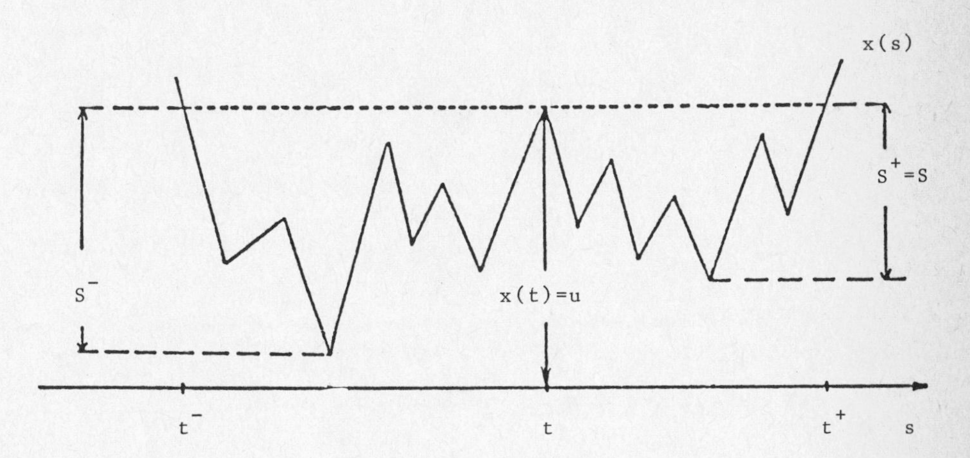

Fig. 2 Definition of RFC–amplitude.

DEFINITION 3: (RFC–amplitude) Let the load process x have a local maximum at t with height u = x(t), and let t^- and t^+ be the times of the last and first down– and upcrossing of the level u before and after t, respectively. With

$$S^- = \max\{u - x(s);\ t^- < s < t\},$$
$$S^+ = \max\{u - x(s);\ t < s < t^+\},$$

the RFC–amplitude, see Fig. 2., at t is

$$S = \min(S^-, S^+).$$

□

Observe that all three definitions give the same amplitudes for a pure sinusoidal function. However, the RFC–wave is the only method that identifies both slowly varying waves and more rapid reversals on top of these.

In this paper we shall present methods to approximate the wavelength and amplitude distribution defined as in Definitions 1–3, when the studied function is a sample path of an ergodic, twice continuously differentiable process. In Sections 2, 3 and 4, we present approximations of the wave–length and amplitude distributions for the RFC–, crest–to–trough and zerocrossing waves, respectively.

2. DISTRIBUTION OF RFC–AMPLITUDE

Now we shall be concerned with the distribution of the RFC–amplitude S, see Definition 3, in the case when the load function is an ergodic twice continuously differentiable process. The presentation, given here, follows mainly [8].

Let x be a sample path of an ergodic load process, and let t_i be the times for its local maxima. The RFC–procedure attaches to each $(t_i, x(t_i))$ an RFC–amplitude S_i, by Definition 3. The ergodic distribution of the RFC–amplitude is then defined as the empirical distribution of S_i, for i = 1,2,..., i.e.

$$F_S(s) = \lim_{T\to\infty} \frac{\#\{t_i \in [-T,T];\ S_i \le s\}}{\#\{t_i \in [-T,T]\}}.$$

Obviously, the value of the RFC–amplitude, originating at a maximum $(t_i, x(t_i))$, is determined by the values of surrounding extremes. More precisely, consider a pair $(t_i, x(t_i))$ and denote by $M_k(t_i)$, $k = 0,\pm1,\pm2,\ldots$, the sequence of local extrema of the function x(s) indexed so that $M_0(t_i) = x(t_i)$ and $M_{-1}(t_i)$, $M_1(t_i)$ are the adjacent local minima, etc. Analogously as for the RFC–distribution, define the ergodic distributions of the $\{M_k\}$–process:

$$F_{M_m,\ldots,M_n}(x_m,\ldots,x_n) = \lim_{T\to\infty} \frac{\#\{t_i \in [-T,T];\ M_k(t_i) \le x_k,\ k=m,\ldots,n\}}{\#\{t_i \in [-T,T]\}}. \tag{1}$$

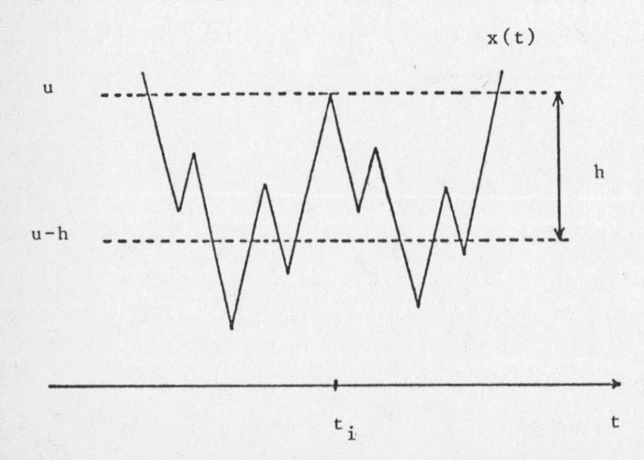

Fig. 3 Illustration of load history with RFC–amplitude greater than h.

By Definition 3, the RFC–amplitude S_i is greater than h if and only if the $M_k(t_i)$–sequence crosses the level M_0–h before it crosses the level M_0 as k goes to plus and minus infinity; see Fig. 3. Consequently, the ergodic distribution of the RFC–amplitudes can be defined using the ergodic properties of the $\{M_k\}$–process. In particular, the conditional distribution of the RFC–amplitude S given the maximum $M_0 = u$, is

$$F_{S|M_0}(h|u) = 1 - P(M_k\text{–process crosses the level } u-h \text{ before it crosses the level } u \text{ as } k \text{ goes to both plus and minus infinity } | M_0 = u)$$

$$= 1 - \sum_{i,j=1}^{\infty} P(M_{-i} \le u-h,\ M_j \le u-h,\ u-h < M_k \le u \text{ for } -i < k < j \mid M_0 = u)$$

$$= 1 - \sum_{i,j=1}^{\infty} P_{ij}. \qquad (2)$$

Obviously, the probabilities P_{ij} in (2) can be easily obtained from the M_k–distributions, given by (1). However, only the marginal density of M_k is known at present, and all joint densities must be approximated by some appropriate method. Lindgren [5] has proposed to approximate the dependence between the extremes of Gaussian ergodic load process by n–step Markov chain with the n–dimensional transition probabilities defined by (1). Two problems need to be addressed. First, to what extent real load sequences obey any Markov rules, and secondly whether for certain standard processes as stationary Gaussian processes, the RFC–distribution is resonably well approximated, if one uses a Markov chain to approximate its process of extremes.

To the first question, no general answer can be offered. However, the class of Markov processes is rich and should cover many types of real load processes. For example, in [16], we have approximated the RFC–distribution, using a one–step Markov chain of extrema, with transition probabilities estimated from real load data; see Fig. 4 from [16], which shows part of a load process. The resulting approximative RFC–distribution was compared to the empirical RFC–distribution and gave very good agreement; see Fig. 5 from [16]. Further, the Markov chain approach, can be extended to cover also processes which switch between different Markov structures, making the class of applicable processes even richer. The second question is discussed in [15], Remark 11.

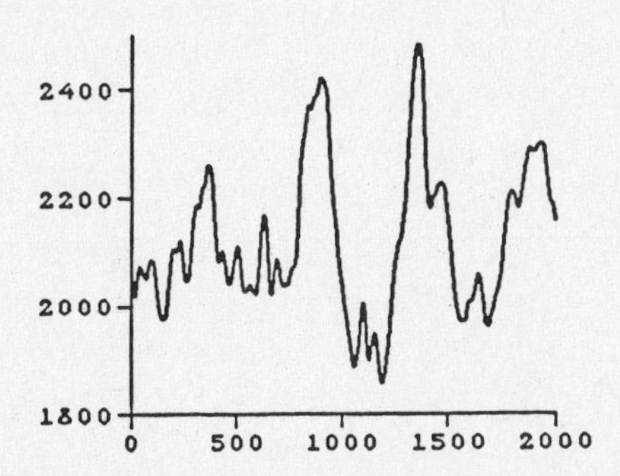

Fig. 4 Truc load data from SAAB–SCANIA in Södertälje; smoothed load record of the first 2000 data, from [16].

In the case when $\{M_k\}$ is an n–step Markov chain, with suitable symmetry properties, the conditional distribution of the RFC–amplitude given the maximum height M_0, from which it is initiated, defined by the sum (2), can be considerably simplified.

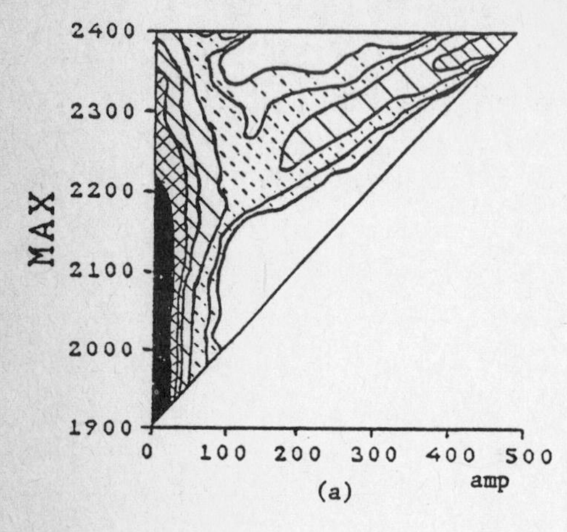

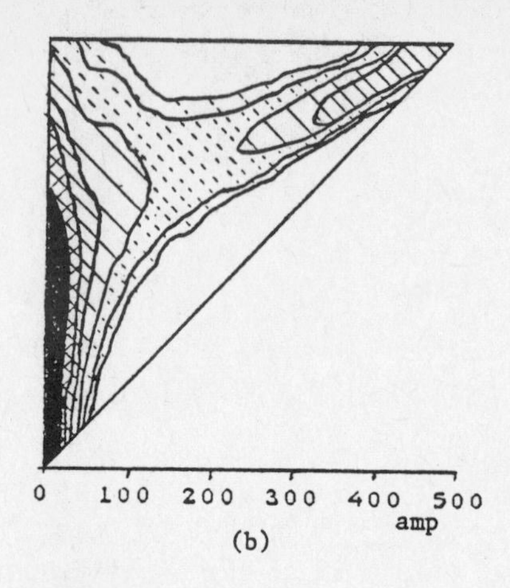

Fig. 5
Isolines of the empirical conditional density of the RFC–amplitude given maximum value estimated from the smoothed load (a), compared with the approximative conditional density of the RFC–amplitude given maximum, based on the one–step Markov assumption (b), from [16].

Above 0.100
0.075 - 0.100
0.050 - 0.075
0.025 - 0.050
0.010 - 0.025
0.005 - 0.010
Below - 0.005

The general case of any $n > 0$ was treated in [15], and here we present the simplest case, one–step Markov version. Consider a one–step Markov chain of extremes $\{M_k\}$. Denote by $p(s_0)$, $u - h \le s_0 \le u$, the conditional probability that M_k, for some $k > 0$, crosses the level $u - h$ before it crosses u, given $M_0 = s_0$. (This probability depends on u and h but for notational convenience we do not write this explicitly.) Since M_k is a one–step Markov process, the processes $\{M_i\}$, $i < 0$, and $\{M_j\}$, $j > 0$, are conditionaly independent given maximum $M_0 = s_0$. Hence the conditional distribution of the RFC–amplitude, given maximum height $M_0 = u$, is simply given by

$$F_{S|M_0}(h|u) = 1 - p(u)^2.$$

To find the probability p(u), let $f_{M_2,M_1|M_0}(s_2, s_1|s_0)$ be the conditional joint density of M_2, M_1 given $M_0 = s_0$. Then, it is an important consequence of the Markov assumption, that $p(s_0)$ can be obtained as the solution of the integral equation

$$p(s_0) = P(M_1 < u-h \mid M_0 = s_0) + \int_{u-h}^{u} p(s_2) f(s_2; s_0) ds_2, \quad s_0 \in [u-h, u], \tag{3}$$

where

$$f(s_2; s_0) = \int_{u-h}^{u} f_{M_2, M_1 \mid M_0}(s_2, s_1 \mid s_0) ds_1.$$

In order to evaluate the conditional distribution $F_{S \mid M_0}$, one has to solve numerically one integral equation (3) for each (u, h)–value.

For stationary Gaussian loads only the marginal density f_{M_k} is known at present, and all joint densities must be approximated by some method. In [15], Section 6, we proposed an approximation for the joint density of n consecutive extremes values and their locations based on the covariance function of the load process. This yields the possibility of approximating the RFC–amplitude distribution when the direct load measurement is difficult or very expensive.

3. DISTRIBUTION OF CREST–TO–TROUGH WAVELENGTH AND AMPLITUDE

In this section we shall present approximations of the distribution of the crest–to–trough wavelength and amplitude (T, H), (see Definition 1), when the function x is the sample path of a stationary, ergodic Gaussian process.

Many attempts have been made to find the exact form of the density of wavelength and amplitude for Gaussian processes. Some results are available for certain processes. However, the analytical methods developed for these special cases are inadequate when applied to more general classes of processes. Thus the problem of finding the (T, H)–density still remains unsolved. The complexity of the problem, is very well illustrated by the following exact formula of the (T, H)–density, for a twice continuously differentiable Gaussian process x,

$$f_{T,H}(t, h) = c \cdot E[x''(0)^- x''(t)^+ I(x', t) \mid x'(0) = x'(t) = 0, x(0) - x(t) = h] \\ f_{x'(0), x'(t), x(0)-x(t)}(0, 0, h), \tag{4}$$

where $x^- = \max(0, -x)$, $x^+ = \max(0, x)$, $c^{-1} = E[x''(0)^- \mid x'(0) = 0] f_{x'(0)}(0)$, and the indicator function I is given by

$$I(x,t) = \begin{cases} 1 \text{ if } x(s) \leq 0 \text{ for all } s, 0 \leq s \leq t, \\ 0 \text{ otherwise.} \end{cases} \tag{5}$$

(The formula (4), given in [17], is an extension of Durbin's formula for the first passage density [2], [13].)

Observe that the expectation in (4) is an infinite dimensional integral and hence is intractable, even though the formula can be used in construction of approximations.

In the following subsections we shall briefly present some of the recently proposed approximations. Subsection 3.1 deals with the approximations of the (T, H)–density, obtained by simplifications of the Slepian model process, and in Subsection 3.2, we present the bounds for the (T, H)–density.

3.1 Approximations based on simplifications of the Slepian model process

Let $x(t)$, $t \geq 0$, be a stationary zero–mean Gaussian process with covariance function r, and assume that its sample paths are a.s. twice continuously differentiable. A sufficient condition [1] for this is that the process be separable and that

$$r^{(4)}(s) = \lambda_4 + o(|\log|s||^{-\alpha}),$$

as $s \to 0$, for some $\alpha > 1$. Further, assume that the covariance function has $\lambda_0 = \mathrm{Var}(x(0)) = \lambda_2 = \mathrm{Var}(x'(0)) = 1$, which is only a matter of scaling.

In order to approximate the densities of the empirical (ergodic) distribution of the crest–to–trough wavelength and amplitude, we have to introduce the Slepian model process ξ for x after a local maximum. This means that the stochastic process $\xi(\cdot)$ is distributed as the long run distribution of $x(\omega, t_k + \cdot)$, when t_k runs over all local maxima of $x(\omega, \cdot)$. Mathematical details about Slepian processes and long run probabilities can be found in Leadbetter et. al. [3], Ch. 10, and Lindgren [6]. We give now a simple representation of the Slepian model process ξ.

Consider a zero–mean Gaussian process Δ, with covariance function

$$\mathrm{Cov}(\Delta(s), \Delta(t)) = \mathrm{Cov}(x(s), x(t) \mid x'(0), x''(0)),$$

and let R be a standard Rayleigh variable, independent of Δ. Then the Slepian model process ξ is given by

$$\xi(s) = -Ra(s) + \Delta(s), \tag{6}$$

where the function a is defined by

$$a(s) = \mathrm{Cov}(x(s), x''(0))/(\mathrm{Var}\, x''(0))^{\frac{1}{2}}.$$

For any continuously differentiable function y, let $T_0(y)$ denote the first zero–upcrossing time of y after zero. Since the ξ–process has a maximum at time zero, and the crest–to–trough wavelength T is a distance in time to the first local minimum of ξ, we have

$$(T,H) \overset{D}{=} (T_0(\xi'), \xi(0) - \xi(T_0)) \tag{7}$$

(where $\overset{D}{=}$ denotes equality in distribution).

In the following we present approximations of the (T, H)–density, obtained by replacing the Slepian process in (7) by a simpler process, for which the density of wavelength and amplitude can be evaluated, see [7], [12].

Let X be a normalised, i.e. $E(X) = 0$ and $\mathrm{Var}(X) = 1$, Gaussian variable from the linear space spanned by the residual process Δ, defined in (6). Consider a decomposition of the residual $\Delta(s)$ into a regression term on X, i.e. the orthogonal projection of $\Delta(s)$ on X, and a new residual $\Delta_1(s)$; then the Slepian model process ξ is

$$\xi(s) = -Ra(s) + Xb(s) + \Delta_1(s) = \xi_1(s) + \Delta_1(s), \tag{8}$$

where $b(s) = \mathrm{Cov}(\Delta(s), X)$, and R, X are independent Rayleigh and Gaussian variables, independent of Δ_1.

In [7], we used ξ_1 as an approximation to ξ and then approximated the wavelength and amplitude T, H by replacing ξ in (7) by its regression curve ξ_1, i.e

$$(T^r, H^r) \overset{D}{=} (T_0(\xi_1'), \xi_1(0) - \xi_1(T_0)). \tag{9}$$

In order to examine the effect of deleting the residual process Δ_1, we shall study the density function of the approximations of (T, H), obtained by replacing the residual process Δ_1 in (8) by a twice continuously differentiable function ω, i.e.

$$\begin{aligned} T_\omega &= T_0(-Ra(\cdot) + Xb(\cdot) + \omega(\cdot)), \\ H_\omega &= -R(a(0) - a(T_\omega)) + X(b(0) - b(T_\omega)) + \omega(0) - \omega(T_\omega). \end{aligned} \tag{10}$$

The (T^r, H^r)–density (9) is then obtained by taking $\omega = 0$ in (10).

Obviously, by (10), (T_ω, H_ω) are functions of (R, X) alone. However since we are intrested in the density of (T_ω, H_ω), we need the inverse mapping, i.e. (R, X) as a function of (T_ω, H_ω). This can be done, using the implicit definition of T_ω, i.e.

$$-Ra'(T_\omega) + Xb'(T_\omega) + \omega'(T_\omega) = 0. \tag{11}$$

Using (11), we can write (10) in a matrix form

$$B(T_\omega)\cdot\begin{bmatrix} R \\ X \end{bmatrix} = \begin{bmatrix} -\omega'(T_\omega) \\ H_\omega - \omega(0) + \omega(T_\omega) \end{bmatrix},$$

where the matrix B is given by

$$B(t) = \begin{bmatrix} -a'(t) & b'(t) \\ a(t)-a(0) & b(0)-b(t) \end{bmatrix}. \tag{12}$$

Now, the density of (T_ω, H_ω) can be obtained by a simple variable transformation and is given in the following lemma.

<u>LEMMA 4:</u> Let (T_ω, H_ω) be the approximation of (T, H) defined by (10). If the determinant $\det B(s) = 0$ only at the finite number of points in any finite interval, the density function of (T_ω, H_ω) is given by

$$f_{T_\omega,H_\omega}(t, h) = (2\pi)^{-\frac{1}{2}} I(t, h; \omega) J(t, h; \omega) p_1(t, h; \omega) \cdot e^{-\|\bar{p}(t,h;\omega)\|^2/2}, \tag{13}$$

where p_1 is the first component of the column vector $\bar{p}$, defined by

$$\bar{p}(t, h, \omega) = B(t)^{-1}\cdot\begin{bmatrix} -\omega'(t) \\ h-\omega(0)+\omega(t) \end{bmatrix}. \tag{14}$$

The Jacobian $J(t, h; \omega)$ and the indicator function $I(t, h; \omega)$ are given by

$$J(t,h;\omega) = \frac{|(-a''(t), b''(t))\cdot\bar{p}(t,h;\omega) + \omega''(t)|}{|\det B(t)|}, \tag{15}$$

$$I(t, h; \omega) = \begin{cases} 1 & \text{if } (-a'(s), b'(s))\cdot\bar{p}(t, h; \omega) + \omega'(s) \le 0 \text{ for all } s,\ 0 \le s \le t, \\ 0 & \text{otherwise,} \end{cases}$$

respectively. □

Since the (T_ω, H_ω)–density is a "smooth" function of (t, h, ω), we can use Fubini's theorem, and obtain the (T, H)–density, by replacing ω by the residual process Δ_1 in (13).

<u>THEOREM 5:</u> Under the assumptions of Lemma 4, the density of the ergodic distribution of crest–to–trough wavelength and amplitude (T, H), is given by

$$f_{T,H}(t, h) = (2\pi)^{-\frac{1}{2}} E_{\Delta_1}\left[I(t, h; \Delta_1) J(t, h; \Delta_1)\cdot p_1(t, h; \Delta_1) \cdot e^{-\|\bar{p}(t,h;\Delta_1)\|^2/2}\right]. \tag{16}$$

□

The expectation (16) is difficult to evaluate exactly, since the indicator $I(t, h; \Delta_1)$ depends on the whole realisation of Δ_1. Even though the formula is a basis for very accurate approximations of the (T, H)–density by approximating the residual process Δ_1, in (16), by some simpler process $\tilde{\Delta}$, which is a function of a finite number of random variables from the space spanned by Δ_1. The $\tilde{\Delta}$–process can be chosen in many different ways, e.g. cosine polynomials, splines etc. Here we shall use the regression curves.

More precisely, let $Y^T = (Y_1,\dots,Y_n)$, be a Gaussian vector in the space spanned by Δ_1. Let $\tilde{\Delta}(s)$ be the orthogonal projection of $\Delta_1(s)$ on Y. Since $\Delta_1(s)$ and Y are jointly Gaussian, $\tilde{\Delta}$ is linear and is of the form

$$\tilde{\Delta}(s) = \bar{b}(s)Y,$$

where $\bar{b}(s)$ is a vector of twice continously differentiable functions.

The **regression approximation** (T_n, H_n) of wavelength and amplitude density (T, H) is obtained by replacing the Δ_1 residual process in (16) by $\bar{b}(\cdot)Y$, i.e.

$$f_{T_n,H_n}(t,h) = (2\pi)^{-\frac{1}{2}}E_Y\left[I(t,h;\bar{b}(\cdot)Y)J(t,h;\bar{b}(\cdot)Y)p_1(t,h;\bar{b}(\cdot)Y)e^{-\|\bar{p}(t,h;\bar{b}(\cdot)Y)\|^2/2}\right], \tag{17}$$

which is an n–dimensional integral.

The approximative wavelength and amplitude (T^r, H^r), (see (9)), is the simplest example of the regression approximation, and corresponds to the choice of regressors Y = 0. Thus the expectation in (17) disappears and we obtain an explicit formula for the (T^r, H^r)–density, viz.

$$f_{T^r,H^r}(t, h) = (2\pi)^{-\frac{1}{2}}J(t, h) \cdot h \cdot p_1(t) \cdot e^{-h^2\|\bar{p}(t)\|^2/2}, \tag{18}$$

for h > 0 and $t \in D$, the set of possible values of T^r, where $p_1(t)$ is the first component of the $\bar{p}$–vector defined by $\bar{p}(t) = \bar{p}(t, 1; 0)$; see (14). In the same way the Jacobian J(t, h) = J(t, h; 0); see (15).

3.2 Bounds for the wavelength and amplitude density

In the previous subsection we presented the regression approximation, which is based on a successively more detailed decomposition of the Slepian model process into one regression term on n suitably chosen random quantities, and one residual process. The wavelength and amplitude density (T, H) was then approximated by the exact density of the wavelength and amplitude for the regression term. By including more random variables in the regression term, one can in general improve the accuracy of the approximation. However, in some applications, one wishes to know exact bounds of the approximations error, even at the cost of some additional numerical work. In this subsection we shall briefly present upper and lower bounds for the (T, H)–density, given in [17].

We turn first to the upper bounds and begin with the exact formula for the (T, H)–density (4), viz.

$$f_{T,H}(t,h) = \int_{-\infty}^{0}\int_{0}^{+\infty} |z_0| z_1 \cdot E[I(x',t) | x'(0) = x'(t) = 0, x(0) - x(t) = h,$$
$$x''(0) = z_0, x''(t) = z_1] f(0,0,h,z_0,z_1) dz_1 dz_0, \tag{19}$$

where the indicator function I is defined by (5), and $f(0, 0, h, z_0, z_1)$ is the density of $(x'(0), x'(t), x(0) - x(t), x''(0), x''(t))$. The upper bounds for the wavelength and amplitude density are obtained by simply overestimating the indicator I. In addition the approximation proposed by Rice & Beer [10], is equivalent to replacing the indicator I, in (19), by one.

A series of upper bounds, proposed in [17], is obtained by replacing the conditional expectation $E[I(x', t) | C]$, in (19), by, successively

1,

$P(x'(S_1) \le 0 | C) = \inf_{0<s<t} P(x'(s) \le 0 | C)$,

$P(x'(S_2) \le 0, x'(S_1) \le 0 | C) = \int_{-\infty}^{0} \inf_{0<s<t} P(x'(s) \le 0 | x'(S_1) = x, C) f_{x'(S_1)|C}(x) dx$, etc.,

where "C" means, given $x'(0) = x'(t) = 0$, $x(0) - x(t) = h$, $x''(0) = z_0$, $x''(t) = z_1$. Observe that S_i, $i = 1,2,\ldots$, are random variables, since they are functions of the values of $x(0) - x(t)$, $x''(0)$, $x''(t)$, $x'(S_1),\ldots,x'(S_{i-1})$. See [17] for a more complete presentation.

It is in general difficult to give a useful lower estimate for the indicator I, and hence formula (19) is not applicable in the construction of lower bounds for the density of (T, H). Therefore we give a second formula, proved in [17], for the (T, H)–density, viz.

$$f_{T,H}(t,h) = \int_{-\infty}^{0}\int_{0}^{+\infty} |z_0| z_1 \cdot \Big[1 - \int_{0}^{t}\int_{0}^{+\infty} k(s, z_0, z_1, z_2, h) \cdot E[I(x', s) | x'(0) = 0,$$
$$x''(0) = z_0, x'(t) = 0, x''(t) = z_1, x'(s) = 0, x''(s) = z_2,$$
$$x(0) - x(h) = h] dz_2 ds\Big]^{+} dz_1 dz_0, \tag{20}$$

where, for a fixed s, the kernel k is given by

$$k(s, z_0, z_1, z_2, h) = z_2 f(0, z_0, 0, z_1, 0, z_2, h),$$

and f is the density of $(x'(0), x''(0), x'(t), x''(t), x'(s), x''(s), x(0)–x(t))$. The indicator function I is given by (5). Like the upper bounds, the lower bounds for the wavelength and amplitude density are obtained by simply overestimating the indicator I in (20); see [17] for more details. In addition the approximation proposed by Lindgren [4], is equivalent to replacing the indicator I in (20) by 1.

4. DISTRIBUTION OF THE ZEROCROSSING WAVELENGTH AND AMPLITUDE

In this section we shall discuss the approximations of the empirical zerocrossing wavelength and amplitude (T^z, H^z) (see Definition 2) distribution when x is a sample path of stationary, ergodic Gaussian process. First, we shall simplify Definition 2, by introducing the half zerocrossing wavelength and amplitidude (T^*, H^*).

DEFINITION 6: The half zerocrossing wavelength T^* and amplitude H^* are the time between the zero upcrossing and the following zero downcrossing and the highest value in this interval, respectively; see Fig. 6. □

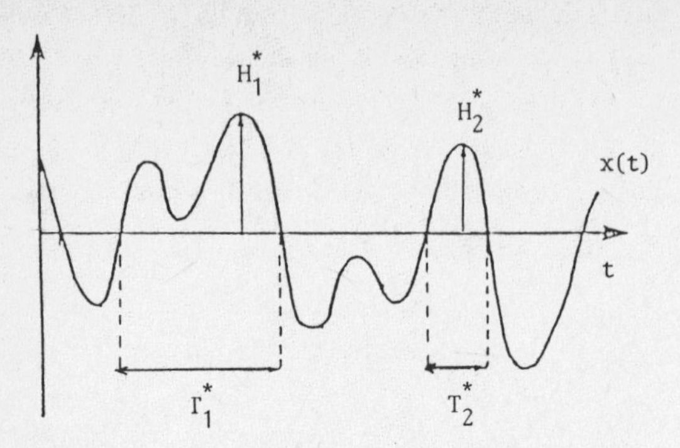

Fig. 6 Definition of half zerocrossing wavelength T^* and amplitude H^*.

In oceanography, one often uses a simple approximation, due to Longuet–Higgins [9], based on an envelope approximation for narrow band Gaussian processes. In that special case, the sample path of the process has usually only one local maximum between a zero up crossing and the following zero down crossing, hence the H^*–amplitude can be approximated by the value of the first local maximum after the zero upcrossing. This is a considerable simplification, leading to an analysis similar to that in Section 3.

Here we shall present the regression approximation of the (T^*, H^*)–density, given in [12], which can be used also for broad band processes.

In order to approximate the ergodic distribution of (T^*, H^*), we have to introduce the Slepian model process ξ_z for x after a zero upcrossing. This means that the stochastic process $\xi_z(\cdot)$ is distributed as the long run distribution of $x(\omega, t_k + \cdot)$, when t_k runs over all zero upcrossings of $x(\omega, \cdot)$. We turn to a simple representation of the Slepian model process ξ_z, similar to (6).

Consider a zero–mean Gaussian process Δ, with covariance function

$$\mathrm{Cov}(\Delta(s), \Delta(t)) = \mathrm{Cov}(x(s), x(t) \mid x(0), x'(0)),$$

and let R be a standard Rayleigh variable, independent of Δ. Then the Slepian model process ξ_z is given by

$$\xi_z(s) = Ra(s) + \Delta(s), \tag{21}$$

where the function a is defined by

$$a(s) = \mathrm{Cov}(x(s), x'(0))/(\mathrm{Var}\, x'(0))^{\frac{1}{2}}.$$

For any continuously differentiable function y, let $T_0(y)$ denote the first zero down–crossing time of y, after zero. Then the ergodic half zerocrossing wavelength and amplitude (T^*, H^*) is given by

$$(T^*, H^*) \overset{D}{=} (T_0(\xi_z), \sup_{0 \le s \le T_0} \xi_z(s)). \tag{22}$$

(The last formula should be compared with (7).)

As in the previous section, one can decompose the Slepian model process ξ_z into one regression term on n suitably chosen random quantities, and one residual process. Then the zerocrossing wavelength and amplitude density (T^*, H^*) is approximated by the exact density of the zerocrossing wavelength and amplitude for the regression term. By adding more random variables in the regression term, one can in general improve the accuracy of the approximation.

More precisely, let $X^T = (X_1,...,X_n)$, be a zero–mean orthonormal Gaussian vector in the space spanned by Δ. Let $\Delta_n(s)$ be the orthogonal projection of $\Delta(s)$ on X. Since $\Delta(s)$ and X are jointly Gaussian, Δ_n is linear and is of the form

$$\Delta_n(s) = \bar{b}(s)X,$$

where $\bar{b}(s) = (b_1(s),...,b_n(s))$ is a vector of continously differentiable functions.

The **regression approximation** (T_n^*, H_n^*) of the wavelength and amplitude density (T^*, H^*) is obtained by replacing in (22) the Slepian model process ξ_z by $Ra(\cdot) + \bar{b}(\cdot)X$, i.e.

$$(T_n^*, H_n^*) = (T_0(Ra(\cdot) + \bar{b}(\cdot)X), \sup_{0 \leq s \leq T_0} Ra(s)+\bar{b}(s)X). \tag{23}$$

The density of (T_n^*, H_n^*) can be derived from density (11), in [12], by taking $\theta = 0$, in (11), and multipling by (2π), which gives the following theorem.

<u>THEOREM 7:</u> Under the general assumptions of this section, if $b_1(s) = 0$ only at a finite number of points in any finite interval, the density (23) of the regression approximation (T_n^*, H_n^*) of the zerocrossing half wave–length and amplitude (T^*, H^*), defined in Definition 6, is given by

$$f_{T_n^*,H_n^*}(t,h) = (2\pi)^{-n/2} \cdot h^{n+1} \int_{\mathbb{R}^{n-1}} I(\bar{y}, t)q(t, \bar{y})^{n+2} \left|\frac{\partial p(t,\bar{y})}{\partial t}\right|$$
$$\cdot e^{-h^2 q(t,\bar{y})^2(1+p(t,\bar{y})^2+\|\bar{y}\|^2)/2} d\bar{y}, \tag{24}$$

for $\bar{y} = (y_1,...,y_{n-1})$, where the functions p and q are defined by

$$p(t, \bar{y}) = -(a(t) + \sum_{i=1}^{n-1} y_i b_{i+1}(t))/b_1(t),$$

$$q(t,\bar{y})^{-1} = \sup_{0 \leq s \leq t} a(s) + p(t, \bar{y})b_1(s) + \sum_{i=1}^{n-1} y_i b_{i+1}(s),$$

respectively. The indicator function $I(\bar{y}, t)$ is given by

$$I(\bar{y}, t) = \begin{cases} 1 \text{ if } a(s) + p(t, \bar{y})b_1(s) + \sum_{i=1}^{n-1} y_i b_{i+1}(s) \geq 0, \text{ for all } s,\ 0 \leq s \leq t, \\ 0 \text{ otherwise}. \end{cases}$$

□

In the following example, we shall derive the density (24) for the simplest case, n = 1. It will indicate the method of proving (24), and illuminate the difference between the regression approximation of crest–to–trough wavelength and amplitude and the zerocrossing ones.

<u>EXAMPLE 8:</u> In this example, we shall derive the density of (T_1^*, H_1^*), i.e

$$T_1^* = T_0(Ra(\cdot) + b_1(\cdot)X_1),$$

$$H_1^* = \sup_{0\le s\le T_1^*} Ra(s) + b_1(s)X_1,$$

where $b_1(s) = \mathrm{Cov}(\Delta(s), X_1)$, and R, X_1 are independent standard Rayleigh and Gaussian variables, respectively.

The density function of (T_n^*, H_n^*) can be determined by using the following simple properties of wavelength and amplitude (T^*, H^*), viz.

$$\begin{aligned} &y(T^*(y)) = 0, \\ &T^*(cy) = T^*(y), \qquad\qquad H^*(cy) = cH^*(y), \end{aligned} \tag{25}$$

which hold for any continuously differentiable function y and constant $c > 0$. Assume that $b_1(t) \neq 0$. Since $R > 0$, using (25) and the definition of (T_1^*, H_1^*), we obtain the following two equations

$$a(T_1^*)+b_1(T_1^*)\cdot X_1/R = 0,$$

$$R\cdot\left[\sup_{0\le s\le T_1^*} a(s) + b_1(s)\cdot X_1/R\right] = H_1^*.$$

Now, we can solve X_1/R and R as functions of T_1^* and H_1^*, i.e.

$$X_1/R = -a(T_1^*)/b_1(T_1^*) = p(T_1^*),$$

$$R = H_1^*/ \sup_{0\le s\le T_1^*} a(s)+b_1(s)p(T_1^*) = H_1^*\cdot q(T_1^*),$$

and the (T_1^*, H_1^*)–density can be obtained by a simple variable transformation giving

$$f(t, h) = (2\pi)^{-\frac{1}{2}}(hq(t))^2|q(t)\cdot\frac{dp(t)}{dt}|\cdot e^{-h^2q(t)^2(1+p(t)^2)/2}, \tag{26}$$

when $h > 0$ and $t \in D$, the set of possible values of t. □

ACKNOWLEDGEMENT I am indebted to professor George Lindgren for discussion, valuable suggestion and comments.

REFERENCES:

[1] Cramèr, H. & Leadbetter, M.R. (1967). Stationary and related stochastic processes. Wiley, New York.

[2] Durbin, J. (1985). The first–passage density of continuous Gaussian process to a general boundary. J. Appl. Prob. 22, pp. 99–122.

[3] Leadbetter, M.R., Lindgren, G. & Rootzèn, H. (1983). Extremes and related properties of random sequences and processes. Springer–Verlag, New York.

[4] Lindgren, G. (1972). Wave–length and amplitude in Gaussian noise. Adv. Appl. Prob. 4, pp. 81–108.

[5] Lindgren, G. (1983). On the use of effect spectrum to determine a fatigue life amplitude spectrum. ITM–Symposium on Stochastic Mechanics. Univ. Lund Statist. Res. Rep. 6, pp. 1–11.

[6] Lindgren, G. (1984). Use and structure of Slepian model processes for prediction and detection in crossing and extreme value theory. Proc. NATO ASI on Statistical Extremes and Applications, Vimeiro 1983. Reidel Publ. Co, pp. 261–284.

[7] Lindgren, G. & Rychlik, I. (1982). Wave characteristic distributions for Gaussian waves – wave–length, amplitude, and steepness. Ocean. Engng. 9, pp. 411–432.

[8] Lindgren, G. & Rychlik, I. (1987). Rain Flow Cycle distributions for fatigue life prediction under Gaussian load processes. Fatigue Fract. Engng Mater. Struct. Vol. 10, No. 3, pp. 251–260.

[9] Longuet–Higgins, M.S. (1983). On the joint distribution of wave periods and amplitudes in a random wave field. Proc. R. Soc. Lond., A 389, pp. 241–258.

[10] Rice, J.R. & Beer, F.P. (1965). On the distribution of rises and falls in a continuous random process. J. Basic Engineering, ASME, Ser. D, 87, pp. 398–404.

[11] Rice, S.O. (1944), (1945). Mathematical analysis of random noise. Bell Syst. Tech. J. 23, pp. 282–332, 24, pp. 46–156.

[12] Rychlik, I. (1987). Regression approximations of wavelength and amplitude distributions. Adv. Appl. Prob. 19, pp. 396–430.

[13] Rychlik, I. (1987). A note on Durbin's formula for the first–passage density. Statistics & Probability Letters 5, pp. 425–428.

[14] Rychlik, I. (1987). A new definition of the rainflow cycle counting method. Int. J. Fatigue 9, No 2, pp. 119–121.

[15] Rychlik, I. (1988). Rain flow cycle distribution for ergodic load processes. SIAM J. Appl. Math., Vol. 48, No. 3, pp. 662–679.

[16] Rychlik, I. (1988). Simple approximations of the rain–flow–cycle distribution for discretized random loads. To appear in Probabilistic Engineering Mechanics, pp. 1–21.

[17] Rychlik, I. (1987). New bounds for the first passage, wave–length and amplitude densities. Univ. Lund Stat. Res. Rep. 1987:9, pp. 1–19.

CHARACTERIZATIONS OF THE EXPONENTIAL DISTRIBUTION BY FAILURE RATE- AND MOMENT PROPERTIES OF ORDER STATISTICS

L. Gajek and U. Gather

Institute of Mathematics, Technical University of Lodź, and Department of Statistics, University of Dortmund

Abstract. This paper gives some characterizations of the exponential distribution. We show that the moments of spacings as well as the failure rates of certain functions of order statistics possess characteristic properties of the exponential distribution.

1. INTRODUCTION

It is the aim of this paper to generalize some characterization theorems for the exponential distribution which are based on the property of identical distributions or identical moments of certain functions of order statistics.

Results of this type are reviewed very well in the work of Galambos, Kotz (1978) and Azlarov and Volodin (1986).

The topic was probably treated first by Puri and Rubin (1970) who proved that, if X, X_1, X_2 are i.i.d. random variables with common Lebesgue density $f(x)$, then X and $|X_1-X_2|$ have the same distribution if and only if X is exponential, that is iff

$$f(x) = \vartheta e^{-\vartheta x},\ x \geq 0, \text{ for some } \vartheta > 0. \tag{1.1}$$

Desu (1971) showed that if $X, X_1,...,X_n$ are i.i.d. random variables, then X and $n\cdot\min(X_1,...,X_n)$ have identical distribution for all $n \in \mathbb{N}$ if and only if X is exponential, a result which goes back to the 1920s when the foundations of extreme value theory were laid down

Key words. Exponential distribution, characterization, order statistics. AMS 1980 subject classifications. 60E05, 62E10, 62E15.

(Galambos and Kotz (1978), p. 18).

Ahsanullah (1976, 1978a, 1978b, 1984) considered

$$D_{i,n} := (n-i)\left(X_{i+1,n}-X_{i,n}\right) \qquad 1\leq i\leq n \tag{1.2}$$

with $D_{0,n} := nX_{1,n}$, $D_{n,n} := 0$

where $X_{1,n}\leq\ldots\leq X_{n,n}$ denote the order statistics of a sample $X_1,\ldots,X_n$ from a distribution with density f(x). He proved that the identical distribution of

$$nX_{1,n} \text{ and } D_{i,n} \text{ for some } 1\leq i<n, \tag{1.3}$$

or of

$$D_{i,n} \text{ and } D_{i+1,n} \text{ for some } 0\leq i<n, \tag{1.4}$$

more generally of

$$D_{i,n} \text{ and } D_{j,n} \text{ for some } 0\leq i<j<n, \tag{1.5}$$

or of

$$X_{j,n}-X_{i,n} \text{ and } X_{j-i,n-i} \text{ for some } 1\leq i<j\leq n, \tag{1.6}$$

characterizes the exponential distribution within the class of all distributions with increasing (decreasing) failure rate. Those are called briefly IFR (DFR) distributions.

In Ahsanullah (1981), (1.4) is weakened to

$$E\left(D_{i,n}\right) = E\left(D_{i+1,n}\right) \text{ for some } 0\leq i<n \tag{1.7}$$

and it is also shown that

$$r_{D_{i,n}}(x) = r_X(x) \quad \forall x\geq 0 \quad \text{for some } 0\leq i<n \tag{1.8}$$

characterizes the exponential distribution. Here $r_Y(y) := \dfrac{g(y)}{1-G(y)}$ denotes the failure rate of a random variable Y with c.d.f. G(y) and Lebesgue density g(y).

In Gather (1988) it is shown that (1.6) for two distinct values j_1 and j_2 of j, $1\leq i\leq j_1<j_2\leq n$, characterizes the exponential distribution without an IFR(DFR) or NBU(NWU) assumption.

In this paper we replace (1.5) and (1.8) by the properties

$$r_{D_{i,n}}(0) = r_{D_{j,n}}(0) \quad \text{for some } 0\leq i<j<n,$$

and

$$r_{D_{i,n}}(0) = r_X(0) \quad \text{for some } 0\leq i<n,$$

and (1.6) by

$$\left.\frac{d^{j-i} r_{X_{j,n}-X_{i,n}}(x)}{dx^{j-i}}\right|_{x=0} = \left.\frac{d^{j-i} r_{X_{j-i,n-i}}(x)}{dx^{j-i}}\right|_{x=0}$$

under the IFR assumption as well as by

$$E\left(X_{j,n}-X_{i,n}\right) = E\left(X_{j-i,n-i}\right) \quad \text{for some } 1 \le i < j \le n,$$

under the NBU assumption.

The following assumption A will be referred to throughout the paper:

A: Let $X, X_1, \ldots, X_n$ be i.i.d. random variables from a distribution with c.d.f. $F(x)$ and Lebesgue density $f_X(x) = f(x)$, $x \in \mathbb{R}$, $\inf\{x\,;\,F(x)>0\} = 0$, and let $F(x)$ be strictly increasing $\forall x>0$.

It is known (see, for example, Krasnosel'skij and Rutickii (1961) or Mitrinovič (1970)) that if $\log(1-F)$ is a concave function on $(0,\infty)$, then there is a nondecreasing right-continuous function r: $\mathbb{R}_+ \to \mathbb{R}$ for which

$$\log\left(1-F(x)\right) = \int_0^x r(t)\, dt,$$

i.e. there is a right-continuous version of the failure rate, provided F is IFR. A similar property holds for the DFR case. Thus we shall assume thoughout the paper that under the IFR(DFR) assumption the failure rate (as well as the density) of the considered random variable X is choosen continuous from the right. Furthermore, if r is monotone, we may define

$$r(0) = \lim_{x \downarrow 0} r(x),$$

where the limit is assumed to be finite.

The same convention concerns failure rates (and densities) of other statistics involved in the paper: they are assumed to be right-continuous on the interval $[0,\infty)$. For the density of $D_{j,n}$, $0 \le j < n$, e.g. this assumption can be fulfilled if the integral in (2.2) below exists as a finite number for all $x \ge 0$.

Conventions of this type are also contained implicitely in the proof of Theorem 2.2 in Ahsanullah (1981).

2. CHARACTERIZATION THEOREMS

Theorem 2.1. Let the assumptions A be fulfilled and let F be IFR (DFR). Then F is exponential with density (1.1) if and only if

$$r_{D_{i,n}}(0) = r_{D_{k,n}}(0) \quad \text{for some } 0 \le i < k < n,\ n \ge 2. \tag{2.1}$$

Proof. The necessity part is trivial and can be derived from the classical result (Ahsanullah (1978b)).

Therefore assume that (2.1) is fulfilled.

From David (1981), p. 11, it is known that $D_{j,n}$, $1 \le j \le n$, has Lebesgue density

$$f_{D_{j,n}}(x) = \frac{n!}{(j-1)!(n-j)!} \int_0^\infty F(u)^{j-1} \cdot \bar{F}\left(u + \frac{x}{n-j}\right)^{n-j+1} \cdot f\left(u + \frac{x}{n-j}\right) \cdot f(u)\, du, \quad x \ge 0. \tag{2.2}$$

Let denote $C_j := \frac{n!}{(j-1)!(n-j)!}$. The densities f, $f_{D_{j,n}}$ and $f_{D_{k,n}}$ are supposed to be continuous from the right. Hence (2.1) yields

$$C_i \int_0^\infty F(u)^{i-1} \bar{F}(u)^{n-i-1} [f(u)]^2\, du = C_k \int_0^\infty F(u)^{k-1} \bar{F}(u)^{n-k-1} [f(u)]^2\, du.$$

Since $C_j^{-1} = \int_0^\infty F(u)^{j-1} \bar{F}(u)^{n-j} f(u)\, du$, this implies

$$\frac{\int_0^\infty F(u)^{i-1} \bar{F}(u)^{n-i}\, r(u)\, f(u)\, du}{\int_0^\infty F(u)^{i-1} \bar{F}(u)^{n-i} f(u)\, du} = \frac{\int_0^\infty F(u)^{k-1} \bar{F}(u)^{n-k}\, r(u)\, f(u)\, du}{\int_0^\infty F(u)^{k-1} \bar{F}(u)^{n-k} f(u)\, du} \tag{2.3}$$

and thus

$$\int_0^\infty F(u)^{i-1} \bar{F}(u)^{n-i}\, r(u)\, f(u)\, du \cdot \int_0^\infty F(u)^{k-1} \bar{F}(u)^{n-k} f(u)\, du$$

$$- \int_0^\infty F(u)^{k-1} \bar{F}(u)^{n-k}\, r(u)\, f(u)\, du \cdot \int_0^\infty F(u)^{i-1} \bar{F}(u)^{n-i} f(u)\, du = 0$$

or

$$0 = \int_0^\infty \int_0^\infty \left[F(u)^{i-1}\overline{F}(u)^{n-i} r(u) F(v)^{k-1}\overline{F}(v)^{n-k} - F(u)^{k-1}\overline{F}(u)^{n-k} r(u) F(v)^{i-1}\overline{F}(v)^{n-i} \right] \cdot f(u)\cdot f(v)\, du dv$$

$$=: \int_0^\infty \int_0^\infty H(u,v)\, du dv$$

$$= \iint\limits_{\{(u,v)\in(0,\infty)^2 \,;\, u\le v\}} H(u,v)\, du dv + \iint\limits_{\{(u,v)\in(0,\infty)^2 \,;\, u>v\}} H(u,v)\, du dv$$

$$= \iint\limits_{u\le v} \Big(\Big[F(u)^{i-1}\, \overline{F}(u)^{n-i}\, r(u)\, F(v)^{k-1}\, \overline{F}(v)^{n-k} - F(u)^{k-1}\, \overline{F}(u)^{n-k}\, r(u)\, F(v)^{i-1}\, \overline{F}(v)^{n-i} \Big] +$$

$$+ \Big[F(v)^{i-1}\, \overline{F}(v)^{n-i}\, r(v)\, F(u)^{k-1}\, \overline{F}(u)^{n-k} - F(v)^{k-1}\, \overline{F}(v)^{n-k}\, r(v)\, F(u)^{i-1}\, \overline{F}(u)^{n-i} \Big] \Big)\, f(u) f(v)\, du dv$$

$$= \iint\limits_{u\le v} \Big(F(u)^{i-1}\ \overline{F}(u)^{n-i}\, F(v)^{k-1}\, \overline{F}(v)^{n-k} \big[r(u) - r(v) \big] +$$

$$+ F(v)^{i-1}\, \overline{F}(v)^{n-i}\, F(u)^{k-1}\, \overline{F}(u)^{n-k} \big[r(v) - r(u) \big] \Big)\, f(u)\, f(v)\, du dv$$

$$= \iint\limits_{u\le v} \big[r(u) - r(v) \big] \Big(F(u)^{i-1}\overline{F}(u)^{n-i} F(v)^{k-1}\overline{F}(v)^{n-k} - F(v)^{i-1}\overline{F}(v)^{n-i} F(u)^{k-1}\overline{F}(u)^{n-k} \Big)\, f(u) f(v)\, du dv$$

$$= \iint\limits_{u\le v} \big[r(v) - r(u) \big] \frac{\overline{F}(u)^n}{F(u)} \frac{\overline{F}(v)^n}{F(v)} \left[\left(\frac{F(u)}{\overline{F}(u)} \right)^i \left(\frac{F(v)}{\overline{F}(v)} \right)^k - \left(\frac{F(v)}{\overline{F}(v)} \right)^i \left(\frac{F(u)}{\overline{F}(u)} \right)^k \right] f(u)\, f(v)\, du dv.$$

And finally

$$0 = \iint\limits_{u\le v} \left[r(v) - r(u)\right] \frac{\bar{F}(u)^n}{F(u)} \frac{\bar{F}(v)^n}{F(v)} \left(\frac{F(u)}{\bar{F}(u)} \cdot \frac{F(v)}{\bar{F}(v)}\right)^i \left[\left(\frac{F(v)}{\bar{F}(v)}\right)^{k-i} - \left(\frac{F(u)}{\bar{F}(u)}\right)^{k-i}\right] f(u)f(v)\, du dv. \quad (2.4)$$

Since $r(x)$ is increasing (decreasing) for all $x>0$, as F is IFR (DFR), it holds $r(v) - r(u) \geq 0$ and also $\frac{F(v)}{\bar{F}(v)} > \frac{F(u)}{\bar{F}(u)}$ for all $0<u<v$.

Hence (2.4) implies that

$$r(u) \equiv \text{const} \quad \forall u>0$$

which is a characterizing property of an exponential distribution with density (1.1). □

Remark 2.1. It is easily seen from the proof of Theorem 2.2 in Ahsanullah (1981) that under the assumptions of our Theorem 2.1 above, X is exponentially distributed with density (1.1), if and only if

$$r_{D_{k,n}}(0) = r_X(0) \quad \text{for some } 1\leq k<n,\ n\geq 2.$$

That is, the involved failure rates need not to be equal for all $x\geq 0$ but only at the point zero.

The next theorem can be seen in the context of the results of Rossberg (1972) and Ahsanullah (1984), who proved that the identical distribution of $X_{i+1,n}-X_{i,n}$ and $X_{1,n-i}$ characterizes the exponential distribution within the class of all continuous distributions and that identical distribution of $X_{j,n}-X_{i,n}$ and $X_{j-i,n-i}$ for some $1\leq i<j\leq n$ (c.f. (1.6) above) within the class of all IFR (DFR) distributions is true iff F is exponential. We replace (1.6) by a failure rate property.

The next theorem will be proven under the condition that the densities of X and $X_{j,n}-X_{i,n}$ are $(j-i-1)$-times continuously differentiable functions on $[0,\infty)$ (the latter one also under the integral sign) although this smoothness condition could be weakened, for example, by considering right derivatives which are continuous from the right.

Theorem 2.2. Let assumption A and the above smoothness condition be fulfilled and let F be IFR (DFR). Then X is exponential iff for some $1\leq i<j\leq n$, $n\geq 2$, the following equality holds

$$\left.\frac{d^{j-i-1}\, r_{X_{j,n}-X_{i,n}}(x)}{dx^{j-i-1}}\right|_{x=0} = \left.\frac{d^{j-i-1}\, r_{X_{j-i,n-i}}(x)}{dx^{j-i-1}}\right|_{x=0}. \quad (2.5)$$

Proof. The necessity part is straightforward. For the opposite direction, assume that (2.5) is fulfilled.

For simplicity we suppose that j-i=:m=2.

The general case with $1 \le m \le n-i$ arbitrary can be treated by the same arguments.

Now the p.d.f. $f_1(x)$ of $X_{i+m,n}-X_{i,n}$ is given by (c.f. David (1981), p. 11)

$$f_1(v) = k_1 \int_0^\infty F(u)^{i-1} [F(u+v) - F(u)]^{m-1} \bar{F}(u+v)^{n-i-m} f(u) f(u+v) du, \quad v \ge 0$$

where

$$k_1 = \frac{n!}{(i-1)!\,(m-1)!\,(n-i-m)!}.$$

The p.d.f. $f_2(v)$ of $X_{m,n-i}$ is given by

$$f_2(v) = k_2 F(v)^{m-1} \bar{F}(v)^{n-i-m} f(v), \quad v \ge 0$$

where

$$k_2 = \frac{(n-i)!}{(m-1)!(n-i-m)!}.$$

Thus, we get

$$f_1'(v) = k_1 \int_0^\infty F(u)^{i-1} f(u) \Big[(m-1)[F(u+v) - F(u)]^{m-2} f(u+v)^2 \bar{F}(u+v)^{n-i-m} + [F(u+v) - F(u)]^{m-1} \cdot$$

$$\cdot (n-i-m) \bar{F}(u+v)^{n-i-m-1} (-1) f(u+v)^2 + [F(u+v) - F(u)]^{m-1} \bar{F}(u+v)^{n-i-m} f'(u+v) f(u+v) \Big] du$$

which yields for m=2 and v=0

$$f_1'(0) = k_1 \int_0^\infty F(u)^{i-1} \bar{F}(u)^{n-i-2} f(u)^3 du.$$

Analogously

$$f_2'(v) = k_2 \Big[(m-1) F(v)^{m-2} f(v)^2 \bar{F}(v)^{n-i-m} + F(v)^{m-1} (n-i-m) \bar{F}(v)^{n-1-m-1} f(v)^2 (-1) +$$

$$+ F(v)^{m-1} \bar{F}(v)^{n-i-m} f'(v) \Big]$$

and with m=2:

$$f_2'(0) = k_2 f(0)^2.$$

Now (2.5) implies

$$k_2 f(0)^2 = k_1 \int_0^\infty F(u)^{i-1} f(u)^3 \bar{F}(u)^{n-i-2} du.$$

Since

$$\frac{k_2}{k_1} = \int_0^\infty F(u)^{i-1} \bar{F}(u)^{n-i} f(u)\, du,$$

$$\int_0^\infty f(0)^2 F(u)^{i-1} \bar{F}(u)^{n-i} f(u)\, du = \int_0^\infty f(u)^2 F(u)^{i-1} \bar{F}(u)^{n-i-2} f(u)\, du.$$

Thus

$$\int_0^\infty \left[\left(\frac{f(0)}{\bar{F}(0)}\right)^2 - \left(\frac{f(u)}{\bar{F}(u)}\right)^2\right] \bar{F}(u)^{n-i} F(u)^{i-1} f(u)\, du = 0$$

which again implies, since $r(u) = \dfrac{f(u)}{\bar{F}(u)}$ is either increasing or decreasing, that

$$r(u) = \text{const} \quad \forall u>0.$$

Hence X is exponentially distributed. □

In the following theorem condition (1.6) is weakened to the equality of the corresponding moments, and that under the weaker assumption of NBU (NWU) distributions.

Theorem 2.3. Let assumption A be fulfilled and let F be NBU (NWU). Then X is exponentially distributed with density (1.1) if and only if

$$E(X_{j,n}) - E(X_{i,n}) = E(X_{j-i,n-i}) \quad \text{for some } 1 \le i < j \le n,\ n \ge 2. \tag{2.6}$$

Proof. Let denote

$V := X_{j-i,n-i}$, $W := X_{j,n} - X_{i,n}$ with c.d.f.'s F_V and F_W,

$$H(z) := H_j(z) = \frac{\int_0^z t^{n-j} (1-t)^{j-i-1}\, dt}{\int_0^1 t^{n-j} (1-t)^{j-i-1}\, dt}, \qquad K(u,x) := \frac{1-F(u+x)}{1-F(u)}, \quad u,x \geq 0.$$

Then as in the proof of the Theorem in Gather (1988),

$$1 - F_W(x) = E\Big(H\big(K(X_{i,n},x)\big)\Big)$$

whereas

$$1 - F_V(x) = H\big(1 - F(x)\big).$$

Now (2.6) implies

$$\int_0^\infty H\big(1 - F(x)\big)\, dx = \int_0^\infty \int_0^\infty H\big(K(y,x)\big)\, f_{X_{i,n}}(y)\, dydx.$$

Hence,

$$\int_0^\infty \int_0^\infty \Big[H\big(1 - F(x)\big) - H\big(K(y,x)\big)\Big]\, f_{X_{i,n}}(y)\, dydx = 0. \tag{2.7}$$

Since H is strictly increasing, we get by the NBU assumption

$$H\big(K(y,x)\big) = H\left(\frac{1 - F(x+y)}{1 - F(y)}\right) \leq H\big(1 - F(x)\big).$$

Thus (2.7) implies

$$K(y,x) \equiv 1 - F(x) \quad \forall x,y \geq 0$$

which characterizes the exponential distribution. □

3. CONCLUDING REMARKS

In the above characterization theorems the sequence of order statistics $X_{k,n}$, $1 \leq k \leq n$, $n \in N$, can be replaced by the corresponding sequence of k-th record values $Y_n^{(k)}$, $1 \le k \le n$, $n=0,1,2,\dots$.

Those are defined by

$$Y_n^{(k)} = X_{L_k(n),L_k(n)+k-1} \quad k \ge 1,\ n=0,1,2,\dots,$$

where

$$L_k(0) := 1$$

$$L_k(n+1) := \min\left\{ j \,;\, X_{L_k(n),L_k(n)+k-1} < X_{j,j+k-1} \right\},$$

c.f. Grudzień (1982), Gajek (1985).

Under assumption A it holds

$$f_{Y_m^{(k)},Y_n^{(k)}}(x_m,x_n) = \frac{k \cdot r_k(x_m) \cdot f(x_n)}{\Gamma(n-m)\,\Gamma(m+1)} \left(1 - F(x_n)\right)^{k-1} \left(R_k(x_m)\right)^m \left(R_k(x_n) - R_k(x_m)\right)^{n-m-1}$$

for $0<x_m<x_n<\infty$,

$$f_{Y_m^{(k)},Y_n^{(k)}} = 0, \quad \text{otherwise},$$

where $R_k(x) = -k \cdot \log\left[1 - F(x)\right]$, $r_k(x) = \frac{d}{dx} R_k(x)$.

If we denote by $D_{n,m}^{(k)} := Y_n^{(k)} - Y_m^{(k)}$, we observe that (c.f. Grudzień (1982), Gajek (1985))

$$f_{D_{n,m}^{(k)}}(z) = \begin{cases} \displaystyle\int_0^\infty k \cdot r_k(u) \left(R_k(u)\right)^m \left(R_k(u+z) - R_k(u)\right)^{n-m-1} \left(1-F(z+u)\right)^{k-1} f(z+u)\,du \left(\Gamma(n-m)\Gamma(m+1)\right)^{-1} & \text{for } 0<z<\infty \\ 0, \quad \text{otherwise} & \end{cases}$$

that is, the distribution of $D_{n,m}^{(k)}$ is of the same structure as that of $X_{j,n} - X_{i,n}$.

Hence, the proofs of the above theorems work also if the order statistics are replaced by the k-th record values.

Acknowledgements. The authors wish to thank Prof. Dr. R.-D. Reiss (Siegen) and Prof. Dr. J. Hüsler (Bern), the organizers of the Oberwolfach conference on Extremvalue Theory. The idea and first draft of this paper originated from discussions during this conference.

The research of U. Gather was partially supported by an award of the Alfried Krupp von Bohlen and Halbach foundation.

REFERENCES

Ahsanullah, M. (1976), On a characterization of the exponential distribution by order statistics, J. Appl. Prob., 13, 818 - 822.

Ahsanullah, M. (1977), A characteristic property of the exponential distribution, Ann. Statist., 5, 580 - 582.

Ahsanullah, M. (1978a), A characterization of the exponential distribution by spacings, J. Appl. Prob., 15, 650 - 653.

Ahsanullah, M. (1978b), On a characterization of the exponential distribution by spacings, Ann. Inst. Statist. Math., 30, A, 163 - 166.

Ahsanullah, M. (1981), On characterizations of the exponential distribution by spacings, Statist. Hefte, 22, 316 - 320.

Ahsanullah, M. (1984), A characterization of the exponential distribution by higher order gap, Metrika, 31, 323 - 326.

Azlarov, T.A. and Volodin, N.A. (1986), Characterization Problems Associated with the Exponential distribution, Springer, New York.

David, H.A. (1981), Order Statistics, Wiley, New York, 2nd. ed.

Desu, M.M. (1971), A characterization of the exponential distribution by order statistics, Ann. Math. Statist., 42, 837 - 838.

Gajek, L. (1985), Limiting properties of difference between the successive k-th record values, Prob., Math. Statist., 5, 221 - 224.

Galambos, J. and Kotz, S. (1978), Characterizations of Probability Distributions. Lecture Notes in Mathematics 675, Springer, Berlin and New York.

Gather, U. (1988), On a characterization of the exponential distribution by properties of order statistics. To appear in Statistics & Probability Letters, 7, No 1.

Grudzień, Z. (1982), Charakteryzacja rozkładów w terminach statystyk rekordowych oraz rozkłady i nomenty statystyk porzadkowych i rekordowych z prób o losowej liczebności, Praca doktorska, UMCS Lublin, Poland.

Karsnosel'skij, M.A. and Rutickii, YA.B. (1961), Convex functions and Orlics spaces. Groningen.

Mitrinovič, D.S. (1970), Analytic Inequalities. Springer, Berlin.

Puri, P.S. and Rubin, H. (1970), A characterization based on absolute differences of two i.i.d. random variables. Ann. Math. Statist., 41, 2113 - 2122.

Rossberg, H.J. (1972), Characterization of the exponential and the Pareto distributions by means of some properties of the distributions which the differences and quotients of order statistics are subject to. Math. Operationsforsch. Statist., 3, 207 - 216.

A CHARACTERIZATION OF THE UNIFORM DISTRIBUTION VIA MAXIMUM LIKELIHOOD ESTIMATION OF ITS LOCATION PARAMETER

Z. Buczolich and G.J. Székely

Abstract. Let $F(x-\Theta)$ be a family of probability density functions with location parameter Θ. If for every ordered sample $x_1 \leqq x_2 \leqq \ldots \leqq x_n$ the midrange $(x_1+x_n)/2$ is a maximum likelihood estimation of Θ then $f(x)$ is the density function of a symmetric uniform distribution. (The converse statement is obvious and well known.)

1. INTRODUCTION.

A classical characterization theorem by Gauss (1809) states that the sample mean $\bar{x} = (1/n) \cdot \sum_{i=1}^{n} x_i$ $(n \geqq 3)$ is a maximum likelihood (ML) estimator of the location parameter Θ of a family of probability density functions $f(x-\Theta)$ if and only if $f(x)$ is a normal distribution with expectation 0 (the unnecessary conditions of this result were gradually dropped by Teicher (1961) and Findeisen (1982 a).) In this paper we give a similar characterization of the symmetric uniform distribution via midrange, the ML estimator of its location parameter. The method of our proof can be generalized to obtain a complete characterization of all convex L-statistics ($L = \sum_{i=1}^{n} a_i x_i$, $a_i \geqq 0$, $\sum_{i=1}^{n} a_i = 1$) that are ML estimators of a location parameter.

In a forthcoming paper we shall show that the following list contains all convex L-statistics that are ML estimators of a location parameter of a probability density function:

Research supported by the Hungarian National Foundation for Scientific Research, Grant No. 1807 and 1808.

(i) $\sum_{i=1}^{n}(x_i/n)$, $n \geq 3$

(ii) $\alpha x_1+(1-\alpha)x_n$, $\alpha \in (0,1)$, $n \geq 3$

(iii) $\alpha x_j+(1-\alpha)x_{j+1}$, $\alpha \in [0,1]$, $n \geq 2$, $j \leq n-1$.

Here the point is that no other convex L-statistics can be an ML estimator of the location parameter. We note that the paper by Findeisen (1982 b) discusses the special case of (iii) when L is the median. In this paper we treat the case (ii) when $\alpha = 1/2$. It is easy to modify our proof for the general case when $\alpha \in (0,1)$ is arbitrary. Then the only difference is that the corresponding uniform density is not necessarily symmetric to the origin, namely $f(x) = 1/t$ if $x \in (-(1-\alpha)t, \alpha t)$ and $f(x) = 0$ otherwise. As a consequence one can see that for an arbitrary positive x_o and real c, a maximum likelihood estimator of the location parameter of the uniform distribution on $(-x_o+c, x_o+c)$ is $\frac{x_1+x_n}{2} - \frac{c}{2x_o}(x_n-x_1)$ (this is not necessarily a convex L-statistics; the coefficients of x_1 and x_n might be negative). Another obvious maximum likelihood estimator of the location parameter is $\frac{x_1+x_n}{2} - c$ (this is not an L-statistics). The point is that both estimators belong to the interval (x_n-x_o-c, x_1+x_o-c).

2. Main result

Take an independent sample from a distribution with density function $f(x-\Theta)$ and arrange the sample elements increasingly: $x_1 \leq x_2 \leq \ldots \leq x_n$. Suppose that $n \geq 3$ and the midrange $(x_1+x_n)/2$ is an ML estimator of Θ, that is

$$(1) \qquad \prod_{i=1}^{n} f(x_i-\Theta) \leq \prod_{i=1}^{n} f(x_i-(x_1+x_n)/2)$$

for all real Θ and $x_1 \leq \ldots \leq x_n$. Using the notation $y_i = x_i-(x_1+x_n)/2$ and $m = -\Theta+(x_1+x_n)/2$ we obtain the inequality

$$(2) \qquad \prod_{i=1}^{n} f(y_i+m) \leq \prod_{i=1}^{n} f(y_i)$$

for all real m and $y_1 \leq y_2 \leq \ldots \leq y_n = -y_1$.

Theorem. Let f be a probability density function ($f \geq 0$ is Lebesque measurable, $\int_{\mathbb{R}} f = 1$) and suppose that (2) holds for every $m \in \mathbb{R}$ and $y_1 \leq \ldots \leq y_n = y_1$. Then there exists an $x_o > 0$ such that $f(x) = \frac{1}{2x_o}$ if $x \in (-x_o, x_o)$ and $f(x) = 0$ if $|x| > x_o$.

3. Lemmas.

In the following we shall denote the Lebesgue measure by λ.

Lemma 1. (Steinhaus). If $A \subset \mathbb{R}$ is measurable and $\lambda(A) > 0$ then the difference set $\{a-a' : a, a' \in A\}$ contains an interval around 0.

For a proof of this lemma we refer to Steinhaus (1920) or Bingham Goldie and Teugels (1987) Theorem 1.1.1.

Lemma 2. Suppose that the set $E \subset R$ has the following properties

(i) if $x, y \in E$ then $(x-y)/2 \in E$ and $(y-x)/2 \in E$;

(ii) there exists an $\varepsilon > 0$ such that $(-\varepsilon, \varepsilon) \subset E$.

Then from $x \in E$ it follows that $(-|x|, |x|) \subset E$.

Proof. Suppose that $x \in E$. We may suppose that $x=1$, otherwise take $(1/x) \cdot E$ instead of E. We have to show that $(-1,1) \subset E$. From property (ii) it follows that $0 \in E$. Thus (i) implies that if $x \in E$ then $x/2 \in E$ and $-x/2 \in E$. Since $1 \in E$ we have $1/2 \in E$ and $-1/2 \in E$. We put $G_n = \{-2^n+1, -2^n+2, \ldots, 2^n-1, 2^n\}$. Suppose that for any $k \in G_n$ we have $k/2^n \in E$. From (i) it follows that for $k \in G_n$ we have $(1-k/2^n)/2 \in E$ and $-(1-k/2^n)/2 \in E$, that is $1/2-k/2^{n+1} \in E$ and $-1/2+k/2^{n+1} \in E$. Since $1 \in E$, by assumption, we proved that for any $k \in G_{n+1}$ we have $k/2^{n+1} \in E$, and thus by induction $k/2^n \in E$ for every n and $k \in G_n$. Suppose that $|y| < 1/2$. We can obviously choose a number z such that $z = k/2^n$ $(k \in G_n)$ and $|z/2-y| < \varepsilon/2$. Then there exists a number v such that $|v| < \varepsilon$ and $(z-v)/2 = y$. By (ii) we have $v \in E$ and hence $y \in E$ by (i). Hence we proved that $(-1/2,1/2) \subset E$. Suppose that we have already showed that $(-1+1/2^n, 1-1/2^n) \subset E$. Then by (i) we have $y = (1+z)/2 \in E$ and $y' = -(1+z)/2 \in E$ for every $(-z)$, $|z| < 1-1/2^n$. Since $(1+1-1/2^n)/2 = 1-1/2^{n+1}$ it is obvious that $(-1+1/2^{n+1}, 1-1/2^{n+1}) \subset E$. Therefore letting $n \to \infty$ we obtain that $(-1,1) \subset E$.

4. Proof of the main result

Since $\int_{\mathbb{R}} f = 1$ there exists an $r > 0$ such that the set $H_r := \{x : f(x) > r\}$ is of positive Lebesgue measure. Lemma 1 implies that the set $\{x-y : x \in H_r,\ y \in H_r\}$ contains an interval around 0. Thus there exists an $\varepsilon_1 > 0$ such that $|x_2 - x_1| < \varepsilon_1$ implies the existence of an $m \in \mathbb{R}$ such that $x_1 + m \in H_r$ $x_2 + m \in H_r$. Therefore if $|x_2 - x_1| < \varepsilon_1$ then we have $\sup\{f^{n-1}(x_1+m) f(x_2+m) : m \in \mathbb{R}\} > r^n > 0$. Choosing $0 > y_1 = y_2 = \dots = y_{n-1} = y > \varepsilon_1/2 =: -\varepsilon_2$ and $y_n = -y$ we infer from (2) that $f^{n-1}(y) f(-y) \geq \sup_m \{f^{n-1}(y+m) f(-y+m)\} \geq r^n$ and hence for $|y| < \varepsilon_2$ we have $f(y) > 0$. We put $E := H_o$. Suppose that $x \in E$ $y \in E$ and $x < y$. Then letting $y_1 = \dots = y_{n-1} = (x-y)/2$, $y_n = (y-x)/2$ and $m = (x+y)/2$ we obtain from (2) that $0 < f^{n-1}(x) f(y) = f^{n-1}((x-y)/2 + (y-x)/2) f((y-x)/2 + (x+y/2)) \leq$ $\leq f^{n-1}((x-y)/2)\ f((y-x)/2)$ and hence $(x-y)/2 \in E$ and $(y-x)/2 \in E$. Using that $f(y) > 0$ for $|y| < \varepsilon_2$ we have $(-\varepsilon_2, \varepsilon_2) \subset E$. Applying Lemma 2 we obtain that if $x \in E$ then $(-|x|, |x|) \subset E$, that is there exists an $x_o \in \mathbb{R} \cup \{+\infty\}$ such that $(-x_o, x_o) = E \setminus (\{-x_o\} \cup \{x_o\})$. Suppose that $x_1 \in (0, x_o)$. Choose a $\delta > 0$ such that $[x_1 - \delta, x_1 + \delta] \subset (0, x_o)$. Since $(x_1 - \delta, x_1 + \delta) \subset E$ and $(-x_1 - \delta, -x_1 + \delta) \subset E$ we can find a number $r > 0$ such that

$$\lambda(H_r \cap (x_1 - \delta, x_1 + \delta)) > (15/16) \cdot 2\delta \tag{3}$$

and

$$\lambda(H_r \cap (-x_1 - \delta, -x_1 + \delta)) > (15/16) \cdot 2\delta. \tag{4}$$

Suppose that $y \in (x_1 - \delta/8, x_1 + \delta/8)$. Then for every $m \in (0, 3\delta/8)$ we have $-y + m \in (-x_1 - \delta, -x_1 + \delta)$ and $y + m \in (x_1 - \delta, x_1 + \delta)$. If for every m either $-y + m \notin H_r$ or $y + m \notin H_r$ then we would obtain that either $\lambda((x_1 - \delta, x_1 + \delta) \setminus H_r) > 3\delta/16$ or $\lambda((-x_1 - \delta, -x_1 + \delta) \setminus H_r) > 3\delta/16$. This contradicts (3) or (4) since $3\delta/16 > 2\delta/16$. Hence for every $y \in (x_1 - \delta/8, x_1 + \delta/8)$ there exists an m such that $-y + m \in H_r$ and $y + m\ H_r$. Letting $y_1 = \dots = y_{n-1} = -y$, $y_n = y$ and $m = m$ we obtain from

(2) that

$$(5) \qquad r^n \leq f^{n-1}(-y+m)f(y+m) \leq f^{n-1}(-y)f(y).$$

Letting $y_1 = \ldots = y_n = 0$ in (2) we have $f^n(m) \leq f^n(0)$ for every $m \in \mathbb{R}$, that is $f(0) \geq f(m)$. Therefore using (5) we obtain that $f(y) \geq$ $\geq r^n/(f(o))^{n-1} > 0$ and $f(-y) \geq (r^n/f(o))^{1/(n-1)} > 0$. (Recall that $f(0) > 0$ because $\int_{\mathbb{R}} f = 1$ and $f(x) \leq f(0)$.) Since x_1 was an arbitrary element of $(0,x_o)$ we proved that the function $g := \log f$ is locally bounded in $(0,x_o)$, and by symmetry in $(-x_o,x_o)$.

Letting $y = y_1 < y_2 = \ldots = y_{n-1} = t < -y = y_n$ and $m=h$ in (2) we obtain (with $g = \log f$)

$$g(y+h)+(n-2)g(t+h)+g(-y+h) \leq$$
$$\leq g(y)+(n-2)g(t)+g(-y),$$

that is

$$(6) \quad g(t+h)-g(t) \leq (1/n-2))(g(y)-g(y+h)+g(-y)-g(-y+h)).$$

If $y < t' := t+h < -y$ then letting $y = y_1 < y_2 = \ldots = y_{n-1} = t' < -y = y_n$ and $m=-h$ we also have

$$(7) \quad \begin{aligned} &g(t)-g(t+h) = g(t'-h)-g(t') \leq \\ &\leq (1/(n-2))\cdot(g(y)-g(y-h)+g(-y)-g(-y-h)). \end{aligned}$$

Suppose that $t \in (-x_o,x_o)$. Choose an $y_o \in (0,x_o)$ and $\delta > 0$ such that $[-y_o-2\delta,-y_o+2\delta] \subset (-x_o,-t-2\delta)$, that is $[y_o-2\delta,y_o+2\delta] \subset (t+2\delta,x_o)$. Then for any $s \in (t-\delta,t+\delta)$, $y \in (y_o-\delta,y_o+\delta)$ and h, $|h| < \delta$ we can apply (6) and (7).

Since g is locally bounded and hence integrable on any compact subset of $(-x_o,x_o)$ we obtain from (6) that

$$(8) \quad \begin{aligned} &g(s+h)-g(s) \leq ((n-2)\delta)^{-1}\cdot \\ &\cdot\left|\int_{y_o}^{y_o+\delta} (g(y)-g(y+h))+(g(-y)-g(-y+h))dy\right| \leq \\ &\leq ((n-2)\delta))-1\left|\int_{y_o}^{y_o+h} g(y)+g(-y)dy\right| + \\ &+ ((n-2)\cdot\delta))^{-1}\left|\int_{y_o+\delta}^{y_o+\delta+h} g(y)+g(-y)dy\right| \leq \\ &\leq ((n-2)\cdot\delta)^{-1}\cdot 4\cdot K\cdot |h| \end{aligned}$$

where K is the bound of g on the compact intervals $[-y_o-2\delta,-y_o+2\delta]$ and $[y_o-2\delta,y_o+2\delta]$. Using (7) we obtain similarly that

$$
\begin{aligned}
& g(s)-g(s+h) \leq \\
& \leq ((n-2)\cdot\delta)^{-1}\cdot\left|\int_{y_o-h}^{y_o} g(y)+g(-y)dy\right| + \\
(9) \quad & + ((n-2)\cdot\delta)^{-1}\left|\int_{y_o+\delta-h}^{y_o+\delta} g(y)+g(-y)dy\right| \leq \\
& \leq ((n-2)\cdot\delta)^{-1}\cdot 4\cdot K\cdot |h|.
\end{aligned}
$$

We thus proved that g is Lipshitz on $(t-\delta,t+\delta)$. Since t was arbitrary we proved that g is a locally Lipshitz function in $(-y_o,x_o)$.

Therefore g is differentiable almost everywhere in $(-x_o,x_o)$. Hence we can choose a set $S\subset(-x_o,0)$, $\lambda(S)=x_o$ such that if $x\in S$ then g is differentiable at x and $-x$. For any $x\in S$ put $y_1=x$, $y_2=\ldots=y_n=-x$ and $m=h$ into (2); then

$$g(x+h)+(n-1)g(-x+h)\leq g(x)+(n-1)g(-x)$$

for every $h\in\mathbb{R}$. Therefore

$$(10) \qquad g'(x)+(n-1)g'(-x)=0.$$

Letting $y_1=y_2=\ldots=y_{n-1}=x$, $y_n=-x$ and $m=h$ in (2) we obtain $(n-1)g(x+h)+g(-x+h)\leq(n-1)g(x)+g(-x)$ for every $h\in\mathbb{R}$. Therefore

$$(11) \qquad (n-1)g'(x)+g'(-x)=0.$$

Since $n\geq 3$ (10) and (11) imply that $g'(x)=0$, $g'(-x)=0$ for every $x\in S$.

The function g is locally Lipschitz in $(-x_o,x_o)$, hence absolutely continuous on any compact subinterval of $(-x_o,x_o)$, that is g is the integral of its derivative there. Since $g'(x)=0$ for almost every $x\in(-x_o,x_o)$ we proved that g is constant on $(-x_o,x_o)$. Thus $f=\exp(g)$ is also constant. We have also showed that $x\notin H_o=E$ for $|x|>x_o$, that is $f(x)=0$ for $|x|>x_o$. From $\int f=1$ it follows that $f(x)=1/(2x_o)$ if $x\in(-x_o,x_o)$ and $f(x)=0^{\mathbb{R}}$ for $|x|>x_o$. This completes the proof.

Finally we note that the classical results we used from real analysis can be found in Saks (1964).

The authors are indebted to professor E. Siebert for calling their attention to the papers by Findeisen.

REFERENCES

Bingham, N.H., Goldie, C.M. and Teugels, J. (1987): Regular Variation, Cambridge Univ. Press.

Findeisen, P. (1982a): Die Charakterisierung der Normalverteilung nach Gauss, Metrika 29, 55-63.

Findeisen, P. (1982b): Charakterisierung der zweiseitigen Exponentialverteilung, Metrika 29 95-102.

Gauss, C.F. (1809): Theoria Motus Corporum Coelestium, Perthes & Besser, Hamburgi (English ed. The Heavenly Bodies Moving Around the Sun in Conic Sections, Dover Pub., New York, 1963).

Saks, S. (1964): Theory of the integral, Dover, New York.

Steinhaus, H. (1920): Sur les distances des points de mesure positive, Fund. Math. 1, 93-104.

Teicher, H. (1961): Maximum likelihood characterization of distributions, Ann. Math. Statist. 32, 1214-1222.

Eötvös University
Mathematical Institute
Budapest, Muzeum krt. 6-8.
H-1088, Hungary

SIMPLE ESTIMATORS OF THE ENDPOINT OF A DISTRIBUTION

SÁNDOR CSÖRGŐ[1] and DAVID M. MASON[2]

University of Szeged and University of Delaware

Abstract. Based on an increasing number of extreme order statistics, Hall (1982) proposed an implicit maximum likelihood type estimator of the endpoint of a distribution when only limited information is available about the behavior of the distribution in the neighborhood of the endpoint. Using continuous time regression methods, we derive simple explicit estimators based on a linear combination of the same number of extreme order statistics. We prove the asymptotic normality of our estimators in very general models and conclude that in the particular models considered by Hall, our explicit estimators inherit all the efficiency and robustness properties of his estimator.

1. INTRODUCTION, MOTIVATION, AND THE RESULTS

Let

$$F(x) = F_\theta(x) = (x-\theta)^{1/a} L_1(x-\theta), \quad \theta \le x < \infty \tag{1.1}$$

be a distribution function, where $-\infty < \theta < \infty$ is the unknown left endpoint of its support, $0 < a < \infty$, and $L_1(y)$, $y \ge 0$, is a function slowly varying at the origin. The problem of estimating the location parameter θ (the threshold parameter) or the completely parallel problem of estimating the upper endpoint of a distribution has been addressed by several authors. Hall (1982) gives the relevant references and comments on them in detail. For an alternative approach and additional references see Smith (1987). Hall (1982) points out, if we only know the very limited information that L_1 is slowly varying then the estimator should be based on a relatively small proportion of small order statistics. Also, when $a > 1/2$ one cannot do better than using a linear combination of a fixed number of small order statistics to estimate θ. In Section 3 of Hall (1982) this problem is basically solved. However, when $0 < a \le 1/2$, which we assume from now on, one can do much better using an increasing number k_n of lower order statistics such that $k_n \to \infty$ and $k_n/n \to 0$ as $n \to \infty$. Assuming that $0 < a \le 1/2$ is *known* and specifying the slowly varying function such that

$$F(x) = c^{-1/a}(x-\theta)^{1/a}\{1 + \mathcal{O}((x-\theta)^{b/a})\} \quad \text{as} \quad x \downarrow \theta, \tag{1.2}$$

where c and b are some positive constants, the left-endpoint version of Theorem 2 of Hall (1982) shows that for maximum likelihood estimates θ_n^H derived under an even more special model and based on

[1] Partially supported by the Hungarian National Foundation for Scientific Research, Grant No. 1808
[2] Partially supported by the Alexander von Humboldt Foundation.

the sequence of the k_n smallest order statistics we have the following: Whenever $0 < a < 1/2$,

$$n^a k_n^{1/2-a}(\theta_n^H - \theta) \xrightarrow{\mathcal{D}} N(0, (1-2a)c^2), \tag{1.3}$$

and when $a = 1/2$

$$(n \log k_n)^{1/2}(\theta_n^H - \theta) \xrightarrow{\mathcal{D}} N(0, c^2), \tag{1.4}$$

provided that

$$k_n = \mathcal{O}(n^{b/(b+1/2)}). \tag{1.5}$$

(Here and in what follows, convergence and order relations are meant as $n \to \infty$ if not specified otherwise.)

The special model of Hall referred to above is when $F = F_\theta$ has a density $f = f_\theta$ satisfying

$$f(x) = 0, \ x \le \theta, \quad \textit{and} \quad f(x) = c^{-1/a}(x-\theta)^{1/a-1}/a \tag{1.6}$$

in a right neighborhood of θ. Hall shows that θ_n^H has certain asymptotic efficiency properties within model (1.6), which also hold under some perturbations of the model.

Although Hall's maximum likelihood estimator θ_n^H is unique, it is implicit because the corresponding likelihood equation cannot be solved in a closed form. The aim of this paper is to derive a simple explicit estimator $\hat{\theta}_n$ based on a linear combination of the smallest k_n observations, which possesses all the desirable properties of Hall's implicit maximum likelihood estimator.

In order to motivate our estimator, let $X_1, X_2, \ldots$ be a sequence of independent and identically distributed random variables (rv's) with distribution function F and for each $n \ge 1$ let $X_{1,n} \le \cdots \le X_{n,n}$ denote the order statistics of the first n rv's. Assume that F has a density f satisfying (1.6) in a right neighborhood of θ. On an appropriate probability space there exists a sequence of Brownian bridges B_n such that for the sample quantile function

$$Q_n(s) = \begin{cases} X_{i,n}, & \text{if } (i-1)/n < s \le i/n, \ i = 1, \ldots n, \\ X_{1,n}, & \text{if } s = 0 \end{cases}$$

we have for each $0 < s \le 1$

$$Q_n(s) \approx \theta + cs^a + cas^{a-1}n^{-1/2}B_n(s) =: Y_n(s; \theta, c). \tag{1.7}$$

For weight functions w, w_0 and w_1 set according to whether $0 < a < 1/2$ or $a = 1/2$

$$\tilde{\theta}_n = \begin{cases} \int_0^{k_n/n} Y_n(s;\theta,c)w(s)ds + w_0(k_n/n)Y_n(k_n/n;\theta,c), \\ \int_{1/n}^{k_n/n} Y_n(s;\theta,c)w(s)ds + w_1(1/n)Y_n(1/n;\theta,c) + w_0(k_n/n)Y_n(k_n/n;\theta,c). \end{cases}$$

We seek weight functions $w^\star, w_0^\star$ and $w_1^\star$ such that $\mathrm{Var}\,\tilde{\theta}_n/c^2$ is minimized among all weight functions w, w_0 and w_1 that satisfy $E\tilde{\theta}_n = \theta$ for all $-\infty < \theta < \infty$ and $0 < c < \infty$.

Applying the continuous time regression methods of Parzen (1961, 1979) to solve this problem, we obtain for $0 < a < 1/2$

$$w^\star(s) = \begin{cases} \left(\frac{k_n}{n}\right)^{2a-1} \frac{(1-a)(1-2a)}{a} s^{-2a}, & 0 < s < k_n/n\ , \\ 0, & \text{otherwise} \end{cases}$$

and

$$w_0^\star(k_n/n) = -(1-2a)/a,$$

and for $a = 1/2$

$$w^\star(s) = \begin{cases} (\log k_n)^{-1} s^{-1}, & 1/n < s < k_n/n, \\ 0, & \text{otherwise,} \end{cases}$$

$$w_0^\star(k_n/n) = -2/\log k_n \quad \text{and} \quad w_1^\star(1/n) = 2/\log k_n.$$

The obtained estimator $\tilde{\theta}_n$ with these weights is the best linear unbiased estimator of θ based on the continuous observation of the Gaussian process Y_n in the interval $(0, k_n/n]$ when $0 < a < 1/2$ and in the interval $[1/n,\ k_n/n]$ when $a = 1/2$. By the relation in (1.7) it is reasonable to expect that the estimator

$$\hat{\theta}_n = \begin{cases} \int_0^{k_n/n} Q_n(s) w^\star(s) ds + w_0^\star(k_n/n) Q_n(k_n/n), & 0 < a < 1/2, \\ \int_0^{k_n/n} Q_n(s) w^\star(s) ds + w_1^\star(1/n) Q_n(1/n) + w_0^\star(k_n/n) Q_n(k_n/n), & a = 1/2, \end{cases}$$

is approximately equal to the best linear unbiased estimator of θ based on a linear combination of the order statistics $X_{1,n}, \ldots, X_{k_n,n}$, say $\hat{\theta}_n^B$. In fact it can be shown that if $E|X_1|^\delta < \infty$ for some $\delta > 0$,

$$\text{MSE}\, \hat{\theta}_n / \text{Var}\, \hat{\theta}_n^B \to 1.$$

The details of showing this are omitted. We remark that, at least for the case $0 < a < 1/2$, the methods in Chernoff, Gastwirth and Johns (1967) give the same optimal weight functions.

Integrating out the integral in $\hat{\theta}_n$ we obtain the following nice closed form for our estimator:

$$\hat{\theta}_n = \begin{cases} k_n^{2a-1} \frac{1-a}{a} \sum_{i=1}^{k_n} (i^{1-2a} - (i-1)^{1-2a}) X_{i,n} - \frac{1-2a}{a} X_{k_n,n}, & 0 < a < 1/2, \\ (\log k_n)^{-1} \{\sum_{i=2}^{k_n} \log(i/(i-1)) X_{i,n} + 2(X_{1,n} - X_{k_n,n})\}, & a = 1/2. \end{cases} \tag{1.8}$$

Having derived our estimator we also drop the fictitious assumption (1.6) about the underlying distribution. It can be easily shown that any of the estimators of the form given in (1.8) is a strongly consistent estimator of θ assuming only that $F(x) = 0$ for all $x \le \theta$ and F is continuous at θ, if $k_n \to \infty$ and $k_n/n \to 0$ as $n \to \infty$.

The purpose of the present paper is to study the asymptotic normality and optimality of our estimator for θ not only in Hall's model (1.2) but in the general model (1.1) assuming only very mild regularity conditions on the slowly varying function.

Since our estimator is a functional of the sample quantile function, it will be more convenient for us to work with the quantile formulations of the models (1.1) and (1.2). Denoting by F^{-1} the left–continuous inverse of F, model (1.1) is equivalent to

$$F^{-1}(s) = \theta + s^a L_2(s)$$

in a right neighborhood of zero, where L_2 is a nonnegative function slowly varying at zero. (See, for instance, Seneta (1976).) As in Hall (1982), model (1.2) is equivalent to

$$F^{-1}(s) = \theta + cs^a\{1 + \mathcal{O}(s^b)\} \quad as \quad s \downarrow 0. \tag{1.9}$$

It is well known that any function L_2 slowly varying at zero can be expressed as

$$L_2(s) = \sigma(s)L(s) = \sigma(s)\exp\Big(\int_s^1 \frac{b(u)}{u}du\Big), \tag{1.10}$$

where $\sigma(\cdot)$ is a function such that $\sigma(s) \to \sigma$ as $s \downarrow 0$, where $0 < \sigma < \infty$ is a constant, and $b(\cdot)$ is a measurable function converging to zero as $s \downarrow 0$.

The most general model in which we can prove the asymptotic normality of our estimator $\hat{\theta}_n$ is when the function $\sigma(s)$ in (1.10) is constantly σ in a right neighborhood of the origin. We point out that if in the general model (1.1), F has a density which is positive in a right neighborhood of zero, then L_2 has a representation such that $\sigma(s) \equiv \sigma$ holds in an appropriate right neighborhood of zero.

To state our first result we need to introduce the following bias term according to whether $0 < a < 1/2$ or $a = 1/2$:

$$\beta_n(a) = \begin{cases} \frac{(1-a)(1-2a)}{a}\left(\frac{k_n}{n}\right)^{2a-1}\int_0^{k_n/n} s^{-a}\{L(s) - L(k_n/n)\}ds, \\ (\log k_n)^{-1}\{2\left(\frac{1}{n}\right)^{1/2}L\left(\frac{1}{n}\right) - 2\left(\frac{k_n}{n}\right)^{1/2}L\left(\frac{k_n}{n}\right) + \int_{1/n}^{k_n/n} s^{-1/2}L(s)ds\}. \end{cases}$$

The proofs of our theorems are postponed until Section 3, following a short section of technical lemmas.

THEOREM 1. *If for all s in a right neighborhood of zero*

$$F^{-1}(s) = \theta + \sigma s^a L(s) \tag{1.11}$$

for some $-\infty < \theta < \infty$, $0 < \sigma < \infty$, and $0 < a \le 1/2$, where L is given in (1.10), then for any sequence of positive integers k_n such that $k_n \to \infty$ and $k_n/n \to 0$, we have for $0 < a < 1/2$

$$\frac{n^a k_n^{1/2-a}}{L(k_n/n)}(\hat{\theta}_n - \theta - \sigma\beta_n(a)) \xrightarrow{\mathcal{D}} N(0, (1-2a)\sigma^2)$$

and for $a = 1/2$

$$\frac{n^{1/2}\log k_n}{\left(\int_{1/n}^{k_n/n} s^{-1}L^2(s)ds\right)^{1/2}}(\hat{\theta}_n - \theta - \sigma\beta_n(1/2)) \xrightarrow{\mathcal{D}} N(0, \sigma^2).$$

As we see the normalizing constants in Theorem 1 depend upon the normalized slowly varying function L. A small restriction of the model in Theorem 1 allows us to get rid of this dependence. Assume that in (1.11) we have

$$\sigma L(s) \to c \quad \text{as} \quad s \downarrow 0, \tag{1.12}$$

where $0 < c < \infty$. Applying the continuous time regression method for the estimation of c in the model (1.6), we obtain the estimator

$$\hat{c}_n = \left(\frac{k_n}{n}\right)^{-a}(X_{k_n,n} - \hat{\theta}_n),$$

where $\hat{\theta}_n$ is given in (1.8). Introducing the bias term

$$\gamma_n(a) = \left(\frac{k_n}{n}\right)^{-a}\left\{\left(\frac{k_n}{n}\right)^a L\left(\frac{k_n}{n}\right) - \beta_n(a)\right\}, \quad 0 < a \leq 1/2,$$

for $\hat{c}_n$, our next result is the following.

THEOREM 2. *If for all s in a right neighborhood of zero*

$$F^{-1}(s) = \theta + \sigma s^a L(s)$$

for some $-\infty < \theta < \infty$, $0 < \sigma < \infty$, and $0 < a \leq 1/2$, where L is given in (1.10), and in addition (1.12) holds with some $0 < c < \infty$, then for any sequence of positive integers k_n such that $k_n \to \infty$ and $k_n/n \to 0$, we have for $0 < a < 1/2$

$$n^a k_n^{1/2-a}(\hat{\theta}_n - \theta - \sigma\beta_n(a)) \xrightarrow{\mathcal{D}} N(0, (1-2a)c^2),$$

for $a = 1/2$

$$(n \log k_n)^{1/2}(\hat{\theta}_n - \theta - \sigma\beta_n(1/2)) \xrightarrow{\mathcal{D}} N(0, c^2),$$

and for all $0 < a \leq 1/2$

$$k_n^{1/2}(\hat{c}_n - c - \sigma\gamma_n(a)) \xrightarrow{\mathcal{D}} N(0, (1-a)^2 c^2).$$

We note that under the conditions of Theorem 2 it can be shown by using Lemmas 2.1 and 2.3 below that $\gamma_n(a) \to 0$ for any $0 < a \leq 1/2$. Thus by the last statement of Theorem 2 we have $\hat{c}_n \xrightarrow{P} c$. Subsequently we get for $0 < a < 1/2$

$$\frac{n^a k_n^{1/2-a}}{\hat{c}_n}(\hat{\theta}_n - \theta - \sigma\beta_n(a)) \xrightarrow{\mathcal{D}} N(0, 1-2a) \tag{1.13}$$

and for $a = 1/2$

$$\frac{(n \log k_n)^{1/2}}{\hat{c}_n}(\hat{\theta}_n - \theta - \sigma\beta_n(1/2)) \xrightarrow{\mathcal{D}} N(0,1). \tag{1.14}$$

Finally, specializing Theorem 2 to Hall's model (1.2) (or (1.9)), we obtain the complete analogues of Hall's results in (1.3) – (1.5) for our explicit estimators.

THEOREM 3. *If for all s in a right neighborhood of zero (1.9) holds for some* $-\infty < \theta < \infty$, $0 < c < \infty$, $0 < a \leq 1/2$ *and* $0 < b < \infty$, *then whenever k_n is a sequence of positive integers such that $k_n \to \infty$ and (1.5) holds, we have for* $0 < a < 1/2$

$$n^a k_n^{1/2-a}(\hat{\theta}_n - \theta) \xrightarrow{\mathcal{D}} N(0, (1-2a)c^2),$$

for $a = 1/2$

$$(n \log k_n)^{1/2}(\hat{\theta}_n - \theta) \xrightarrow{\mathcal{D}} N(0, c^2),$$

and for all $0 < a \leq 1/2$

$$k_n^{1/2}(\hat{c}_n - c) \xrightarrow{\mathcal{D}} N(0, (1-a)^2 c^2).$$

Of course, under the conditions of Theorem 3 we have (1.13) and (1.14) with the bias term $\sigma\beta_n(a)$ deleted.

Since in Hall's model (1.9) our estimators have exactly the same asymptotic distributional properties as his maximum likelihood type estimators, we conclude that our simple estimators enjoy all asymptotic efficiency and robustness properties described in his Theorem 3 and the comments that follow it.

When $a < 1/2$ is unknown, Hall shows in his Theorem 6 that a modified form of his implicit estimator is still meaningful under (1.9) and its asymptotic behavior is the same as that of the original except that the limiting variance increases by a factor of $(a^{-1} - 1)^2$. In our case a natural idea would be to plug one of the many explicit consistent estimators of a in the formula for $\hat{\theta}_n$, for example a version of the celebrated Hill estimator. However, a satisfactory solution to the problem is much more involved. The joint continuous time regression estimators for the endpoint, the index and the leading constant can be derived using the methods of Kinderman and LaRiccia (1985), but their analysis will require a separate effort and will be considered elsewhere. We mention here that the referee pointed out to us that if one is not interested in efficiency the methods of Hosking and Wallis (1987) could be adapted to construct an estimator of θ when a is unknown. Of course in such a case one can always use the minimum observation.

We should point out that when $a = 1/2$, Polfeldt (1970) was the first to consider an estimator of θ based on a linear combination of the smallest k_n order statistics. It appears to us that the continuous time regression method has been applied to a nonstandard estimation problem ($a = 1/2$) for the first time in the present paper.

Finally, we note that in the problem of estimating the upper endpoint θ, i.e., when $1 - F(x) = (\theta - x)^{1/a} L_1(\theta - x)$, $\quad -\infty < x \leq \theta$, our estimator takes the form according to whether $0 < a < 1/2$ or $a = 1/2$

$$\hat{\theta}_n = \begin{cases} k_n^{2a-1} \frac{1-a}{a} \sum_{i=1}^{k_n} (i^{1-2a} - (i-1)^{1-2a}) X_{n+1-i,n} + \frac{1-2a}{a} X_{n+1-k_n,n}, \\ (\log k_n)^{-1} \{\sum_{i=2}^{k_n} \log(i/(i-1)) X_{n+1-i,n} - 2(X_{n,n} - X_{n+1-k_n})\}. \end{cases}$$

All the above results hold for this estimator of the upper endpoint with the obvious changes in the notation.

2. TECHNICAL LEMMAS

In the following four lemmas $L(s)$, $0 < s < 1$, denotes any function slowly varying at zero. The proofs of the first two lemmas are found in deHaan (1970), the third is proved in S. Csörgő, Horváth and Mason (1986), and the fourth is proved in S. Csörgő and Mason (1986).

LEMMA 2.1. *If* $\beta > -1$, *then*

$$\lim_{s\downarrow 0} \int_0^s u^\beta L(u)du/(s^{\beta+1}L(s)) = \frac{1}{1+\beta},$$

while if $\beta < -1$, *then there exists a* $0 < \delta < 1$ *such that*

$$\lim_{s\downarrow 0} \int_s^\delta u^\beta L(u)du/(s^{\beta+1}L(s)) = \frac{1}{|1+\beta|}.$$

LEMMA 2.2. *For all* $0 < \tau \leq \gamma < \infty$,

$$\lim_{s\downarrow 0} \sup_{\tau s \leq t \leq \gamma s} L(t)/L(s) = 1$$

and

$$\lim_{s\downarrow 0} \inf_{\tau s \leq t \leq \gamma s} L(t)/L(s) = 1.$$

LEMMA 2.3. *Let* k_n *be any sequence of positive integers such that* $k_n \to \infty$ *and* $k_n/n \to 0$. *Then for any* $0 < \beta < \infty$, *we have*

$$\lim_{n\to\infty} \{(k_n/n)^\beta L(k_n/n)\}/\{(1/n)^\beta L(1/n)\} = \infty.$$

LEMMA 2.4. *Let* k_n *be any sequence of positive integers such that* $k_n \to \infty$ *and* $k_n/n \to 0$. *Then*

$$\lim_{n\to\infty} L(1/n)/\int_{1/n}^{k_n/n} s^{-1}L(s)ds = 0$$

and

$$\lim_{n\to\infty} L(k_n/n)/\int_{1/n}^{k_n/n} s^{-1}L(s)ds = 0.$$

In the following two lemmas $L(s)$, $0 < s < 1$, denotes any nonnegative function slowly varying at zero. These lemmas are completely analogous to Lemmas 5 and 6 in S. Csörgő and Mason (1986), respectively, and their proofs go along the same lines. Therefore we omit the proofs here.

LEMMA 2.5. *For any* $0 < a < 1/2$ *set*

$$d^2(s) = \int_0^s \int_0^s (u \wedge v - uv) u^{-a-1} v^{-a-1} L(u) L(v) du dv.$$

Then

$$\lim_{s \downarrow 0} d^2(s) / (s^{1-2a} L^2(s)) = 2/((1-a)(1-2a)).$$

LEMMA 2.6. *Setting*

$$d_n^2(k_n) = \int_{1/n}^{k_n/n} \int_{1/n}^{k_n/n} (u \wedge v - uv) u^{-3/2} v^{-3/2} L(u) L(v) du dv,$$

we have

$$\lim_{n \to \infty} d_n^2(k_n) / \int_{1/n}^{k_n/n} s^{-1} L^2(s) ds = 4.$$

3. PROOFS

M. Csörgő, S. Csörgő, Horváth and Mason (1986) have constructed a probability space $(\Omega, \mathcal{A}, P)$ carrying a sequence $U_1, U_2, \ldots$ of independent rv's uniformly distributed on (0,1) and a sequence of Brownian bridges $B_n(s)$, $0 \le s \le 1$, $n = 1, 2, \ldots$, such that if $U_{1,n} \le \cdots \le U_{n,n}$ denote the order statistics of $U_1, \ldots, U_n$ and

$$U_n(s) = \begin{cases} U_{i,n}, & \text{if } (i-1)/n < s \le i/n,\ i = 1, \ldots, n, \\ U_{1,n}, & \text{if } s = 0 \end{cases}$$

is the corresponding quantile function, then for any fixed number $0 \le \nu < 1/2$ we have

$$\sup_{1/n < s < 1} n^{\nu} \frac{|n^{1/2}(U_n(s) - s) - B_n(s)|}{s^{1/2-\nu}} = \mathcal{O}_p(1). \tag{3.1}$$

This follows immediately from Theorem 2.1 in the above paper.

Since for each n we have according to whether $0 < a < 1/2$ or $a = 1/2$

$$\hat{\theta}_n \stackrel{\mathcal{D}}{=} \begin{cases} k_n^{2a-1} \frac{1-a}{a} \sum_{i=1}^{k_n} (i^{1-2a} - (i-1)^{1-2a}) F^{-1}(U_{i,n}) - \frac{1-2a}{a} F^{-1}(U_{k_n,n}), \\ (\log k_n)^{-1} \{ \sum_{i=2}^{k_n} \log(i/(i-1)) F^{-1}(U_{i,n}) + 2(F^{-1}(U_{1,n}) - F^{-1}(U_{k_n,n})) \}, \end{cases} \tag{3.2}$$

where $\stackrel{\mathcal{D}}{=}$ denotes equality in distribution, from now on we work without loss of generality on the above space $(\Omega, \mathcal{A}, P)$ and, with a slight abuse of notation, will denote by $\hat{\theta}_n$ the right side of the distributional equality (3.2). Notice that, with this convention, when $0 < a < 1/2$

$$\hat{\theta}_n = \left(\frac{k_n}{n}\right)^{2a-1} \frac{(1-a)(1-2a)}{a} \int_0^{k_n/n} s^{-2a} F^{-1}(U_n(s)) ds - \frac{1-2a}{a} F^{-1}(U_n(k_n/n)), \tag{3.3}$$

and when $a = 1/2$

$$\hat{\theta}_n = (\log k_n)^{-1}\{\int_{1/n}^{k_n/n} s^{-1}F^{-1}(U_n(s))ds + 2(F^{-1}(U_n(1/n)) - F^{-1}(U_n(k_n/n)))\}. \tag{3.4}$$

The proofs of the theorems require the following lemmas. Throughout these lemmas $0 < a \leq 1/2$ and we let Q denote the nonnegative function

$$Q(s) = s^a L(s) = s^a \exp\left(\int_s^1 \frac{b(u)}{u} du\right), \quad 0 < s \leq \delta, \tag{3.5}$$

for some $0 < \delta < 1$, where $L(\cdot)$ is given in (1.10), i.e., $b(\cdot)$ is a real valued measurable function such that $b(s) \to 0$ as $s \downarrow 0$.

For $0 < s \leq \delta$ write

$$\Delta_n(s) = |n^{1/2}\{Q(U_n(s)) - Q(s)\} - as^{a-1}B_n(s)L(s)|.$$

LEMMA 3.1. *Under condition (3.5), for any $0 < a \leq 1/2$, $0 \leq \nu < 1/2$ and sequence of positive integers k_n such that $k_n \to \infty$ and $k_n/n \to 0$, we have for all $1/n \leq s \leq k_n/n$*

$$\begin{aligned}\Delta_n(s) &= \mathcal{O}_p(n^{-\nu})L(s)s^{a-1/2-\nu} + \mathcal{O}_p(1)s^{a-2}L(s)n^{1/2}|U_n(s) - s|^2 \\ &\quad + o_p(1)s^{a-1}L(s)n^{1/2}|U_n(s) - s|,\end{aligned}$$

where the $\mathcal{O}_p(n^{-\nu})$ and $\mathcal{O}_p(1)$ rv's depend only on n and the $o_p(1)$ rv depends only on k_n/n.

Proof. For any $1/n \leq s \leq k_n/n$, we have

$$\begin{aligned}\Delta_n(s) &\leq |n^{1/2}\{(U_n(s))^a - s^a\}L(s) - as^{a-1}B_n(s)L(s)| \\ &\quad + n^{1/2}|L(U_n(s)) - L(s)|(U_n(s))^a,\end{aligned}$$

which by applying a two-term Taylor expansion in the first term is

$$\begin{aligned}&\leq as^{a-1}L(s)|n^{1/2}(U_n(s) - s) - B_n(s)| \\ &\quad + n^{1/2}\frac{a(1-a)}{2}(\xi_n(s))^{a-2}L(s)(U_n(s) - s)^2 \\ &\quad + n^{1/2}|L(U_n(s)) - L(s)|(U_n(s))^a \\ &=: \Delta_n^{(1)}(s) + \Delta_n^{(2)}(s) + \Delta_n^{(3)}(s),\end{aligned}$$

where

$$s \wedge U_n(s) \leq \xi_n(s) \leq s \vee U_n(s). \tag{3.6}$$

Note that

$$\begin{aligned}\Delta_n^{(1)}(s) &\leq as^{a-1/2-\nu}L(s) \sup_{1/n \leq t \leq 1} |n^{1/2}(U_n(t) - t) - B_n(t)|/t^{1/2-\nu} \\ &= \mathcal{O}_p(n^{-\nu})L(s)s^{a-1/2-\nu}\end{aligned}$$

by (3.1).

For any $n \geq 1$ and $1 < \rho < \infty$ set

$$A_n^{(1)}(\rho) = \{s/\rho < U_n(s) < \rho s \quad \text{for} \quad 1/n < s < 1 - 1/n\}.$$

Observe that on the event $A_n^{(1)}(\rho)$, for any $1/n \leq s \leq k_n/n$,

$$\Delta_n^{(2)}(s) \leq \frac{a(1-a)}{2} \rho^{2-a} s^{a-2} L(s) n^{1/2} (U_n(s) - s)^2.$$

Also on the event $A_n^{(1)}(\rho)$, for any $1/n \leq s \leq k_n/n$ and large enough n,

$$\begin{aligned} \Delta_n^{(3)}(s) &\leq n^{1/2} \rho^a s^a | \exp\Big(\int_{U_n(s)}^1 \frac{b(u)}{u} du\Big) - \exp\Big(\int_s^1 \frac{b(u)}{u} du\Big)| \\ &= n^{1/2} \rho^a s^a \exp(\eta_n(s)) | \int_s^{U_n(s)} \frac{b(u)}{u} du |, \end{aligned}$$

where $\eta_n(s)$ lies in the interval determined by

$$\int_{U_n(s)}^1 \frac{b(u)}{u} du \quad \text{and} \quad \int_s^1 \frac{b(u)}{u} du.$$

This last expression is in turn

$$\leq \rho^{a+1} s^{a-1} L(\xi_n^\star(s)) n^{1/2} |U_n(s) - s| \sup_{0 < u \leq k_n/n} |b(u)|,$$

where $\xi_n^\star(s)$ is like $\xi_n(s)$ in (3.6). Since we are on the event $A_n^{(1)}(\rho)$, this last bound is

$$\leq \rho^{a+1} s^{a-1} L(s) n^{1/2} |U_n(s) - s| \sup_{s/\rho \leq t \leq \rho s} (L(t)/L(s)) \sup_{0 < u \leq k_n/n} |b(u)|,$$

which since $b(u) \to 0$ as $u \downarrow 0$, and by Lemma 2.2, is

$$= o(1) s^{a-1} L(s) n^{1/2} |U_n(s) - s|.$$

On the other hand, it is well known that

$$\lim_{\rho \to \infty} \liminf_{n \to \infty} P\{A_n^{(1)}(\rho)\} = 1.$$

(See for instance Rényi (1973).) Thus we see from the above inequalities that the proof of the lemma is complete.

Set

$$D_n(a, k_n) = \begin{cases} (k_n/n)^{1/2-a} L(k_n/n), & \text{if } 0 < a < 1/2, \\ \left(\int_{1/n}^{k_n/n} s^{-1} L^2(s) ds\right)^{1/2}, & \text{if } a = 1/2. \end{cases}$$

LEMMA 3.2. *Under the conditions of Lemma 3.1, for any* $0 < a < 1/2$,

$$\begin{aligned}\frac{n^{1/2}}{D_n(a,k_n)}&\int_0^{k_n/n}\{Q(U_n(s))-Q(s)\}s^{-2a}ds\\&=\frac{a}{D_n(a,k_n)}\int_0^{k_n/n}B_n(s)s^{-a-1}L(s)ds+o_p(1),\end{aligned}\tag{3.7}$$

and

$$\begin{aligned}\frac{n^{1/2}}{D_n(1/2,k_n)}&\int_{1/n}^{k_n/n}\{Q(U_n(s))-Q(s)\}s^{-1}ds=\\&=\frac{1/2}{D_n(1/2,k_n)}\int_{1/n}^{k_n/n}B_n(s)s^{-3/2}L(s)ds+o_p(1).\end{aligned}\tag{3.8}$$

Proof. We will first show that for every $0 < a \le 1/2$ and for Δ_n of Lemma 3.1 we have

$$\frac{1}{D_n(a,k_n)}\int_{1/n}^{k_n/n}\Delta_n(s)s^{-2a}ds=o_p(1).\tag{3.9}$$

Applying Lemma 3.1 we see that for any $0 < \nu < 1/2$ the left side here is

$$\begin{aligned}&=(D_n(a,k_n))^{-1}O_p(n^{-\nu})\int_{1/n}^{k_n/n}L(s)s^{-a-1/2-\nu}ds\\&+(D_n(a,k_n))^{-1}O_p(1)\int_{1/n}^{k_n/n}s^{-a-2}L(s)n^{1/2}|U_n(s)-s|^2ds\\&+(D_n(a,k_n))^{-1}o_p(1)\int_{1/n}^{k_n/n}s^{-a-1}L(s)n^{1/2}|U_n(s)-s|ds\\&=:S_n^{(1)}+S_n^{(2)}+S_n^{(3)}.\end{aligned}$$

First consider $S_n^{(1)}$. Assume $0 < a < 1/2$ and choose ν such that $a + 1/2 + \nu < 1$. In this case

$$\begin{aligned}S_n^{(1)}&\le(D_n(a,k_n))^{-1}O_p(n^{-\nu})\int_0^{k_n/n}L(s)s^{-a-1/2-\nu}ds\\&=O_p(k_n^{-\nu})\int_0^{k_n/n}L(s)s^{-a-1/2-\nu}ds/(L(k_n/n)(k_n/n)^{-a+1/2-\nu}),\end{aligned}$$

which by Lemma 2.1 equals $O_p(k_n^{-\nu}) = o_p(1)$. Now assume that $a = 1/2$. We have

$$\begin{aligned}S_n^{(1)}&=O_p(n^{-\nu})\int_{1/n}^{k_n/n}L(s)s^{-1-\nu}ds/\Big(\int_{1/n}^{k_n/n}L^2(s)s^{-1}ds\Big)^{1/2}\\&\le O_p(n^{-\nu})\int_{1/n}^{\delta_0}L(s)s^{-1-\nu}ds/\Big(\int_{1/n}^{k_n/n}L^2(s)s^{-1}ds\Big)^{1/2}\\&=O_p(1)\frac{\int_{1/n}^{\delta_0}L(s)s^{-1-\nu}ds}{L(1/n)(1/n)^{-\nu}}\Big(\frac{L^2(1/n)}{\int_{1/n}^{k_n/n}L^2(s)s^{-1}ds}\Big)^{1/2}\end{aligned}$$

for any $0 < \delta_0 \leq \delta$ if n is large enough. Applying Lemma 2.1 to the first ratio, with an appropriate choice of δ_0, and applying Lemma 2.4 to the second ratio and with the slowly varying function L^2, we see that the last term is $o_p(1)$. Therefore

$$S_n^{(i)} \xrightarrow{P} 0, \tag{3.10}$$

holds for $i = 1$.

For the cases $i = 2, 3$ we need the fact that there exists a constant $0 < K < \infty$ such that for any $n \geq 1$ and $1/n \leq s \leq 1 - 1/n$,

$$E|U_n(s) - s|^2 \leq Ks(1-s)/n \tag{3.11}$$

and

$$E|U_n(s) - s| \leq K(s(1-s))^{1/2}/n^{1/2}. \tag{3.12}$$

These inequalities follow easily from the well-known formula for the variance of a uniform order statistic (cf. David (1970) p. 28). By taking expectations under the integral sign and applying (3.11) and (3.12) we see that

$$S_n^{(2)} = O_p(1)(D_n(a, k_n))^{-1} n^{-1/2} \int_{1/n}^{k_n/n} s^{-a-1} L(s) ds$$

and

$$S_n^{(3)} = o_p(1)(D_n(a, k_n))^{-1} \int_{1/n}^{k_n/n} s^{-a-1/2} L(s) ds.$$

The proof that both $S_n^{(2)}$ and $S_n^{(3)}$ are $o_p(1)$ rv's now proceeds much like the proof given above, based on Lemmas 2.1 and 2.4. Thus we have established (3.9) for any $0 < a \leq 1/2$. This is equivalent to (3.8) when $a = 1/2$.

To complete the proof of (3.7) we need to show that for $0 < a < 1/2$,

$$S_n^{(4)} = \frac{n^{1/2}}{D_n(a, k_n)} \int_0^{1/n} \{Q(U_n(s)) - Q(s)\} s^{-2a} ds = o_p(1) \tag{3.13}$$

and

$$S_n^{(5)} = \frac{a}{D_n(a, k_n)} \int_0^{1/n} B_n(s) s^{-a-1} L(s) ds = o_p(1). \tag{3.14}$$

Choose any $\rho > 1$. Note that since Q is nondecreasing, on the event $A_n^{(2)}(\rho) = \{U_n(1/n) \leq \rho/n\}$ we have

$$\begin{aligned} S_{n,1}^{(4)} &= n^{1/2}(D_n(a, k_n))^{-1} \int_0^{1/n} Q(U_n(s)) s^{-2a} ds \\ &\leq (\rho^a/(1-2a)) n^{1/2-a} L(\rho/n)(1/n)^{1-2a}/D_n(a, k_n) \\ &= (\rho^a/(1-2a)) L(\rho/n)(1/n)^{1/2-a}/((k_n/n)^{1/2-a} L(k_n/n)), \end{aligned}$$

which, since L is slowly varying at zero, is

$$= \mathcal{O}(1)L(1/n)(1/n)^{1/2-a}/((k_n/n)^{1/2-a}L(k_n/n)).$$

Lemma 2.3 implies that the last term converges to zero. Since

$$\lim_{\rho\to\infty}\liminf_{n\to\infty} P\{A_n^{(2)}(\rho)\} = 1,$$

we see that

$$S_{n,1}^{(4)} \xrightarrow{P} 0. \tag{3.15}$$

The fact that

$$S_{n,2}^{(4)} = n^{1/2}(D_n(a,k_n))^{-1}\int_0^{1/n} Q(s)s^{-2a}ds \to 0$$

is proven similarly. Thus (3.15) and (3.16) imply that (3.13) holds.

On the other hand, observe that

$$E|S_n^{(5)}| \le a(D_n(a,k_n))^{-1}\int_0^{1/n} s^{-a-1/2}L(s)ds. \tag{3.16}$$

This upper bound can be shown to be $o(1)$ by use of Lemmas 2.1 and 2.3. This finishes the proof of (3.7) and hence that of the lemma.

LEMMA 3.3. *Under the conditions of Lemma 3.1, for any* $0 < a < 1/2$,

$$\frac{(\frac{k_n}{n})^{1-2a}}{D_n(a,k_n)}n^{1/2}\{Q(U_n(\frac{k_n}{n})) - Q(\frac{k_n}{n})\} = a(\frac{k_n}{n})^{-1/2}B_n(\frac{k_n}{n}) + o_p(1) \tag{3.17}$$

and whenever $s_n = 1/n$ *or* k_n/n,

$$\frac{n^{1/2}}{D_n(1/2,k_n)}\{Q(U_n(s_n)) - Q(s_n)\} = \frac{1/2}{D_n(1/2,k_n)}(s_n)^{-1/2}B_n(s_n)L(s_n) + o_p(1) \tag{3.18}$$

and

$$\frac{1/2}{D_n(1/2,k_n)}(s_n)^{-1/2}B_n(s_n)L(s_n) = o_p(1). \tag{3.19}$$

Proof. First consider (3.17). Applying Lemma 3.1 and facts (3.11) and (3.12), we see that for any $0 < \nu < 1/2$ the left side of (3.17) is

$$\begin{aligned}
&= \frac{(k_n/n)^{1/2-a}}{L(k_n/n)}\{\mathcal{O}_p(n^{-\nu})L(\frac{k_n}{n})(\frac{k_n}{n})^{a-1/2-\nu} + \mathcal{O}_p(1)(\frac{k_n}{n})^{a-2}L(\frac{k_n}{n})n^{1/2}|U_n(\frac{k_n}{n}) - \frac{k_n}{n}|^2 \\
&\quad + o_p(1)(\frac{k_n}{n})^{a-1}L(\frac{k_n}{n})n^{1/2}|U_n(\frac{k_n}{n}) - \frac{k_n}{n}|\} \\
&= \frac{(k_n/n)^{1/2-a}}{L(k_n/n)}L(k_n/n)(k_n/n)^{a-1/2}\{\mathcal{O}_p(k_n^{-\nu}) + \mathcal{O}_p(k_n^{-1/2}) + o_p(1)\}.
\end{aligned}$$

This is $o_p(1)$ and hence (3.17) is valid. Similarly, Lemma 3.1, the facts in (3.11) and (3.12) and Lemma 2.4 imply (3.18) for both choices $s_n = 1/n$ and $s_n = k_n/n$. Hence we only have to show

(3.19). But we see that the expectation of the modulus of the left side of (3.19) is less than or equal to

$$L(s_n)/\Big(\int_{1/n}^{k_n/n} s^{-1}L^2(s)ds\Big)^{1/2},$$

which by Lemma 2.4 converges to zero. The proof of Lemma 3.3 is also complete.

Proof of Theorem 1. First we consider the case $0 < a < 1/2$. Using (1.11) and the integral representation of $\hat{\theta}_n$ given in (3.3), an elementary computation shows that

$$\begin{aligned}\frac{n^a k_n^{1/2-a}}{L(k_n/n)}(\hat{\theta}_n - \theta - \sigma\beta_n(a)) &= \frac{\sigma(1-a)(1-2a)}{a}\,\frac{n^{1/2}}{D_n(a,k_n)}\int_0^{k_n/n} s^{-2a}\{Q(U_n(s)) - Q(s)\}ds \\ &\quad - \frac{\sigma(1-2a)}{a}\,\frac{(k_n/n)^{1-2a}n^{1/2}}{D_n(a,k_n)}\{Q(U_n(k_n/n)) - Q(k_n/n)\},\end{aligned}$$

which by Lemmas 3.2 and 3.3 is

$$\begin{aligned}&= \sigma(1-2a)\Big\{\frac{1-a}{D_n(a,k_n)}\int_0^{k_n/n} B_n(s)s^{-a-1}L(s)ds - \big(\frac{k_n}{n}\big)^{-1/2}B_n\big(\frac{k_n}{n}\big)\Big\} + o_p(1) \\ &=: N_n(0;a,k_n) + o_p(1).\end{aligned} \tag{3.20}$$

The rv's $N_n(0;a,k_n)$ are normally distributed and have mean zero for each $n \geq 1$. Moreover,

$$\begin{aligned}&E(N_n(0;a,k_n))^2 \\ &= \sigma^2(1-2a)^2\Big\{\frac{(1-a)^2}{(k_n/n)^{1-2a}L^2(k_n/n)}\,d^2(k_n/n) + \big(\frac{k_n}{n}\big)^{-1}\frac{k_n}{n}\big(1-\frac{k_n}{n}\big) \\ &\quad - \frac{2(1-a)(1-\frac{k_n}{n})}{(k_n/n)^{1-a}L(k_n/n)}\int_0^{k_n/n} s^{-a}L(s)ds\Big\},\end{aligned}$$

which by Lemmas 2.5 and 2.1 converges to

$$\sigma^2(1-2a)^2\Big\{\frac{2(1-a)^2}{(1-a)(1-2a)} + 1 - 2\Big\} = \sigma^2(1-2a).$$

This proves the theorem when $0 < a < 1/2$.

Now let $a = 1/2$. Again an elementary computation shows that by (1.11) and the integral representation of $\hat{\theta}_n$ given in (3.4) we have

$$\begin{aligned}&\frac{n^{1/2}\log k_n}{\Big(\int_{1/n}^{k_n/n} s^{-1}L^2(s)ds\Big)^{1/2}}(\hat{\theta}_n - \theta - \sigma\beta_n(1/2)) \\ &= \frac{\sigma n^{1/2}}{D_n(1/2,k_n)}\Big\{\int_{1/n}^{k_n/n}\{Q(U_n(s)) - Q(s)\}s^{-1}ds + 2\{Q(U_n(1/n)) - Q(1/n)\} \\ &\quad - 2\{Q(U_n(k_n/n)) - Q(k_n/n)\}\Big\},\end{aligned}$$

which by Lemmas 3.2 and 3.3 is

$$\begin{aligned}&= \frac{\sigma}{2D_n(1/2,k_n)}\int_{1/n}^{k_n/n} B_n(s)s^{-3/2}L(s)ds + o_p(1) \\ &=: N_n(0;1/2,k_n) + o_p(1).\end{aligned} \tag{3.21}$$

Again, the rv's $N_n(0;1/2,k_n)$ are normal with zero mean for each $n \geq 1$. Moreover,

$$E(N_n(0;1/2,k_n))^2 = \frac{\sigma^2}{4} d_n^2(k_n) / \int_{1/n}^{k_n/n} s^{-1} L^2(s) ds \to \sigma^2$$

by Lemma 2.6, and the theorem is completely proven.

Proof of Theorem 2. The statements for the estimate $\hat{\theta}_n$ follow directly from Theorem 1. The statement for the estimate $\hat{c}_n$ follows from the representations of $\hat{\theta}_n - \theta - \sigma\beta_n(a)$ given in (3.20) and (3.21) and from Lemma 3.3 by the same methods as used in the proof of Theorem 1.

Proof of Theorem 3. Calculus shows that in the model (1.9) and under the condition (1.5) on k_n, $n^a k_n^{1/2-a} \beta_n(a) \to 0$ for any $0 < a < 1/2$, $(n \log k_n)^{1/2} \beta_n(1/2) \to 0$ and $k_n^{1/2} \gamma_n(a) \to 0$ for any $0 < a \leq 1/2$. Therefore the theorem follows from Theorem 2.

REFERENCES

1. CHERNOFF, H., GASTWIRTH, J.L. and JOHNS, M.V., Jr. (1967). Asymptotic distribution of linear combinations of order statistics with applications to estimation. *Ann. Math. Statist.* 38, 52-72.

2. CSÖRGŐ, M., CSÖRGŐ, S., HORVÁTH, L. and MASON, D. M. (1986). Weighted empirical and quantile processes. *Ann. Probab.* 14, 31-85.

3. CSÖRGŐ, S., HORVÁTH, L. and MASON, D. M. (1986). What portion of the sample makes a partial sum asymptotically stable or normal? *Probab. Th. Rel. Fields* 72, 1-16.

4. CSÖRGŐ, S. and MASON, D. M. (1986). The asymptotic distribution of sums of extreme values from a regularly varying distribution. *Ann. Probab* 14, 974-983.

5. DAVID, H. A. (1970). *Order Statistics.* Wiley, New York.

6. DE HAAN, L. (1970). *On Regular Variation and its Application to the Weak Convergence of Sample Extremes,* Tract No. 32, Mathematical Centre, Amsterdam.

7. HALL, P. (1982). On estimating the endpoint of a distribution. *Ann. Statist.* 10, 556-568.

8. HOSKING, J.R.M. and WALLIS, T.J. (1987). Parameter and quantile estimation for the generalized pareto distribution. *Technometrics* 29, 339–349.

9. KINDERMAN, R. P. and LARICCIA, V. N. (1985) Closed form asymptotically efficient estimators based upon order statistics. *Statist. Probab. Letters* 3, 29-34.

10. PARZEN, E. (1961). Regression analysis of continuous parameter time series. *Proc. Fourth Berkeley Symp. Math Statist. Probab., Vol 1,* 469-489.

11. PARZEN, E. (1979). Nonparametric statistical data modeling. *J. Amer. Statist. Assoc* 74, 105-121.

12. POLFELDT, T. (1970). Asymptotic results in non-regular estimation. *Skand. Aktuarietidskr.* Supplement 1-2.

13. RÉNYI, A. (1973). On a group of problems in the theory of order statistics. *Selected Transl. in Math. Statist. and Probab. Amer. Math. Soc. 13,* 289-298. (Translated from *Magyar Tud. Akad. Mat. Fiz. Oszt. Közl.* 18 (1968), 23-30.)

14. SENETA, E. (1976). *Regularly Varying Functions.* Lecture Notes in Math. No. 508, Springer-Verlag, Berlin.

15. SMITH, R. L. (1987). Estimating tails of probability distributions. *Ann. Statist.* 15, 1174-1207.

BOLYAI INSTITUTE
UNIVERSITY OF SZEGED
ARADI VÉRTANÚK TERE 1
H-6720 SZEGED
HUNGARY

DEPARTMENT OF MATHEMATICAL SCIENCES
UNIVERSITY OF DELAWARE
NEWARK, DELAWARE 19716
U.S.A.

ASYMPTOTIC NORMALITY OF HILL'S ESTIMATOR

Jan BEIRLANT and Jozef L. TEUGELS

Katholieke Universiteit Leuven, Belgium

Abstract. Hill's estimator has been shown to be very useful in tail and quantile estimation. In an earlier paper the authors found a broad class of underlying distributions such that Hill's estimator is asymptotically normal. In this note the domain of attraction of the normal law is further specified.

1. Introduction

Let $X_1, X_2, \ldots, X_n$ be a sample of size n from a distribution function (d.f.)F with $F(0) = 0$, survival function $\bar{F} = 1\text{-}F$ and $\inf\{x : F(x) = 1\} = \infty$. The ordered sample is denoted by $X_{(1)}, X_{(2)}, \ldots, X_{(n)}$. Hill (1975) introduced the estimator

$$H_{m,n} = \frac{1}{m} \sum_{i=1}^{m} \log X_{(n-i+1)} - \log X_{(n-m)} \tag{1}$$

as a maximum likelihood estimator for the tail of a distribution. This estimator is an empirical version of the mean residual life function of the log-transformed data :

$$e(x) = \left(\int_{e^x}^{\infty} \bar{F}(y)\frac{dy}{y}\right)/\bar{F}(e^x).$$

In fact

$$H_{m,n} = \hat{e}_n\,(\log X_{(n-m)}) = \int_{X_{(n-m)}}^{\infty} \{1\text{-}\hat{F}_n(y)\}\, \frac{dy}{y} \Big/ (1\text{-}\hat{F}_n(X_{(n-m)}))$$

AMS Subject Classification : 26A12,60F05

Keywords and Phrases : Tail estimation, quantile estimation, de Haan theory.

where $\hat{F}_n$ is the empirical d.f. based on the sample.

From this it follows that $H_{m,n}$ should be useful in tail estimation.

In an earlier paper (1987) the authors solved the domain of attraction problem for Hill's estimator when $n \to \infty$ and m stays fixed, assuming that F belongs to C↑ the class of distributions on $\mathbb{R}^+$ for which F is continuous and strictly increasing in a neighborhood of ∞.

Define $k(u) = \log F^{\leftarrow}(1-u^{-1})$, $u \geq 1$ (with $f^{\leftarrow}(t) = \inf\{x : f(x) \geq t\}$ for any function f).

It was shown that

$$\mathcal{L}(H_{m,n}/g(\tfrac{n}{m})) \longrightarrow \begin{Bmatrix} \mathcal{L}(\overline{\omega}_m) \\ \text{resp.} \\ \mathcal{L}(Y_m) \end{Bmatrix} \quad (n \to \infty)$$

with $\overline{\omega}_m = \frac{1}{m}\sum_{i=1}^{m} \omega_i$ (ω_i being iid exponentials with means one) and

$$E(\exp(-\lambda Y_m)) = (m!)^{-1} \int_0^\infty e^{-v} v^m \left\{1 - \frac{1}{v}\int_0^\infty e^{-w}\left(\frac{mw\ell}{\lambda} + v^{-\ell}\right)^{-1/\ell} dw\right\}^m dv \; (\lambda > 0, \ell > 0)$$

iff

$$(2) \qquad (k(xu) - k(x))/_{g(x)} \longrightarrow \begin{Bmatrix} \log u \\ \text{resp.} \\ \frac{u^\ell - 1}{\ell} \; (\ell > 0) \end{Bmatrix} \quad (x \to \infty),\ u > 0.$$

with g slowly varying (resp. regularly varying with index ℓ). We will refer to the first case in (2) as the case "$\ell = 0$". Relation (2) can be written as

$$k(xu) - k(x) \sim k_\ell(u) g(x) \quad (x \to \infty),\ u > 0,$$

with $k_\ell(u) = \int_1^u v^{\ell-1}\, dv$.

It was shown that in case $\ell = 0$ sequences $(c_n)_n$ and $(d_n)_n$ can be found such that for appropriate $m = m_n = o(n)$

$$(H_{m,n} - c_n)/_{d_n}$$

converges to a normal law as $m,n \to \infty$.

Comparable results can be found in Lô (1987) and Csörgő and Mason (1984).
In this note this last result is extended to a broader class of distributions.

In the literature on regular variation (see e.g. Bingham, Goldie and Teugels (1987)) the class of functions satisfying relation (2) with $\ell = 0$ is denoted with Π (with auxiliary function g) (de Haan (1970)). If a function belongs to Π, its generalized inverse belongs to the so-called class Γ. The next proposition extends this fact. Read $(1+\ell u)^{1/\ell}$ as e^u when $\ell = 0$.

PROPOSITION 1. *Let* $f : \mathbb{R} \longrightarrow (0,\infty)$ *be non-decreasing and right-continuous, for which there exists a measurable function* $h : \mathbb{R} \to (0,\infty)$ *such that*

$$(3) \qquad f(x + uh(x))/f(x) \longrightarrow (1+\ell u)^{1/\ell} \quad (x \to \infty), \forall u > -\tfrac{1}{\ell} (\ell \geq 0).$$

Then

$$(4) \qquad (f^{\leftarrow}(xv) - f^{\leftarrow}(x)/h(f^{\leftarrow}(x)) \longrightarrow k_\ell(v) \quad (x \to \infty), \forall v > 0,$$

and $h \circ f^{\leftarrow}$ *is regularly varying with index* ℓ.
Conversely, (4) *implies* (3) *with auxiliary function* $h \circ (f^{\leftarrow} \circ f)$. *Moreover, relation* (3) *holds locally uniformly in* u. □

This result is a generalisation of a result of de Haan (1973). (see also Theorem 3.10.2 and 3.10.4 in Bingham et al. (1987)), who considered the case $\ell = 0$. The proofs in Bingham et al. (1987) can be extended to the present general situation without any problem.

Along the same lines as the representation theorem for the class Γ, one derives the following generalisation.

PROPOSITION 2. *A non-decreasing and right-continuous function* $f : \mathbb{R} \to (0,\infty)$ *satisfies* (3)

iff

$$f(x) = \exp\left\{\eta(x) + \int_0^x \frac{dt}{h(t)}\right\}$$

where $\eta(x) \to d \in \mathbb{R}$ *and* $\dfrac{h(x + uh(x))}{h(x)} \longrightarrow 1 + u\ell$ *as* $x \to \infty$. □

In the next section the previous result will be used with $f = 1/\bar{F} \circ \exp$ (, and hence $f^{\leftarrow} = k$), $h \circ k = g$ (, and hence $h = g \circ f$).

In what follows we make use of the following results from Beirlant and Teugels (1987). Let

$q_n = \frac{m}{n-1}$, $p_n^2 = \frac{m(n-m-1)}{(n-1)^3}$. Then

$$E\left[\exp it\left[\frac{H_{m,n} - c_n}{d_n}\right]\right] = I_1 \int_{-q_n/p_n}^{(1-q_n)/p_n} I_2(z)\, I_3(t,z)dz$$

where $I_1 \longrightarrow \frac{1}{\sqrt{2\pi}}$, and $I_2(z) \longrightarrow e^{-z^2/2}$ locally uniformly (l.u.) in z.

Moreover,

$$\text{(5)} \qquad \log I_3(t,z) = -it\frac{c_n}{d_n} + $$

$$m\log\left\{1 + \frac{itg(\frac{n}{m})}{md_n(q_n+p_n z)} \int_0^\infty \bar{F}\left[e^{vg(\frac{n}{m})}\bar{F}^{i}(q_n+p_n z)\right] e^{ivtg(\frac{n}{m})/md_n} dv\right\}$$

$$= -it\frac{c_n}{d_n} + m\log\left\{1 + it\,\frac{g(\frac{n}{m})}{md_n} \int_0^\infty e^{ivt\frac{g(\frac{n}{m})}{md_n}} \frac{\bar{F}\left[e^{vg(\frac{n}{m})+k(\frac{1}{q_n+p_n z})}\right]}{\bar{F}\left[e^{k(\frac{1}{q_n+p_n z})}\right]} dv\right\}.$$

2. Conditions for degenerate or normal limit law

We state the following generalisations of Theorem 4 in Beirlant and Teugels (1987). Throughout, let $F \in C\uparrow$, $f(x) = 1/\bar{F}(\exp x)$, $(x > 0)$, and $k(u) = f^{i}(u) = \log F^{i}(1-u^{-1})$ $(u > 1)$.

<u>THEOREM 1.</u> *Assume*

$$(k(xu) - k(x))/g(x) \longrightarrow k_\ell(u) \quad (x \to \infty) \quad \forall u > 0$$

for some regularly varying function g *with index* $\ell \in [0,1)$.

If $n \to \infty$, $m = m_n \to \infty$, $m_n = o(n)$, *then* $H_{m,n}/g(\frac{n}{m}) \xrightarrow{P} 1/(1-\ell)$.

<u>THEOREM 2.</u> *Assume*

$$(k(xu) - k(x))/g(x) \longrightarrow k_\ell(u) \quad (x \to \infty) \quad \forall u > 0$$

for some regularly varying function g *with index* $\ell \in [0,\frac{1}{2})$.

If $m \to \infty$, $m_n \to \infty$, $m_n = o(n)$, *and*

$$(6) \qquad \sqrt{m_n}\left\{\int_0^\infty \frac{\bar F\left[e^{ug(\frac{n}{m})}\ \bar F^{i}(q_n+p_n z)\right]}{(q_n+p_n z)}\,du - \frac{1}{1-\ell}\right\} \longrightarrow A \in \mathbb{R}$$

l.u. in z, then

$$\sqrt{m_n}\left\{\frac{H_{m,n}}{g(\frac{n}{m})} - \frac{1}{1-\ell}\right\} \xrightarrow{\mathcal{D}} \mathcal{N}\left[A, \frac{1}{(1-\ell)^2(1-2\ell)}\right].$$

We only prove the result of Theorem 2. As in Beirlant and Teugels (1987) it follows from (5) that if $\ell \in [0,\frac{1}{2})$

$$\log I_3(t,z) = -it\frac{c_n}{d_n} + \frac{it}{d_n} g(\tfrac{n}{m})\, J_1(n,m,z)$$

$$-\frac{t^2}{2}\,\frac{g^2(\frac{n}{m})}{m d_n^2}\left\{2J_2(n,m,z) - J_1^2(n,m,z)\right\} + O(m^{-2}d_n^{-3})$$

where

$$J_j(n,m,z) = \frac{1}{q_n+p_n z}\int_0^\infty u^{j-1}\,\bar F(e^{ug(\frac{n}{m})}\ F^{i}(1-q_n-p_n z))du \quad (j=1,2).$$

Choosing $c_n = \frac{1}{(1-\ell)} g(\frac{n}{m})$, $d_n = g(\frac{n}{m})/\sqrt{m}$

$$\log I_3(t,z) = it\sqrt{m}\,\{J_1 - \tfrac{1}{1-\ell}\} - \tfrac{t^2}{2}\,\{2J_2 - J_1^2\} + O(m^{-2}d_n^{-3}).$$

Similarly as in Teugels (1981) it then only remains to show that

$$J_j(n,m,z) \longrightarrow \int_0^\infty u^{j-1}(1+\ell u)^{-1/\ell}\,du = \begin{cases} {}^1/1-\ell & , j=1 \\ {}^1/(1-\ell)(1-2\ell) & , j=2 \end{cases}$$

l.u. in z as $n \to \infty$, $m_n = o(n)$.

First, using Proposition 1 we find that

$$\frac{\bar F(e^{ug(\frac{n}{m})}\ \bar F^{i}(q_n+p_n z))}{q_n+p_n z} = \frac{\bar F(e^{k(\frac{1}{q_n+p_n z}) + uh(k(\frac{n}{m}))})}{\bar F(e^{k(\frac{1}{q_n+p_n z})})}$$

$$\longrightarrow (1+\ell u)^{1/\ell},\ u>0$$

l.u. in z as $n \to \infty$, $m_n = o(n)$.

With the help of Proposition 2 one derives the following Potter(1942) - type bounds (see also Bingham et al. (1987), p. 25) :

$$\frac{\bar{F}\left(e^{k\left(\frac{1}{q_n+p_n z}\right) + u(h\circ k)\left(\frac{n}{m}\right)}\right)}{\bar{F}\left(e^{k\left(\frac{1}{q_n+p_n z}\right)}\right)} \le A\,(1+\ell u)^{-\frac{1}{B\ell}}$$

with $\frac{1}{\ell} > B > 1$, if n is large enough.

Hence the dominated convergence theorem can be applied and for any $T > 0$

$$\int_0^\infty u^{j-1}\ \frac{\bar{F}\left(e^{ug\left(\frac{n}{m}\right)}\right)\ \bar{F}^i\,(q_n \pm p_n T)}{q_n \mp p_n T}\ du$$

$$\longrightarrow \int_0^\infty u^{j-1}(1+\ell u)^{-1/\ell}du$$

if $\ell \in [0,1)$ when $j = 1$ and if $\ell \in [0,\frac{1}{2})$ when $j = 2$.

Remarks

1. Under additional assumptions it is possible to simplify condition (6) in the statement of Theorem 2. E.g. Beirlant and Willekens (1988) considered the relations

   (ΓR_1) $\qquad f(x + uh(x))/f(x) = e^u(1 + O(r(x)))\quad (x \to \infty)\quad$ l.u. in $u \in \mathbb{R}$

   and

   (ΓR_2) $\qquad f(x + uh(x))/f(x) - e^u(1 + m(u)r(x))\quad (x \to \infty)\quad$ l.u. in $u \in \mathbb{R}$

   where $\qquad \lim_{x\to\infty} r(x + uh(x))/r(x) = \exp(\gamma u)\quad$ (for some $\gamma \le 0$)

$$\text{and} \qquad m(u) = \begin{cases} c\gamma^{-1}(e^{\gamma u}-1) - \frac{a}{\gamma}\int_0^u (e^{\gamma v}-1)dv & \text{if } \gamma < 0 \\ cu - au^2/2 & \text{if } \gamma = 0 \end{cases}$$

Hence in case $\ell = 0$, when (ΓR_1) is assumed, condition (6) can be replaced by

$$\sqrt{m_n}\, r(k(\tfrac{n}{m})) \longrightarrow 0 \tag{7}$$

and the conclusion of Theorem 2 holds with $A = 0$. When (ΓR_2) is assumed, condition (6) can be replaced by

$$\sqrt{m}\, r(k(\tfrac{n}{m})) \longrightarrow B \tag{8}$$

and the conclusion of Theorem 2 holds with

$$A = B \int_0^\infty m(t)e^{-t}dt = \frac{B}{(1-\gamma)}(a-c).$$

2. As in Beirlant and Teugels (1987) applications to tail and quantile estimation can be stated.

THEOREM. 3. *Under the conditions of Theorem* 1

$$\sup_{x \geq \bar{F}^i(\frac{n}{m})} \left| \frac{n}{m}\bar{F}(x) - \left[1 + \ell\frac{\log(x/X_{(n-m)})}{g(\frac{n}{m})}\right]^{-\ell} \right| \xrightarrow{P} 0$$

and

$$\left\{\log \bar{F}^i(\tfrac{m}{n}t) - \log X_{(n-m)}\right\} \frac{1}{H_{m,n}} \xrightarrow{P} -\log t/(1-\ell), \quad \textit{for every } t < \frac{n}{m}.$$

References.

Beirlant J. and Teugels J.L. (1987). Asymptotics of Hill's estimator. *Th. Probability Appl.* 31, 463-469.

Beirlant J. and Willekens E. (1988). Rapid variation with remainder and rates of convergence. Memorandum COSOR 88-07 Eindhoven University of Technology.

Bingham N.H., Goldie C.M. and Teugels J.L. (1987). *Regular Variation.* Encyclopedia of Math. and its Applic. **27**, Cambridge University Press.

Csörgő S. and Mason D. (1984). Central limit theorems for sums of extreme values. *Math. Proc. Camb. Phil. Soc.* **98**, 547-558.

De Haan L. (1970). *On regular variation and its application to the weak convergence of sample extremes.* Math. Centre Tract 32, Amsterdam.

Hill B.M. (1975). A simple general approach to inference about the tail of a distribution. *Ann. Statist.* **3**, 1163-1174.

Lô G.S. (1987). Asymptotic behavior of Hill's estimate and applications. *Journal Appl. Probability* **23**, 922-936.

Potter H.S.A. (1942). The mean value of Dirichlet series II.*Proc. London Math. Soc.* (2) **47**, 1-19.

Teugels J.L. (1981). Limit theorems on order statistics, *Ann. Probability* **9**, 868-880.

Extended Extreme Value Models and Adaptive Estimation of the Tail Index

R.-D. Reiss
Department of Mathematics, University of Siegen

Abstract. Classical extreme value models are families of limit distributions of sample maxima. Now, consider expansions of length two where limit distributions are the leading terms. Such expansions define extended extreme value models.

We will study the asymptotic performance of an adaptive estimator of the scale parameter α in an extended Gumbel model, thus also getting an estimator of the tail index $1/\alpha$ in a model of Pareto type distributions. Under the present conditions the new estimator is asymptotically superior to those given in literature.

1. Introduction

L. Weiss (1971) studied a model of densities belonging to a neighborhood of a generalized Pareto type II density. It was proved by Weiss that the variational distance between the joint distribution of the k(n) smallest order statistics and the corresponding k(n)-dimensional extreme value distribution tends to zero as the sample size n tends to infinity. This approximation and a quick estimator of the shape parameter in the extreme value model was used to define an estimator of the tail index of the original density.

The problem of estimating the tail index was also investigated by Hill (1975) who studied an estimator related to the maximum likelihood (m.l.) estimator in the parametric extreme value model.

Falk (1985) combined both approaches, namely, the use of m.l. estimators and the approximation of the nonparametric model of joint distributions of extremes by the parametric extreme value model.

An alternative estimation procedure based on the subsample method

---

Key words.Tail index, adaptive procedure.
AMS 1980 subject classifications.Primary, 62E20, 62G30.

(in other words: annual maxima or Gumbel method) was studied by Reiss (1987). Now the estimator of the tail index is based on the maxima of $k(n)$ subsamples each having size $m(n) = [n/k(n)]$.

We give a short outline of the estimation of the tail index of a density belonging to a neighborhood of a Pareto density. Assume that

$$f(x) = \alpha^{-1}x^{-(1+1/\alpha)}(1 + g(x)), \quad \alpha > 0 \tag{1.1}$$

with $|g(x)| \le Cx^{-\delta/\alpha}$ for sufficiently large x, $0 < \delta \le 1$ and $C > 0$. Moreover, we will include a scale parameter into our considerations. Hill's estimator is given by

$$\alpha^{**}_{k,n} = k^{-1} \sum_{i=1}^{k} \log(X_{n-i+1:n}/X_{n-k:n}) \tag{1.2}$$

where $X_{n:n} \ge \ldots \ge X_{n-k:n}$ are the $k+1$ upper extremes. It was proved by Falk (1985) that

$$\left| P\left\{k^{1/2}\alpha^{-1}(\alpha^{**}_{k,n} - \alpha) \le t\right\} - \Phi(t)\right| = 0\left((k/n)^{\delta}k^{1/2} + k^{-1/2}\right) \tag{1.3}$$

where Φ denotes the standard normal d.f. For a related result we also refer to Häusler and Teugels (1985).

The estimator $\alpha^{*}_{k,n}$ constructed by Reiss (1987) is given by

$$\alpha^{*}_{k,n} = \hat{\alpha}_k(\log M_m^{(1)},\ldots,\log M_m^{(k)}) \tag{1.4}$$

where $n = km$, $M_m^{(i)}$ are independent maxima of a sample of size m and $\hat{\alpha}_k$ is the m.l. estimator of a sample of size k of the scale parameter α in the Gumbel model with unknown location and scale parameter. It was proved that

$$\left| P\left\{(k\pi^2/6)^{1/2}\alpha^{-1}(\alpha^{*}_{k,n} - \alpha) \le t\right\} - \Phi(t)\right| = 0\left((k/n)^{\delta}k^{1/2} + k^{-1/2}\right) \tag{1.5}$$

which shows the superiority of $\alpha^{*}_{k,n}$ compared to $\alpha^{**}_{k,n}$ under the present conditions. A related comparison within the parametric framework was carried out by Hüsler and Tiago de Oliveira (1986). We also refer to Smith (1987).

From (1.3) and (1.5) we see that in both cases there is a trade-off between the following two requirements:

(a) k has to be large to gain efficiency,

(b) k has to be small enough to get asymptotic normality of the estimator.

An illuminating description of this situation can be found in Csörgö, Deheuvels and Mason (1985). Assume that f satisfies the stronger condition

$$f(x) = \alpha^{-1} x^{-(1+1/\alpha)} \left(1 - cx^{-\delta/\alpha} + o(x^{-\delta/\alpha})\right) \quad \text{as } x \uparrow \infty. \tag{1.6}$$

It can be seen that the performance of estimators of α based on k upper extremes depends on the proper balance between the variance and the bias: If k increases then the variance decreases and the bias increases (and vice versa). The bias depends on the constants c and δ being unknown to the statistician.

Under condition (1.6) an optimal sequence $k^*(n)$ is of the form

$$k^*(n) = \lambda(c)\, n^{2\delta/(2\delta+1)}. \tag{1.7}$$

Using an estimator of $k^*(n)$, based on consistent estimators of c and δ, Hall and Welsh (1985) obtained an adaptive version of the Hill estimator.

The next step beyond the efforts described above suggests itself. The bias prevents the use of estimators based on a larger number k(n) of upper extremes and, thus, an increase of efficiency. On the other hand, the bias term is estimable under a slightly stronger condition. So an improved estimation procedure can be established by using the estimated bias as a correction term.

The present considerations should be regarded as a pilot study showing that the method works asymptotically. We believe that further progress can be achieved w.r.t. the following two aspects:

(a) extension and modification of the model,

(b) construction of efficient estimators.

2. Models

If (1.1) holds for $\delta = 1$ [e.g. Fréchet, Pareto and Cauchy densities satisfy this condition] then Hill's estimator and the estimator in (1.4) are sufficiently accurate. The choice of k(n) should be made according to the insight provided by asymptotic theory and the experience gained from simulations. If δ is small and the constant c is not close to zero, then one should not waste one's time any longer with the procedures in (1.2) and (1.4) because they are inaccurate for small and moderate sample sizes. In this situation it is a challenging problem to find new sufficiently accurate procedures under acceptable conditions.

We will take another look at the problem where in contrary to (1.1) the model includes another nuisance parameter. Assume that for some $\alpha > 0$ and $x \geq x_0 > 0$,

$$f(x) = \alpha^{-1}x^{-(1+1/\alpha)}\left(1 - cx^{-\delta/\alpha} + h(x)\right) \tag{2.1}$$

where $c \geq 0$, $0 < \delta < \rho \leq 1$, $|h(x)| \leq Lx^{-\rho/\alpha}$ and $L > 0$. Condition (2.1) implies that f satisfies (1.1) and (1.6). Under this stronger condition the estimators in (1.2) and (1.4) merely have the performance as under condition (1.1).

Under condition (2.1) sample maxima belong to the domain of attraction of a Fréchet distribution with parameter $1/\alpha$. More precisely, it is well known that $m^{-\alpha}M_m$ is asymptotically distributed according to the Fréchet d.f.

$$G_{1,1/\alpha}(x) = \exp(-x^{-1/\alpha}), \quad x > 0. \tag{2.2}$$

We will measure the distance between distributions w.r.t. the Hellinger distance. Notice that we use the the same symbol for the distribution and its d.f. Let F_i be a distribution with Lebesgue density f_i. The Hellinger distance between F_1 and F_2 is defined by

$$H(F_1,F_2) = \left[\int(f_1(x)^{1/2} - f_2(x)^{1/2})^2 dx\right]^{1/2}. \tag{2.3}$$

The following theorem, concerning an expansion of length 2 of distributions of maxima, is taken from Reiss (1989), Theorem 5.2.11. We will consider expansions of length 2 of the following form: For $x > 0$ define

$$G^{(0,\sigma)}_{1,1/\alpha,\tau}(x) = G_{1,1/\alpha}(x/\sigma)\left(1 + m^{-\delta}\vartheta\ (x/\sigma)^{-(1+\delta)/\alpha}\right) \tag{2.4}$$

with $\tau = (m,\vartheta,\delta)$. If $\vartheta \geq 0$ and

$$\vartheta\delta^{\delta} \leq m^{\delta}, \tag{2.5}$$

then $G^{(0,\sigma)}_{1,1/\alpha,\tau}$ is a d.f.

Theorem 1. Let F be a d.f. with density $x \to \sigma^{-1}f(x/\sigma)$ for some $\sigma > 0$ where f satisfies condition (2.1). Then, with $\vartheta = c/(1+\delta)$,

$$H(F^m, G^{(0,m^{\alpha}\sigma)}_{1,1/\alpha,\tau}) = O\left(m^{-\min(2\delta,\rho)}\right). \tag{2.6}$$

The leading term of the expansion is a Fréchet d.f. with scale parameter $m^{\alpha}\sigma$. If this Fréchet d.f. is taken as an approximation of F^m then the remainder term is of order $m^{-\delta}$. We assume $c \geq 0$ in condition (2.1) to obtain d.f.'s in (2.4) for sufficiently large m. Of course, this condition leads to a rather artificial model; we made this assumption to avoid further technicalities. If $c < 0$ in condition (2.1) then (2.6) still holds w.r.t. the variational distance. However, the Hellinger distance is only defined for distributions. An extension of (2.6) to the case $c < 0$ can be achieved by redefining the expansion in (2.4) in an appropriate way so that one has only to deal with d.f.'s. Notice that the Fréchet d.f.'s are special cases of (2.4) with $c = 0$ so that, in fact, the model obtained by expansions of length 2 extends the classical Fréchet model.

Consider the following d.f.'s with location parameter μ and scale parameter α:

$$G_{3,\tau}^{(\mu,\alpha)}(x) = G_3((x-\mu)/\alpha)\left(1 + m^{-\delta}\vartheta \exp[-(1+\delta)(x-\mu)/\alpha]\right) \tag{2.7}$$

where again $\tau = (m,\vartheta,\delta)$, and

$$G_3(x) = \exp(-e^{-x}) \tag{2.8}$$

is the standard Gumbel d.f. with density g_3. If ξ is a random variable with d.f. $G_{1,1/\alpha,\tau}^{(0,\sigma)}$ then $\eta = \log \xi$ is a random variable with d.f. $G_{3,\tau}^{(\log \sigma,\alpha)}$. Thus the log function transforms the extended Fréchet model to an extended Gumbel model with unknown location and scale parameter. In this sequel, write $G_{3,\tau} = G_{3,\tau}^{(0,1)}$.

3. Extended Gumbel Model: Estimation of Scale Parameter

Let $Y_{1:k} \leq \ldots \leq Y_{k:k}$ be the order statistics of k i.i.d. random variables with common d.f. $G_{3,\tau}^{(\mu,\alpha)}$. Notice that $\mu + \sigma G_{3,\tau}^{-1}$ is the quantile function (q.f.) of $G_{3,\tau}^{(\mu,\alpha)}$. Check that

$$|G_{3,\tau}^{-1}(q) - G_{3,\tau}^{*}(q)| = O(m^{-2\delta}) \tag{3.1}$$

where

$$G_{3,\tau}^{*}(q) = G_3^{-1}(q) - m^{-\delta}\vartheta(-\log q)^{\delta}. \tag{3.2}$$

Given the values z_i we take order statistics of the form

$$Z_{i,k} = Y_{[kG_3(z_i)]:k}. \tag{3.3}$$

Notice that

$$G^*_{3,\tau}(G_3(z)) = z-m^{-\delta}\vartheta\exp(-\delta z). \tag{3.4}$$

From Theorem 4.2.4 in Reiss (1989) and (3.1)-(3.4) one easily gets

$$\sup_t \left| P\left\{(k^{1/2}/\alpha)\left(Z_{i,k}-(\mu+\alpha(z_i-m^{-\delta}\vartheta\exp(-\delta z_i)))\right) \le t\right\} - P\{X_i \le t\}\right|$$
$$= O(k^{-1/2} + k^{1/2}m^{-2\delta} + m^{-\delta}) \tag{3.5}$$

where X_i is a normal r.v. with expectation equal to zero and variance

$$\sigma_i^2 = G_3(z_i)(1-G_3(z_i))/g_3(z_i)^2.$$

If $z_i < z_{i+1}$ then normalized order statistics as given in (3.4) are asymptotically jointly distributed like normal r.v.'s X_i with expectation equal to zero and covariances

$$\sigma_{i,j} = G_3(z_i)(1-G_3(z_j))/g_3(z_i)g_3(z_j) \tag{3.6}$$

for $i \le j$ (see Reiss (1989), Theorem 4.5.3). This implies that

$$\sup_t \left| P\left\{\left(k^{1/2}/\alpha\sigma\right)\left(\frac{Z_{2,k}-Z_{1,k}}{\Delta} - \alpha\left(1 + \frac{m^{-\delta}\vartheta}{\Delta}\exp(-\delta z_1)[1-\exp(-\delta\Delta)]\right)\right) \le t\right\}\right.$$
$$\left. - P\{X_0 \le t\}\right| = O(k^{-1/2} + k^{1/2}m^{-2\delta} + m^{-\delta}) \tag{3.7}$$

where $\Delta = z_2-z_1$, X_0 is a standard normal r.v., and

$$\sigma^2 = \left(q_1(1-q_1))/h_1^2-2q_1(1-q_2)/h_1h_2+q_2(1-q_2)/h_2^2\right)/(z_2-z_1)^2 \tag{3.8}$$

with $h_i = g_3(z_i)$ and $q_i = G_3(z_i)$.

If the term of order $O(m^{-\delta})$ on the left-hand side of (3.7) is omitted then the term $k^{1/2}m^{-2\delta}$ has to be replaced by $k^{1/2}m^{-\delta}$ thus restricting the choice of k. Now we consider an adaptive estimation procedure starting with the statistic

$$\frac{Z_{2,k}-Z_{1,k}}{\Delta} - \alpha\,\frac{m^{-\delta}\vartheta}{\Delta}\exp(-\delta z_1)[1-\exp(-\delta\Delta)]. \tag{3.9}$$

In this sequel, let $z_{i+1}-z_i = \Delta$ for $i = 1,2,3$. Notice that (3.7) suggests

$$\frac{(Z_{3,k}-Z_{2,k})-(Z_{2,k}-Z_{1,k})}{\Delta(1-\exp(-\delta\Delta))} \tag{3.10}$$

as an estimator of $-\alpha m^{-\delta}\vartheta\Delta^{-1}\exp(-\delta z_1)[1-\exp(-\delta\Delta)]$ leading to the statistic

$$\frac{Z_{2,k}-Z_{1,k}}{\Delta} + \frac{(Z_{3,k}-Z_{2,k})-(Z_{2,k}-Z_{1,k})}{\Delta(1-\exp(-\delta\Delta))} \tag{3.11}$$

as an estimator of α. Finally, (3.10) motivates the use of

$$W_k^* = \frac{(Z_{4,k}-Z_{3,k})-(Z_{3,k}-Z_{2,k})}{(Z_{3,k}-Z_{2,k})-(Z_{2,k}-Z_{1,k})}$$

as an estimator of $\exp(-\delta\Delta)$. If it is known that $\delta_0 \le \delta \le \rho$ then we may take

$$W_k = \min(\max(W_k^*,\exp(-\rho\Delta)),\exp(-\delta_0\Delta)) \tag{3.12}$$

instead of W_k^*. Thus, we have to investigate the performance of

$$\hat{\alpha}_k = \frac{Z_{2,k}-Z_{1,k}}{\Delta} + \frac{Z_{3,k}-2Z_{2,k}+Z_{1,k}}{\Delta(1-W_k)} \tag{3.13}$$

as an estimator of α. In this sequel, it is understood that $k \equiv k(n)$ and $m \equiv m(n)$ are such that $k \to \infty$, $m \to \infty$, and $k^{1/2}m^{-2\delta} \to 0$ as $n \to \infty$. We will prove that the sequence $k^{1/2}(\hat{\alpha}_k-\alpha)$ is stochastically bounded.

Theorem 2. (i) If $\vartheta k^{1/2}m^{-\delta} = O(n^0)$ then

$$k^{1/2}(\hat{\alpha}_k-\alpha) = O_p(n^0). \tag{3.14}$$

(ii) If $\delta_0 < \delta < \rho$ and $\vartheta k^{1/2}m^{-\delta} \to \infty$ as $n \to \infty$ then $k^{1/2}(\hat{\alpha}_k-\alpha)$ is asymptotically normal with expectation equal to zero (and a variance that may be deduced from (3.16)).

Proof. I. Let X_i, $i = 1,\dots,4$, be jointly normally distributed r.v.'s with mean vector zero and covariances as given in (3.6). Rewriting $k^{1/2}(\hat{\alpha}_k-\alpha)$ and applying Theorem 4.5.3 in Reiss (1989) we obtain

$$P\{k^{1/2}(\hat{\alpha}_k-\alpha) \le t\}$$

$$= P\left\{\frac{X_2-X_1}{\Delta} + \frac{X_3-2X_2+X_1}{\Delta(1-V_k)} + \left(\frac{1-\exp(-\delta\Delta)}{1-V_k}-1\right)\frac{k^{1/2}m^{-\delta}\kappa(1)}{\Delta(1-\exp(-\delta\Delta))} \le t\right\} \tag{3.15}$$

$$= O(k^{-1/2} + k^{1/2}m^{-2\delta} + m^{-\delta})$$

uniformly in t where V_k is defined like W_k with W_k^* replaced by

$$V_k^* = \frac{X_4-2X_3+X_2+k^{1/2}m^{-\delta}\kappa(2)}{X_3-2X_2+X_1+k^{1/2}m^{-\delta}\kappa(1)},$$

and

$$\kappa(i) = -\alpha\vartheta\exp(-\delta z_i)[1-\exp(-\delta\Delta)]^2.$$

Now, (i) is immediate.

II. Observe that

$$V_k^* = \frac{\exp(-\delta\Delta) + (X_4-2X_3+X_2)k^{-1/2}m^{\delta}/\kappa(1)}{1 + (X_3-2X_2+X_1)k^{-1/2}m^{\delta}/\kappa(1)}.$$

Hence, $P\{V_k^* = V_k\} \to 1$ as $n \to \infty$, and V_k may asymptotically be replaced by V_k^* in (3.15). Moreover,

$$\left(\frac{1-\exp(-\delta\Delta)}{1-V_k^*} - 1\right) \frac{k^{1/2}m^{-\delta}\kappa(1)}{\Delta(1-\exp(-\delta\Delta))}$$

$$= \frac{X_4-2X_3+X_2 - \exp(-\delta\Delta)(X_3-2X_2+X_1)}{\Delta[1-\exp(-\delta\Delta)]^2} (1 + o_p(n^0))$$

which also yields that

$$\frac{1-\exp(-\delta\Delta)}{1-V_k^*} = 1 + o_p(n^0).$$

Thus, $k^{1/2}(\hat{\alpha}_k-\alpha)$ is asymptotically distributed like

$$\frac{X_2-X_1}{\Delta} + \frac{X_4-2X_3+X_2 + (1-2\exp(\delta\Delta))(X_3-2X_2+X_1)}{\Delta[1-\exp(-\delta\Delta)]^2} \tag{3.16}$$

which proves (ii). ∎

4. Estimation of the Tail Index

Given $n = km$ let again $M_m^{(i)}$, $i = 1,\ldots,k$, be independent maxima of a sample of size m having the common d.f. F^m. If F has the density $x \to \sigma^{-1}f(x/\sigma)$ for some $\sigma > 0$ where f satisfies condition (2.1) then, according to Theorem 1,

$$\log M_m^{(1)},\ldots,\log M_m^{(k)} \tag{4.1}$$

may be treated like independent r.v.'s with common d.f. $G_{3,\tau}^{(\log(m^{\alpha}\sigma),\alpha)}$ within an error bound of order $O(k^{1/2}m^{-\min(2\delta,\rho)})$. Let $\tilde{Y}_{1:k} \le \ldots \le$

$\tilde{Y}_{k:k}$ be the order statistics of $\log M_m^{(1)},\ldots,\log M_m^{(k)}$. Moreover, let $\tilde{\alpha}_{k,n}$ be defined like $\hat{\alpha}_k$ with $Y_{i:k}$ replaced by $\tilde{Y}_{i:k}$. The following result is immediate from Theorems 1 and 2.

Theorem 3. Consider $k \equiv k(n)$ such that $k \to \infty$ and $k^{1/2}(k/n)^{\min(2\delta,\rho)} \to 0$ as $n \to \infty$. Assume that $0 < \delta_0 \le \delta \le \rho$. Under condition (2.1) we obtain:

(i) If $ck^{(2\delta+1)/2\delta}/n = O(n^0)$ then

$$k^{1/2}(\tilde{\alpha}_{k,n}-\alpha) = O_p(n^0). \tag{4.2}$$

(ii) If $\delta_0 < \delta < \rho$ and $ck^{(2\delta+1)/2\delta}/n \to \infty$ as $n \to \infty$ then $k^{1/2}(\tilde{\alpha}_{k,n}-\alpha)$ is asymptotically normal with expectation equal to zero (and a variance that may be deduced from (3.16)).

Let us compare this result with those indicated in the introduction. To simplify the argument we assume that $1/2 \le \delta \le \rho = 1$. In (1.3) and (1.5) the choice of the estimators was restricted by the requirement that $k = o(n^{2\delta/(2\delta+1)})$; within the new framework this condition is weakened to $k = o(n^{2/3})$ leading to estimators $\tilde{\alpha}_{k,n}$ of a higher accuracy in case of $\delta < 1$.

References

Csörgo, S., Deheuvels, P. and Mason, D. (1985). Kernel estimates of the tail index of a distribution. Ann. Statist. 13, 1050-1078.

Falk, M. (1985). Uniform convergence of extreme order statistics. Habilitation Thesis, University of Siegen.

Hall, P. and Welsh, A. H. (1985). Adaptive estimates of parameters of regular variation. Ann. Statist. 13, 331-341.

Häusler, E. and Teugels, J. L. (1985). On asymptotic normality of Hill's estimator for the exponent of regular variation. Ann. Statist. 13, 743-756.

Hill, B. M. (1975). A simple approach to inference about the tail of a distribution. Ann. Statist. 3, 1163-1174.

Hüsler, J. and Tiago de Oliveira, J. (1986). The usage of the largest observations for parameter and quantile estimation for the Gumbel distribution; an efficiency analysis. Preprint.

Reiss, R.-D. (1987). Estimating the tail index of the claim size distribution. Blätter der DGVM 18, 21-25.

Reiss, R.-D. (1989). Approximate Distributions of Order Statistics (With Applications to Nonparametric Statistics). Springer Series in Statistics. New York: Springer.

Smith, R. L. (1987). A theoretical comparison of the annual maximum and threshold approaches to extreme value analysis. Report No. 53, University of Surrey.

Weiss, L. (1971). Asymptotic inference about a density function at an end of its range. Nav. Res. Logist. Quart. 18, 111-114.

ASYMPTOTIC RESULTS FOR AN EXTREME VALUE ESTIMATOR OF THE AUTOCORRELATION COEFFICIENT FOR A FIRST ORDER AUTOREGRESSIVE SEQUENCE

W. P. McCormick and G. Mathew

University of Georgia, Athens, GA and Southwest Missouri State University, Springfield, MO

Abstract. For an AR(1) process with positive or bounded innovations, an estimate of the autocorrelation coefficient based on an extreme value statistic is proposed. Asymptotic properties are investigated. In particular, an asymptotic essentially nonparametric confidence interval for the autocorrelation coefficient is derived.

1. INTRODUCTION

In this paper we study the asymptotic properties of an extreme value estimator of the autocorrelation coefficient for the first-order autoregressive model

$$X_n = \theta X_{n-1} + \varepsilon_n, \quad n=1, 2, \ldots \tag{1.1}$$

with $0 < \theta < 1$.

Our interest is in the case where the errors have a distribution whose support is bounded in some way. It is the boundedness of the support which gives rise to the possibility of an extreme value estimate for θ and in certain cases ourestimate converges to the parameter at a faster rate than the traditional sample autocorrelation coefficient.

We consider two cases. In the first case we consider positive innovations appearing in (1.1). Such sequences have been considered by Collings (1975) and Tavares (1978) to describe the input process for dams. We further mention the important work of Lawrence and Lewis (1977) and Gaver and Lewis (1980) wherein autoregressive models were introduced having a marginal gamma distribution. With regard to these models, we mention that for a first-order autoregressive exponential process, Raftery (1980) considered the limiting distribution of the maximum likelihood estimate for θ.

In the case that $\varepsilon_n \geqq 0$ in (1.1), we have that

$$X_n/X_{n-1} \geqq \theta, \quad n \geqq 1$$

which suggests

$$W_n = \min_{1 \leqq k \leqq n} X_k/X_{k-1} \tag{1.2}$$

Key words. AR(1) processes, regular variation, weak limit. AMS 1980 subject classification. Primary 62M10, Secondary 60F05.

can serve as an estimate of θ. In fact under our assumptions on the distribution of the ε_n, θ is the left endpoint of the distribution of X_n/X_{n-1} so that W_n is a natural estimate for this parameter.

In the second case we consider errors with distribution having bounded symmetric support. In this case we show that, under a tail balancing condition, two extreme value estimates of the same stochastic order arise and we discuss an optimum way to combine them. Examples where the innovation sequence has compact support may be found in McKenzie (1985).

For the case of bounded support with $-\infty<x_0 = \inf\{x:F(x)>0\}$ and $x_1 = \sup\{x:F(x)<1\}<\infty$, we have in the symmetric case $(x_0 = -x_1)$ in which we can assume $x_1 = 1$, that

$$\frac{X_n}{X_{n-1}} - \frac{1}{|X_{n-1}|} < \theta < \frac{X_n}{X_{n-1}} + \frac{1}{|X_{n-1}|}, \ n \geq 1.$$

We are not requiring $\theta > 0$ in this case, but only that $|\theta| < 1$. From these relations, we consider the two sequences of estimators

$$T_{1,n} = \bigvee_{k=1}^{n} \left[\frac{X_k}{X_{k-1}} - \frac{1}{|X_{k-1}|}\right] \tag{1.3}$$

and

$$T_{2,n} = \bigwedge_{k=1}^{n} \left[\frac{X_k}{X_{k-1}} + \frac{1}{|X_{k-1}|}\right]. \tag{1.4}$$

Section 2 contains asymptotic results for the estimates for θ given above.

In order to obtain our limit law results, we need to impose a regularity condition on the tail of the distribution of the errors; otherwise, the distribution of X_k/X_{k-1} need not belong to a domain of minimum attraction. In this paper we require F to be regularly varying with index α at its left endpoint. It then follows that our limit law results depend on α, both in the norming constants and in the limiting distribution. In practice the value of α would be unknown. In section 4 we show how α may be estimated by a statistic depending on an increasing number of the small order statistics. We then use this estimate to obtain an essentially nonparametric confidence interval for θ. Our method uses a Renyi-type representation for the order statistics valid for dependent observations. This representation is developed in Section 3.

2. LIMITING DISTRIBUTIONS

In this section we present asymptotic results for the estimates introduced in Section 1. Let $\{X_n\}$ be the AR(1) process given by

$$X_n = \theta X_{n-1} + \varepsilon_n, \tag{2.1}$$

where $\varepsilon_n \sim F$ which satisfies

$$\lim_{t\to 0} \frac{F(tx)}{F(t)} = x^{\alpha}, \ x > 0 \tag{2.2}$$

and

$$\int_0^\infty x^\beta dF(x) < \infty \text{ for some } \beta > \alpha. \tag{2.3}$$

Under assumption (2.3) it is easy to check that

$$X_0 = \sum_{j=0}^{\infty} \theta^j \varepsilon_{-j} \tag{2.4}$$

exists almost surely and its distribution provides the invariant probability measure for the Markov chain defined in (2.1). We consider the limiting distribution of the estimate in (1.2) under two models. The stationary AR(1) process in (2.1) with X_0 given in (2.4), so that

$$X_n = \sum_{j=0}^{\infty} \theta^j \varepsilon_{n-j} \tag{2.5}$$

and the nonstationary process $\tilde{X}_n$ which satisfies the difference equation in (2.1) with $\tilde{X}_0 = 0$, that is,

$$\tilde{X}_n = \sum_{j=0}^{n-1} \theta^j \varepsilon_{n-j}. \tag{2.6}$$

For the proof of the following result, we refer to Davis and McCormick (1988).

<u>Theorem 2.1</u>. Let $W_n = \bigwedge_{k=1}^{n} X_k/X_{k-1}$ where X_k is given in (2.5) and $\tilde{W}_n = \bigwedge_{k=2}^{n} \tilde{X}_k/\tilde{X}_{k-1}$ with $\tilde{X}_k$ given in (2.6). Then if F satisfies (2.2) and (2.3), we have

$$\lim_{n\to 0} P\{a_n^{-1}(W_n-\theta)c_\alpha \leq x\}$$

$$= \lim_{n\to\infty} P\{a_n^{-1}(\tilde{W}_n-\theta)c_\alpha \leq x\} = 1-\exp\{-x^\alpha\},\ x > 0,$$

where $a_n = F^{\leftarrow}(\frac{1}{n}) = \inf\{x{:}F(x) \geq n^{-1}\}$ and $c_\alpha = (EX_0^\alpha)^{1/\alpha}$ with X_0 given in (2.4).

<u>Remark 1</u>. By (2.2) we have that the norming constants $a_n^{-1} = n^{1/\alpha}L(n)$ where L is a slowly varying function. Thus if the index of regular variation in (2.2) is smaller than 2, the estimate W_n or $\tilde{W}_n$ converges at a faster rate than the least squares estimate,

$$\hat{\theta}_n = \sum_{t=1}^{n-1} (X_t-\bar{X})(X_{t+1}-\bar{X}) / \sum_{t=1}^{n} (X_t-\bar{X})^2$$

for which $\sqrt{n}(\hat{\theta}_n-\theta)$ has a limiting normal distribution provided the ε_n have a finite variance.

<u>Remark 2</u>. Theorem 2.1 above is derived in Davis and McCormick as a Corollary of a more general limiting result for point process. However, a direct proof utilizing a result of O'Brien (1987) for the limiting distribution of the maximum of a stationary Markov chain is possible. An outline of this approach is as follows.

Under the assumption that the error distribution has nontrivial absolutely continuous component with respect to Lebesque measure and has compact support, it follows from Doob (1953) p. 197 that the stationary Markov Chain (X_{n-1},ε_n) is strong mixing, which in turn implies that the sequence $\{\varepsilon_k/X_{k-1}\}$ is strong mixing. Since strong mixing implies the AIM(c_n) condition in O'Brien where $P(\frac{\varepsilon_k}{X_{k-1}} \le c_n) = \frac{\tau}{n}$, then by Theorem 2.1 of that paper, in order to conclude

$$\lim_{n\to\infty} P\{\bigwedge_{k=1}^{n} \frac{\varepsilon_k}{X_{k-1}} \le c_n\} = 1-e^{-\tau},$$

it suffices to check that

$$\lim_{n\to\infty} P\{\bigwedge_{k=2}^{p_n} \frac{\varepsilon_k}{X_{k-1}} \ge c_n \,|\, \frac{\varepsilon_1}{X_0} < c_n\} = 1$$

where we can take $p_n = O(\log n)$. A straightforward computation checks this last condition.

For the case that the support of the error distribution is [-1,1] we require the balancing condition that for some $\alpha > 0$ and all $x > 0$

$$\lim_{t\downarrow 0} \frac{F(-1+tx)}{F(-1+t)+1-F(1-t)} = qx^{\alpha} \tag{2.7}$$

and

$$\lim_{t\downarrow 0} \frac{1-F(1-tx)}{F(-1+t)+1-F(1-t)} = px^{\alpha} \tag{2.8}$$

where $0 \le p, q$ and $p + q = 1$. Again we refer to Davis and McCormick for the proof of the following result.

<u>Theorem 2.2</u>. Let $T_{1,n}$ and $T_{2,n}$ be given as in (1.3) and (1.4) and $\tilde{T}_{i,n}$ be the same as $T_{i,n}$ with $\tilde{X}_k$ in place of X_k in the definition of $\tilde{T}_{i,n}$, i=1, 2. Then if F satisfies (2.7) and (2.8) we have

$$\lim_{n\to\infty} P\{a_n^{-1}(T_{1,n}-\theta) \le -x, a_n^{-1}(T_{2,n}-\theta) > y\}$$

$$= \lim_{n\to\infty} P\{a_n^{-1}(\tilde{T}_{1,n}-\theta) \le -x, a_n^{-1}(\tilde{T}_{2,n}-\theta) > y\}$$

$$= \exp\{-c_1 x^{\alpha} - c_2 y^{\alpha}\}$$

where a_n satisfies $F(-1 + a_n) +1-F(1-a_n) \sim \frac{1}{n}$ and

$$c_1 = pE(X_0^+)^{\alpha} + q(X_0^-)^{\alpha}$$

and

$$c_2 = E(X_0)^{\alpha} - c_1$$

with X^+ and X^- denoting positive and negative part of X respectively.

Remark. The linear combination a T_{1n} + bT_{2n} with a + b = 1 which minimizes the asymptotic mean square error is a natural choice of a point estimate of θ based on the $T_{i,n}$. Direct calculation yield

$$a = \frac{\Gamma(1+\frac{2}{\alpha})c_2^{-2/\alpha}+\Gamma^2(1+1/\alpha)(c_1c_2)^{-1/\alpha}}{\Gamma(1+\frac{2}{\alpha})(c_1^{-2/\alpha}+c_2^{-2/\alpha})+2\Gamma^2(1+\frac{1}{\alpha})(c_1c_2)^{-1/\alpha}}$$

3. A REPRESENTATION FOR SMALL ORDER STATISTICS

Let $X_1, X_2, \ldots, X_n$ be iid random variables with Exp(1) distribution, that is, $P(X_i \le x) = 1 - \exp(-x)$, $x > 0$. Then the Renyi representation for the order statistics $X_{1,n} < X_{2,n} \ldots < X_{n,n}$ states that on a common probability space there exist iid random variables Z_i with distribution Exp(1) such that

$$X_{i,n} = \frac{Z_1}{n} + \frac{Z_2}{n-1} + \ldots + \frac{Z_i}{n-i+1}, \quad 1 \le i \le n. \tag{3.1}$$

This useful representation can be exploited to yield asymptotic results for the order statistics from a sample with distribution belonging to the minimum domain of attractions of a Type II distribution. For example, this was done in deHann (1981) where an essentially nonparametric confidence interval for the unknown left endpont of a distribution function was derived. In this section we derive an analogous representation to (3.1) for dependent observations. This representation is then applied in the same way as in deHann's result to obtain an essentially nonparametric confidence interval for θ, the correlationcoefficient in (2.1).

Consider a stationary sequence X_n, $n \ge 1$ with common continuous distribution F having a finite left endpoint μ and such that for some $\alpha > 0$

$$\lim_{t \to 0} \frac{F(tx+\mu)}{F(t+\mu)} = x^{\alpha}, \quad x > 0. \tag{3.2}$$

Let $X_{1,n} < \ldots < X_{n,n}$ denote the ordered values of $X_1, \ldots, X_n$. Then the main result of this section is to show that under suitable mixing conditions on the X_n, we have for a sequence r_n which tends to infinity and which plays a role in our mixing conditions, there exists random variables Y_i and Z_i, $1 \le i \le n$ defined on the same probability space satisfying the following condition

$$(Y_1, Y_2, \ldots, Y_n) \overset{d}{=} (X_{1,n}, X_{2,n}, \ldots X_{n,n}) \tag{3.3}$$

and for any $\delta > 0$ there exists $n_0 = n_0(\delta)$ such that $n \ge n_0$ and $1 \le k \le r_n$

$$(1-\delta)\{[\sum_1^k Z_j - \delta]^{(1-\delta)} \wedge [\sum_1^k Z_j - \delta]^{(1+\delta)}\} \le (\frac{Y_k - \mu}{a_n})^{\alpha} \tag{3.4}$$

$$\le (1+\delta)\{[\sum_1^k Z_j + \delta]^{(1-\delta)} \vee [\sum_1^k Z_j + \delta]^{(1+\delta)}\}$$

on a set Ω_n with $P(\Omega_n) \to 1$, where $F(a_n + \mu) = \frac{1}{n}(1+o(1))$ as $n \to \infty$, and where the Z_i are iid Exp(1).

We remark that property (3.3) asserts that the Y_k provide a representation for the order statistics $X_{k,n}$, while property (3.4) exhibits a relationship between the represented order statistics and sums of iid Exp(1) random variables. We begin our program by showing how to define the Y_k as functions of the Z_k. To that end define

$$H_1(y) = P(X_{1,n} \leq y) \tag{3.5}$$

and for $k \geqq 1$

$$H_{k+1}((x_1, \ldots, x_k),y) = P\{X_{k+1,n} \leq x_k+y | X_{1,n} = x_1, \ldots, X_{k,n} = x_k\} \tag{3.6}$$

with $x_1 < \ldots < x_k$ and $y > 0$.

Let $\leftarrow$ denote the functional inverse and define

$$\psi_1(\cdot) = (-\ln[1 - H_1(\cdot)])^{\leftarrow} \tag{3.7}$$

and

$$\psi_k(\underline{x},\cdot) = (-\ln[1 - H_k(\underline{x},\cdot)])^{\leftarrow}, \ k \geq 2 \tag{3.8}$$

with $\underline{x} = (x_1, \ldots, x_{k-1})$.

Now for Z_k iid Exp(1) random variables, set

$$Y_k = Y_{k-1} + \psi_k(\underline{Y}_{k-1},Z_k), \ k \geq 2 \text{ and } Y_1 = \psi_1(Z_1) \tag{3.9}$$

where $\underline{Y}_k = (Y_1, \ldots, Y_k)$.

Lemma 3.1. For the Y_k defined in (3.9) we have

$$(Y_1, \ldots, Y_n) \stackrel{d}{=} (X_{1,n}, \ldots, X_{n,n}) \tag{3.10}$$

Proof. For a fixed n we show (3.10) by induction on k the number of statistics which may be represented by the Y_i. For k=1, in order to show that $Y_1 \stackrel{d}{=} X_{1,n}$, observe

$$P(Y_1 \leqq y) = P(\psi_1(Z_1) \leqq y) = P(Z_1 \leqq -\ln(1-H_1(y))) = H_1(y).$$

Now assume $(Y_1, \ldots, Y_{k-1}) \stackrel{d}{=} (X_{1,n}, \ldots, X_{k-1,n})$. Then for k, we have

$$P(Y_1 \leqq y_1, \ldots, Y_k \leqq y_k)$$

$$= \int_{-\infty}^{y_{k-1}} \ldots \int_{-\infty}^{y_1} P(Y_k \leqq y_k | Y_1 = x_1, \ldots, Y_{k-1} = x_{k-1}) P(\underline{Y}_{k-1} \varepsilon d\underline{x})$$

$$= \int_{-\infty}^{y_{k-1}} \ldots \int_{-\infty}^{y_1} P(\psi_k(\underline{x},Z_k) \leqq y_k - x_{k-1}) P(\underline{Y}_{k-1} \varepsilon d\underline{x})$$

$$= \int_{-\infty}^{y_{k-1}} \ldots \int_{-\infty}^{y_1} P(Z_k \leqq -\ln[1-H_k(\underline{x},y_k-x_{k-1})]) P(\underline{Y}_{k-1} \varepsilon d\underline{x})$$

$$= \int_{-\infty}^{y_{k-1}} \cdots \int_{-\infty}^{y_1} P(X_{k,n} \leqq y_k | X_{1,n} = x_1, \ldots, X_{k-1} = x_{k-1})$$

$$P(X_{1,n} \varepsilon dx_1, \ldots X_{k-1,n} \varepsilon dx_{k-1})$$

$$= P(X_{1,n} \leqq y_1, \ldots, X_{k,n} \leqq y_k).$$

Thus (3.10) follows.

If the underlying sample $X_1, \ldots, X_n$ is taken from the Exp(1) distribution, then it is easy to check that the ψ_k defined in (3.8) are given by $\psi_k(\underline{x},y) = \frac{y}{n-k+1}$. Thus by (3.9) we see that Lemma 3.1 has the Renyi representation (3.1) as a special case. In order for the representation in (3.9) to be useful, we need to obtain tractable approximations to the functions defined in (3.7) and (3.8). This is a computationally involved task. For the reader who is interested in seeing the complete details, we refer to the technical report of this paper. Here, we will present only a sketch of the proofs. Moreover in giving the outline of the argument, we do not always give a complete statement of the hypotheses needed for the results. However, we do give a complete statement of the main theorem.

As a first step in establishing (3.4) we consider stationary sequences having an Exp(1) marginal. This, of course, can always be arranged via the transformation

$$\tilde{X}_n = -\ln(1-F(X_n)),\ n \geq 1 \tag{3.11}$$

where F is the common marginal distribution of the X_n. In view of the regularity of F near its left endpoint under the assumption in (3.2), we obtain the following relationship between the small order statistics of (X_n) and $(\tilde{X}_n)$.

Step 1. Let $\tilde{X}_{k,n}$ denote the order statistics for the $\tilde{X}_k$, $1 \leqq k \leqq n$. Then if F satisfies (3.2), we have for any $\delta > 0$ there exists $n_o = n_o(\delta)$ and $c = c(\delta)$ such that on the set $(X_{r_n,n} - \mu < c)$

$$T_1(n\tilde{X}_{k,n}) \leq \left(\frac{X_{k,n}-\mu}{a_n}\right)^{\alpha} \leq T_2\ (n\tilde{X}_{k,n}),\ 1 \leq k \leq r_n, \tag{3.12}$$

where

$$T_1(z) = (1-\delta)[z^{(1-\delta)} \wedge z^{(1+\delta)}] \text{ and } T_2(z) = (1+\delta[z^{(1-\delta)} \vee z^{(1+\delta)}] \tag{3.13}$$

and $F(a_n + \mu) = \frac{1}{n}$.

The proof of (3.12) relies on the relation

$$-\ln(1-x) \sim x \text{ as } x \to 0$$

and the bounds for a regularly varying functionF satisfying (3.2)

$$(1-\delta)[(\tfrac{x}{y})^{(\alpha-\delta)} \wedge (\tfrac{x}{y})^{(\alpha+\delta)}] \leq \frac{F(x+\mu)}{F(y+\mu)} \leq (1+\delta)[(\tfrac{x}{y})^{(\alpha-\delta)} \vee (\tfrac{x}{y})^{(\alpha+\delta)}]$$

where $0 \leq x, y \leq c = c(\delta)$.

Our next simplification is to restrict attention to essentially m-dependent sequences. More precisely suppose $X_k^{(n)}$, $k \geq 1$, $n \geq 1$ is an array, which is stationary in each row, and having marginal distribution $F(x) = 1 - \exp(-x)$, $x > 0$. We further assume that the n-th row is m_n-dependent. Let $X_{k,n}$, $1 \le k \le n$ denote the first n terms of the n-th row arranged in increasing order. Let

$$H_1^{(n)}(y) = P(X_{1,n} \le y) \text{ and } H_k^{(n)}(\underline{x}, y) = P(X_{k,n} \le X_{k-1}+y | X_{1,n}=x_1, \ldots, X_{k-1,n}=x_{k-1})$$

with $\underline{x} = (x_1, \ldots, x_{k-1})$ and define the functions

$$\psi_1^{(n)}(\cdot) = (-\ln[1-H_1^{(n)}(\cdot)])^{\leftarrow} \text{ and } \psi_k^{(n)} = (-\ln[1-H_k^{(n)}(\underline{x},\cdot)])^{\leftarrow},\ k \ge 2 \qquad (3.14)$$

Step 2. Under appropriate conditions on m_n as defined above and r_n the number of small order statistics under consideration, we show that the functions $\psi_k^{(n)}$ defined in (3.14) satisfy

$$\frac{y-c_{1,n}}{n} \le \psi_k^{(n)}(\underline{x},y) \le \frac{y+c_{2,n}}{n} \quad 1 \le k \le r_n \qquad (3.15)$$

with $\underline{x} = (x_1, \ldots, x_{k-1})$ and $0 < c_{i,n} = o(r_n^{-1})$ as $n \to \infty$, i=1, 2 provided $x_1 \le \ldots \le x_{r_n}$ are sufficiently small.

The proof of (3.15) uses a blocking argument and the assumed Exp(1) marginal distribution. Thus, we see that the dependence in the array forces us to deviate slightly from the result in the iid Exp(1) case, namely, $\psi_k(\underline{x},y) = \frac{y}{n-k+1}$.

Having established the important relations in (3.15) for the Exp(1) case, we then apply the result for a general array $X_k^{(n)}$, $k \ge 1$, $n \ge 1$ which is stationary in each row with marginal distribution F_n satisfying (3.2). We suppose that the sequence is m_n-dependent in the n-th row and let $X_{1,n}, X_{2,n}, \ldots X_{n,n}$ denote the ordered values for the first n terms of that row.

Step 3. We show that under suitable conditions on m_n and r_n, there exists a representation

$$(Y_1, Y_2, \ldots Y_n) \stackrel{d}{=} (X_{1,n}, X_{2,n}, \ldots X_{n,n})$$

such that on a set Ω_n with $P(\Omega_n) \to 1$ as $n \to \infty$ and for any $\delta > 0$ and all n sufficiently large

$$T_1\left(\sum_1^k Z_j - \delta\right) \le \left(\frac{Y_k - \mu}{a_n}\right)^{\alpha} \le T_2\left(\sum_1^k Z_j + \delta\right), \quad 1 \le k \le r_n \qquad (3.16)$$

where T_i, i=1,2 is defined in (3.13) and $F_n(a_n+\mu) = \frac{1}{n}$.

To show (3.16) we first consider the transformed array $\tilde{X}_k^{(n)} = -\ln(1-F_n(X_k^{(n)}))$ which has Exp(1) marginal. By Lemma 3.1 we have the representation for the ordered $\tilde{X}_k^{(n)}$ values given by

$$\tilde{Y}_k = \tilde{Y}_{k-1} + \tilde{\psi}_k^{(n)}(\underline{\tilde{Y}}_{k-1}, Z_k),\ k \ge 2$$

and

$$\tilde{Y}_1 = \tilde{\psi}_1^{(n)}(Z_1).$$

By Step 2 we have that (3.15) holds for the $\tilde{\psi}_k^{(n)}$ which, in turn, yields

$$\sum_1^k Z_j - kc_{1,n} \leq n\tilde{Y}_k \leq \sum_1^k Z_j + kc_{2,n}, \quad 1 \leq k \leq r_n. \tag{3.17}$$

Finally, a representation Y_k, $1 \leq k \leq n$ for the $X_{k,n}$ is obtained via $Y_k = F_n^{-1}(1 - \exp(-\tilde{Y}_k))$. Step 1 may then be applied to the (Y_k) and $(\tilde{Y}_k)$ to yield

$$T_1(n\tilde{Y}_k) \leq \left(\frac{Y_k - \mu}{a_n}\right)^\alpha \leq T_2(n\tilde{Y}_k), \quad 1 \leq k \leq r_n. \tag{3.18}$$

Thus by (3.17), (3.18) and the property $c_{i,n} r_n \to 0$ as $n \to \infty$, we obtain (3.16).

The rational behind working with an array of m_n-dependent sequences is the simplicity of the dependence and that we can adequately approximate a sequence obtained as a function of an AR(1) sequence this way. However, before we can ascribe a representation like (3.16) to the small order statistics of a stationary sequence (X_k), we need to know that closeness of an approximating m_n-dependent sequence $(X_k^{(n)})$ to (X_k) implies closeness of their respective ordered values.

<u>Step 4</u>. Let (a_k) and $(\tilde{a}_k)$, $1 \leq k \leq n$ be real numbers. Then if (b_k) and $(\tilde{b}_k)$ are the values (a_k) and $(\tilde{a}_k)$ arranged in increasing order of magnitude we have

$$\max_{1 \leq k \leq n} |b_k - \tilde{b}_k| \leq \max_{1 \leq k \leq n} | a_k - \tilde{a}_k | \tag{3.19}$$

The above can be seen as follows. Let $(i(1), i(2), \ldots i(n))$ and $(\tilde{i}(1), \tilde{i}(2), \ldots \tilde{i}(n))$ be the permutations on $\{1, 2, \ldots, n\}$ defined by

$$b_k = a_{i(k)} \text{ and } \tilde{b}_k = \tilde{a}_{\tilde{i}(k)}.$$

Clearly if $\max_{1 \leq k \leq n} |b_k - \tilde{b}_k| = b_k - \tilde{b}_k$ with $\tilde{i}(k) = i(k)$ then (3.19) holds. So assume $\tilde{i}(k) \neq i(k)$. If $\tilde{i}(k) \notin \{i(1), \ldots, i(k-1)\}$, then $b_k - \tilde{b}_k \leq a_{\tilde{i}(k)} - \tilde{a}_{\tilde{i}(k)}$ so that in this case (3.19) holds. Finally, if $\tilde{i}(k) \in \{i(1), \ldots, i(k-1)\}$, then there must exist a $1 \leq j \leq k-1$ such that $\tilde{i}(j) \notin \{i(1), \ldots, i(k-1)\}$, which implies $b_k - \tilde{b}_k \leq a_{\tilde{i}(j)} - \tilde{a}_{\tilde{i}(j)}$. A similar argument shows that (3.19) holds if $\max_{1 \leq k \leq n} |b_k - \tilde{b}| = \tilde{b}_k - b_k$. Hence Step 4 is shown.

We complete our program of preliminary results by showing that closeness of two random vectors is not destroyed by taking an equivalent version.

<u>Step 5</u>. Suppose $(X_1, \ldots, X_n)$ and $(Y_1, \ldots, Y_n)$ are such that $P(\max_{1 \leq k \leq n} |X_k - Y_k| < \varepsilon) = 1$. Then if $(X_1^*, \ldots, X_n^*) \stackrel{d}{=} (X_1, \ldots, X_n)$, there exists $(Y_1^*, \ldots, Y_n^*)$ with $(Y_1^*, \ldots, Y_n^*) \stackrel{d}{=} (Y_1, \ldots, Y_n)$ such that $P(\max_{1 \leq k \leq n} |X_k^* - Y_k^*| < \varepsilon) = 1$. The domain of the X_k^* may need to be enlarged.

To see Step 5 suppose $\underline{X}^* = (X_1^*, \ldots, X_n^*)$ is defined on (Ω, F, Q) where $F = \sigma(\underline{X}^*)$. We enlarge Ω to $\tilde{\Omega} = \Omega \times \mathbb{R}^n$ and let $\tilde{F} = F \times B(\mathbb{R}^n)$. We extend $\underline{X}^*$ to the bigger domain $\tilde{\Omega}$ by defining $\underline{X}^*(\omega, \underline{y}) = \underline{X}^*(\omega)$. Now define $\underline{Y}^*$ on $\tilde{\Omega}$ by $\underline{Y}^*(\omega, \underline{y}) = \underline{y}$. Finally, define $\tilde{Q}$ on $\tilde{F}$ by $\tilde{Q}([\underline{X}^* \in A] \cap [\underline{Y}^* \in B]) = P([\underline{X} \in A] \cap [\underline{Y} \in B])$. Then $\underline{Y}^*$ is our desired representation.

Theorem 3.1. Let $\{X_k, k \geq 1\}$ be a stationary sequence and suppose an approximating array $\{X_k^{(n)}, k\geq 1, n\geq 1\}$ exists for which for any $\varepsilon > 0$

$$\lim_{n\to\infty} P\{ \max_{1\leq k\leq n} |X_k - X_k^{(n)}| > \varepsilon a_n\} = 0$$

where a_n is given by $F_n(a_n + \mu) = \frac{1}{n}$ and F_n satisfying (3.2) is the marginal distribution in the n-th row.

We suppose that the array satisfies the following local and global mixing conditions. For each n the $X_k^{(n)}$, $k\geq 1$ are m_n - dependent. Let $V_j^n = a_n^{-1}(X_j^{(n)} - \mu)$. Then for a given sequence $r_n \to \infty$, we suppose that a sequence $p_n \to \infty$ can be found so that

$$\beta_{1,n} = n \sum_{j=2}^{p_n} P(V_1^n < r_n^{2/\alpha}, V_j^n < r_n^{2/\alpha}) = o(1) \text{ as } n \to \infty$$

and

$$\max\left(\frac{p_n r_n^4}{n}, \frac{m_n r_n^4}{p_n}, a_n r_n^{2/\alpha}\right) = o(1) \text{ as } n \to \infty.$$

Furthermore, we require

$$\beta_{2,n} = m_n r_n \sup P(V_t^n < r_n^{2/\alpha} | V_{i_1}^n = x_1, \ldots, V_{i_k}^n = x_k) = o(1) \text{ as } n \to \infty,$$

where the supremum is over all indices $1\leq i_1 \neq \ldots \neq i_k \neq t \leq n$ and $0<x_1<\ldots<x_k<r_n^{1/\alpha}$ and $1\leq k\leq r_n$.

Let $X_{k,n}$, $1\leq k\leq n$ be the order statistics for X_k, $1\leq k\leq n$. Then there exists $(Y_1, \ldots, Y_n) \stackrel{d}{=} (X_{1,n}, \ldots, X_{n,n})$ such that for any $\delta > 0$ on the set $(Y_{r_n} - \mu < c = c(\delta)) \cap \Omega_n$ with $P(\Omega_n) \to 1$ as $n \to \infty$, we have for all n sufficiently large

$$(1-\delta)\left[\left(\sum_1^k Z_j - \delta\right)^{(1-\delta)} \wedge \left(\sum_1^k Z_j - \delta\right)^{1+\delta}\right] \leq \left(\frac{Y_k - \mu}{a_n}\right)^\alpha$$

$$\leq (1+\delta)\left[\left(\sum_1^k Z_j + \delta\right)^{(1-\delta)} \vee \left(\sum_1^k Z_j + \delta\right)^{(1+\delta)}\right], \quad 1\leq k\leq r_n \tag{3.20}$$

where the Z_j are iid Exp(1) random variables.

The proof of Theorem 3.1 follows from the preceding steps by noting that under the mixing conditions for the array, we have by Step 3 that a representation like (3.20) holds for the small order statistics for the array. Then (3.20) itself holds by Steps 4 and 5.

4. CONFIDENCE INTERVAL FOR THE AUTOCORRELATION COEFFICIENT.

In this section we apply the representation given in Theorem 3.1 to the problem of obtaining a confidence interval for θ in the model given in (2.1). The connection to Theorem 3.1 is that if X_n is the process defined by the difference equation in (2.1) with $\varepsilon_n \sim F$ satisfying

$$\lim_{t\to 0} \frac{F(xt)}{F(t)} = x^\alpha, \; x > 0 \tag{4.1}$$

and $X_o = \sum_{j=0}^{\infty} \theta^j \varepsilon_{-j}$, then the sequence $Y_k = X_k/X_{k-1}$, $k \geq 1$, is a stationary sequence having distribution function H which has θ as its left endpoint and is regularly varying there with index α. In fact we have the following result which can be shown by a straightforward argument.

Lemma 4.1. Let $H(x) = P\{X_n/X_{n-1} \leq x\}$. Then if F satisfies (4.1) and additionally that

$$\int_0^{\infty} x^{\beta} dF(x) < \infty \text{ for some } \beta > \alpha, \tag{4.2}$$

we have

$$\lim_{t\to 0} \frac{H(\theta+tx)}{F(t)} = x^{\alpha} EX_n^{\alpha}. \tag{4.3}$$

To obtain an approximating array for the sequence Y_k, define $Y_k^n = X_k^{m_n}/X_{k-1}^{m_n}$ where $X_k^m = \sum_{i=0}^{m} \theta^i \varepsilon_{k-i}$ and

$$m_n = \left[\frac{-2(1+\alpha)}{\alpha \ln \theta} \ln n\right]. \tag{4.4}$$

Lemma 4.2. For any $\varepsilon > 0$

$$\lim_{n\to\infty} n\, P\{|Y_k - Y_k^n| > \varepsilon\, a_n\} = 0$$

where $F(a_n) = \frac{1}{n}$.

Proof. By the regular variation of F in (4.1), it can be checked that for $m \geq \frac{4}{\alpha}$,

$$E\left(\left[\sum_{1}^{m} \varepsilon_i\right]^{-1}\right) < \infty. \tag{4.5}$$

Therefore, since after some calculation we have

$$|Y_j - Y_j^n| \leq \theta^{-m}\left[\sum_{1=0}^{m-1} \varepsilon_{j-1-i}\right]^{-1}\left(2+\varepsilon_j \theta^{-2m}\left[\sum_{i=m}^{2m-1} \varepsilon_{j-1-i}\right]^{-1}\right) \sum_{i=m_n+1}^{\infty} \theta^i \varepsilon_{j-1},$$

we have by taking expectations in the above and using (4.5) that

$$E|Y_j - Y_j^n| \leq c\theta^{m_n} \tag{4.6}$$

where the constant $c = c(\theta,\alpha)$ does not depend on j. Thus by (4.6) and Chebychev's inequality, we have

$$mP\{|Y_j - Y_j^n| > \varepsilon\, a_n\} \leq c\, \varepsilon^{-1}\, n a_n^{-1}\, \theta^{m_n} = o(1) \text{ as } n \to \infty$$

by choice of m_n and where we have used the fact that $a_n \sim n^{1/\alpha} U(\frac{1}{n})$ where U is a slowly varying function.

Lemma 4.3. For the array $Y_k^n = X_k^{m_n}/X_{k-1}^{m_n}$ with m_n given in (4.4), we suppose that the error distribution F has density f which is regularly varying with index $\alpha - 1$ at zero.

Furthermore, we suppose that (4.2) is satisfied for every $\beta > 0$, that is, $E\,\varepsilon_1^\beta < \infty$ for all $\beta > 0$. Then there exists a $\gamma > 0$ such that if $p_n = o(n^\gamma)$ and r_n is chosen to tend to infinity sufficiently slowly so as to satisfy

$$\max\left(\frac{p_n r_n^4}{n}, \frac{m_n r_n^4}{p_n}, a_n r_n^{2/\alpha}\right) = o(1) \text{ as } n \to \infty,$$

where $F(a_n) = \frac{1}{n}$, we have that the local and global mixing conditions for the array $(Y_k^n,\ k\geq 1,\ n\geq 1)$ described in Theorem 3.1 are satisfied.

Proof. Rather then present the details of the proof which are technical and straightforward, we highlight the main steps. Letting $H_n(x) = P(Y_k^n \leq x)$, we can check that

$$\lim_{t\to 0} \frac{H_n(\theta+tx)}{H_n(\theta+t)} = x^\alpha,\ x > 0,$$

so that H_n is taking the place of F_n in Theorem 3.1. Next if d_n is chosen so that

$$H_n(\theta+d_n) \sim \frac{1}{n} \text{ as } n \to \infty \tag{4.7}$$

then using Lemma 4.1, it follows that

$$d_n \sim c_\alpha^{-1} a_n \quad \text{as } n \to \infty, \tag{4.8}$$

where $c_\alpha = (EX_o^\alpha)^{1/\alpha}$ and $F(a_n) = \frac{1}{n}$. By (4.8) we can check that $\beta_{1,n}$ and $\beta_{2,n}$ tend to zero with a_n given in (4.8) appearing in the definitions of $\beta_{1,n}$ and $\beta_{2,n}$. Note that the a_n in Theorem 3.1 corresponds to our d_n given in (4.7). Utilizing the m_n-dependence of the $Y_k{}^n$, we find after calculation that for some $\varepsilon > 0$

$$n \sum_{j=2}^{p_n} P(a_n^{-1}(Y_1^n-\theta) < r_n^{2/\alpha},\ a_n^{-1}(Y_j^n-\theta) < r_n^{2/\alpha})$$

$$\leq n \sum_{j=2}^{m_n} P(a_n^{-1}(Y_n^n-\theta) < r_n^{2/\alpha},\ a_n^{-1}(Y_j^n-\theta) < r_n^{2/\alpha})$$

$$+ np_n[P(a_n^{-1}(Y_n^n-\theta) < r_n^{2/\alpha}]^2$$

$$\leq m_n r_n^{4(1+\varepsilon)} n^{-\varepsilon} + r_n^{4(1+\varepsilon)} p_n n^{-1}. \tag{4.9}$$

Thus provided $p_n = O(n^\gamma)$ for γ sufficiently small, we have by (4.9) that $\beta_{1,n} = o(1)$ as $n \to \infty$. In checking that $\beta_{2,n}$ tends zero, the only difficulty relates to variables in the conditioning with an index coming after t. To handle such terms one can check that for any fixed k and all n sufficiently large

$$P(Y_1^n < a_n x \mid Y_2^n = y_2 a_2, \ldots, Y_k^n = y_k a_n) \leq \frac{c}{n} \frac{EX_0^{k\alpha}}{EX_0^{(k-1)\alpha}}. \tag{4.10}$$

But the term in the right hand side of (4.10) tends to zero as $n \to \infty$ provided $k = k_n$ tends to infinity sufficiently slowly. We remark that in establishing (4.10) we use the additional assumptions of a regularly varying density f and existence of all moments of ε_1.

Theorem 4.1. Let $Y_k = X_k/X_{k-1}$ and suppose the hypothesis of Lemma 4.3 holds. Then on a set Ω_n such that $P(\Omega_n) \to 1$ as $n \to \infty$, we have for any $\delta > 0$ and all $n \geq n_0$

$$T_1(\sum_1^k Z_j - \delta) \leq \left(\frac{Y_{k,n} - \theta}{a_n}\right)^\alpha c_\alpha^\alpha$$

$$\leq T_2(\sum_1^k Z_j + \delta), \quad 1 \leq k \leq r_n, \tag{4.11}$$

where $Y_{k,n}$ equals the k-th order statistic for the Y_k and where T_i, i=1, 2 are given in (3.13) and r_n tends sufficiently slowly to infinity and the Z_j are iid Exp(1).

Proof. Since the $\{Y_k^n\}$ satisfy the hypothesis of Theorem 3.1 under the conditions given in Lemma 4.3, we have by Thorem 3.1, Lemma 4.2 and (4.8), that (4.11) holds on the set $\Omega_n = (Y_{r_n,n} - \theta < c) \cap \Lambda_n$ with $P(\Lambda_n) \to 1$ as $n \to \infty$. It is easy to check that for any $c > 0$

$$P(Y_{r_n,n} - \theta < c) \to 1 \text{ as } n \to \infty$$

provided r_n tends to infinity sufficiently slowly. Thus $P(\Omega_n) \to 1$ as $n \to \infty$ proving the theorem.

We now apply the results to obtain a confidence interval for θ. Let

$$U_n = (Y_{1,n} - \theta)/(Y_{2,n} - Y_{1,n}). \tag{4.12}$$

Corollary 4.1. Under the assumptions of Theorem 4.1, we have

$$\lim_{n\to\infty} P(U_n \leq u) = P(Z_1^{1/\alpha}/[(Z_1+Z_2)^{1/\alpha} - Z_1^{1/\alpha}] \leq u) \tag{4.13}$$

where U_n is given in (4.12) and Z_i are independent Exp(1) random variables.

Proof. Using the representation in (4.11) for the $Y_{k,n}$ it can be checked that on a set Ω_n with $P(\Omega_n) \to 1$ and for any $\delta > 0$ that for all n sufficiently large

$$\frac{Z_1^{1/\alpha}}{[(Z_1 + Z_2)^{1/\alpha} - Z_1^{1/\alpha}]} - \delta \leq U_n \leq \frac{Z_1^{1/\alpha}}{[(Z_1 + Z_2)^{1/\alpha} - Z_1^{1/\alpha}]} + \delta. \tag{4.14}$$

Since δ is arbitrary, (4.13) is immediate from (4.14).

Corollary 4.2. Assuming the hypothesis of Theorem 4.1 to be satisfied, we have that

$$\log\left(\frac{Y_{r_n,n} - Y_{3,n}}{Y_{3,n} - Y_{2,n}}\right)/\log r_n \xrightarrow{P} \frac{1}{\alpha} \text{ as } n \to \infty \tag{4.15}$$

provided $r_n \to \infty$ sufficiently slowly so that (4.11) holds.

Proof. Define functions

$$S_1(x,y) = \left(\frac{T_1(y-\delta)}{T_2(x+\delta)}\right)^{1/\alpha} \text{ and } S_2(x,y) = \left(\frac{T_2(y+\delta)}{T_1(x-\delta)}\right)^{1/\alpha}$$

Then by (4.11) we have that on a set Ω_n with $P(\Omega_n) \to 1$

$$S_1\left(\sum_1^3 Z_j, \sum_1^{r_n} Z_j\right) \le \frac{Y_{r_n,n}^{-\theta}}{Y_{3,n}^{-\theta}} \le S_2\left(\sum_1^3 Z_j, \sum_1^{r_n} Z_j\right). \tag{4.16}$$

Letting

$$Q_{1,n} = \frac{Y_{r_n,n}^{-\theta}}{Y_{3,n}^{-\theta}} \text{ and } Q_{2,n} = \frac{Y_{2,n}^{-\theta}}{Y_{3,n}^{-\theta}}$$

we have

$$\frac{Y_{r_n,}-Y_{3,n}}{Y_{3,n}-Y_{2,n}} = (Q_{1,n}-1)(1-Q_{2,n})^{-1},$$

so that

$$\log \frac{Y_{r_n,n}-Y_{3,n}}{Y_{3,n}-Y_{2,n}}/\log r_n = \frac{\log(Q_{1,n}-1)}{\log r_n} - \frac{\log(1-Q_{2,n})}{\log r_n}. \tag{4.17}$$

But as in Corollary 4.2, $Q_{2,n}$ has a limiting distribution so that the second term in (4.17) is asymptotically neglibible in probability.

Next observe that for y large we have

$$\log S_1(x,y) = \frac{1}{\alpha}\log\left(\frac{1-\delta}{1+\delta}\right) + \frac{(1-\delta)}{\alpha}\log(y-\delta)$$
$$- \log\{(x+\delta)^{(1-\delta)/\alpha} \vee (x+\delta)^{(1+\delta)/\alpha}\} \tag{4.18}$$

and

$$\log S_2(x,y) = \frac{1}{\alpha}\log\left(\frac{1+\delta}{1-\delta}\right) + \frac{1+\delta}{\alpha}\log(y+\delta)$$
$$- \log\{(x-\delta)^{(1-\delta)/\alpha} \wedge (x-\delta)^{(1+\delta)/\alpha}\}. \tag{4.19}$$

Thus since by the Strong Law of Large Numbers we have $\sum_1^{r_n} Z_j = (1+o(1))r_n$ a.s. as $n \to \infty$, we have by (4.16), (4.18), and (4.19) that on Ω_n

$$\frac{1-\delta}{\alpha} < \frac{\log(Q_{1,n}-1)}{\log r_n} < \frac{1+\alpha}{\alpha}.$$

Letting $n \to \infty$ and then $\delta \to 0$ in the above yields the corollary.

Our pivot for obtaining a confidence interval for θ now follows from Corollaries 4.1 and 4.2 exactly as in deHaan (1981). Namely we have that if

$$R_n = \frac{Y_{2,n}^{-\theta}}{Y_{1,n}^{-\theta}}$$

then by Corollary 4.1

$$\lim_{n\to\infty} P\{\alpha \log R_n \le x\} = 1 - e^{-x}.$$

Furthermore, letting

$$S_n = \log \frac{Y_{r_n,n}-Y_{3,n}}{Y_{3,n}-Y_{2,n}} / \log r_n$$

we have by Corollary 4.2 and the above that

$$\frac{\log R_n}{S_n} \Rightarrow 1 - e^{-x}$$

so that $\frac{\log R_n}{S_n}$ is a pivot for θ.

Acknowledgement. The authors thank Ishwar Basawa for suggesting this problem.

REFERENCES

Collings, P.S. (1975). Dams with autoregressive inputs. J. Appl. Prob., 12, 533-541.

Davis, R.A. and McCormick. W.P. (1988). Estimation for First-Order Autoregressive Processes with Positive or Bounded Innovations.

Doob, J.L. (1953). Stochastic Process. Wiley, New York.

Gaver, D.P. and Lewis, P.A.W. (1980). First-order autoregressive gamma sequences and point processes. Adv. Appl. Prob. 12, 727-745.

Haan, L de (1981). Estimation of the minimum of a function using order statistics. J. Amer. Statist. Assoc. 76, 467-469.

Lawrance, A.J. and Lewis, P.A.W. (1977). An exponential moving-average sequence and point processes. J. Appl. Prob. 14, 98-113.

McCormick, W.P. and Mathew, G. (1988). Asymptotic Results for an extreme value estimator of the autocorrelation coefficient for an AR(1) sequence. T.R. no. 90 Statistics Dept. Univ. of Ga. Athens, Ga.

McKenzie, E. (1985). An autoregressive process for beta random variables. Management Science, 31, 988-997.

O'Brien, G.L. (1987). Extreme Values for Stationary and Markov Sequences. Ann. of Probab. 15, 281-291.

Raftery, A.E. (1980). Estimation efficace pour un processus autoregressif exponential a densite discontinue. Pub. Inst. Stat. Univ. Paris 25, 65-91.

Tavares, L.V. (1978). Firm outflow from multiannual reserviors with skew and autocorrelated inflows. J. of Hydrology, 93-112.

THE SELECTION OF THE DOMAIN OF ATTRACTION OF AN EXTREME VALUE DISTRIBUTION FROM A SET OF DATA

E. Castillo*, J. Galambos** and J.M. Sarabia*
*University of Cantabria (Santander-Spain) and **Temple University (Philadelphia-USA)

Abstract. Methods for determining from a set of data the type of the extreme value distribution that attracts the population distribution are developed. The methods are based on the fact that the cumulative distribution functions of the three classical types for the maximum, when drawn on a Gumbel probability paper, exhibit different curvatures. Both a quick "visual selection method" and a more accurate "fit of attraction test" are discussed. The test statistics are location and scale invariant. The asymptotic results leading to the tests are of independent interest. In order to make our proposed test applicable for small sample sizes, the distribution of our major test statistic is tabulated for Gumbel, uniform, exponential and Cauchy parents via Monte Carlo simulation. Finally, our methods are demonstrated by reevaluating a published set of data.

1. INTRODUCTION

Given a set of data for which the extremes have a tendency, we would like to determine the actual form of the (asymptotic) extreme value distribution to be applied. More precisely, assume that the data $X_1,X_2,...,X_n$ from a population with distribution function F(x) suggest that the maximum $Z_n=\max(X_1,X_2,...,X_n)$, when properly normalized, converges in distribution to one of the extreme value distributions, whose types are given by

$$H_{1,\gamma}(x) = \exp(-x^{-\gamma}), \quad x>0, \quad \text{(Frechet type)} \tag{1.1}$$

$$H_{2,\gamma}(x) = \exp[-(-x)^{\gamma}], \quad x<0, \quad \text{((reversed) Weibull)} \tag{1.2}$$

and

$$H_{3,0}(x) = \exp[-\exp(-x)], \quad \text{all } x, \quad \text{(Gumbel)} \tag{1.3}$$

where γ is some positive constant. The problem is to determine the actual type from among these three distribution functions that attracts F(x). Since (1.1)-(1.3) are not exact, but only limit distributions for (a linearly normalized) Z_n, customary goodness of fit tests are not appropriate.

Keywords : Domains of attraction, probability paper, test of hypothesis
AMS 1980 Subject classifications. Primary 62E20, Secondary 62G30.

There are a number of suggestions in the literature for the solution of this problem, some of which, however, are applicable only with very large sample size, while others ignore the asymptotic character of the model and identify F(x) with one of the limiting types (1.1)-(1.3). Our goals are therefore somewhat different from earlier investigations (see Pickands (1975), Hill (1975), Otten and Monfort (1978), Galambos (1980), Tiago de Oliveira (1981), Mason (1982), Du Mouchel (1983), Davis and Resnick (1984), Gomes (1984) and the relevant entries in this proceedings).

It should be noted that the total set of data should not be utilized in a proper and effective method of analyzing the behaviour of Z_n. Clearly, if $F(x)=H_{3,0}(x)$ for $x \geq A=-\log\log 2$, and $F(x)=G(x)$ otherwise, where G(x) is an arbitrary distribution function with $G(-\log\log 2)=1/2$, then F(x) is continuous, Z_n ultimately behaves as if the data came from $H_{3,0}(x)$, but below the median, that is, 50% of the data, all observations are irrelevant in regard to Z_n (by changing the value of A in this example, one can see that a general method should not use a positive percentage of the data). That is, if $X_{1:n} \leq X_{2:n} \leq \ldots \leq X_{n:n}$ $(=Z_n)$ are the order statistics of the X_j, $1 \leq j \leq n$, any general method should rely on $X_{n-k:n}$, $0 \leq k \leq M$, only, where $M/n \to 0$ as $n \to +\infty$. Fortunately, recent studies show (see Janssen (1988)) that such a set of upper order statistics contain all information concerning the limiting distribution of Z_n, when properly normalized.

It has been observed many times in the literature that, on a Gumbel probability paper, that is, when the scale of one coordinate axis is in $H_{3,0}$ units, $H_{3,0}(x)$ is a straight line, while $H_{1,\gamma}(x)$ is concave and $H_{2,\gamma}(x)$ is convex. We shall demonstrate that, in the upper tail, F(x) imitates the shape of that limiting distribution that attracts it. Hence, one can easily detect a wrong selection just by looking at the tails of the empirical distribution functions on a Gumbel probability paper. We shall carry this observation further, and develop a test of significance based on the curvature of the upper tail of the empirical distribution function. The methods will be demonstrated by reevaluating a set of published data.

As in the preceding paragraphs, we concentrate on the maximum Z_n. However, one can easily transform any statement of the present paper to the lower tail of F(x) and of the corresponding limiting distributions when the minimum is of interest.

2. THE PROPOSED METHOD

The method presented below has an appealing geometrical idea. We represent the three types of limiting distributions (1.1) through (1.3) by their von Mises form

$$H_c(x;\lambda,\delta) = \exp\{-[1+c(x-\lambda)/\delta]_+^{-1/c}\}\,, \qquad (2.1)$$

where $[y]_+=y$ if $y\geq 0$ and equals zero otherwise, and c is a parameter going through the real numbers including c=0 in which case (2.1) is to be interpreted as lim H_c as $c\to 0$. We thus have that H_c of (2.1) is of the type of $H_{1,\gamma}$, $H_{2,\gamma}$ or $H_{3,0}$ depending on whether c>0, c<0 or c=0.

Now we plot $H_c(x;\lambda,\delta)$ on Gumbel probability paper. That is, we use the variables

$$\eta = -\log(-\log p)\ ,\ \xi = x \tag{2.2}$$

where p is the value of the cumulative distribution function and x is the value of the random variable under study. This leads to the family of curves

$$\eta = \{\log[1+c(\xi-\lambda)/\delta]\}/c\ ,\quad 1+c(\xi-\lambda)/\delta \geq 0\ , \tag{2.3}$$

which becomes the straight line

$$\eta = (\xi-\lambda)/\delta \quad (c=0) \tag{2.4}$$

as $c\to 0$ (use Taylor´s expansion). In the remaining cases, we see from the shapes of the logarithmic curves (or differentiate twice) that, on Gumbel probability paper, the reversed Weibull distributions are convex and the Frechét types are concave curves. Now, in Theorem 1 of Section 4 we prove that the upper tail of F(x) has the same shape as H_c if F(x) is in the domain of attraction of H_c(for the maximum). This leads to the following visual method of selecting the domain of attraction which F(x) belongs to : Plot the empirical distribution function on Gumbel probability paper. Check the upper tail (but never the whole empirical distribution function) and conclude that F is in the domain of attraction of $H_{1,\gamma}$, $H_{2,\gamma}$ or $H_{3,0}$ depending on whether the inspected upper tail is a line, concave, or convex, respectively (see the last section for an actual application).

The visual method can be expanded into a test of significance. By the discussion that led to the visual method we have that, in order to test that F belongs to the domain of attraction of a particular H_c, we can test whether the upper tail of the empirical distribution function is convex, concave or a straight line. This, in turn, reduces to testing the magnitude of the relative slopes of two straight lines fitted to disjoint groups of upper extremal order statistics. The specific method we propose is as follows.

Let $n_1<n_2<n_3<n_4<n$ be integers such that $n_1/n\to 1$ as $n\to\infty$ Otherwise, each n_j may depend on n or may be predetermined. Let S_{12} and S_{34} be the slopes of the least squares straight lines fitted, on Gumbel probability paper, to the empirical distribution function $F_n(x)$ with $n_1\leq nx\leq n_2$ and $n_3\leq nx\leq n_4$ respectively. That is,

$$S_{ij} = \frac{m(ij)\,\theta_{11}(ij) - \theta_{10}(ij)\ \theta_{01}(ij)}{m(ij)\,\theta_{20}(ij) - \theta_{10}^{2}(ij)}\ , \tag{2.5}$$

where $m(ij)=n_j-n_i+1$,

$$\theta_{10}(ij) = \sum_{k=n_i}^{n_j} (-\log\{-\log[(k-0.5)/n]\})\ , \tag{2.6}$$

$$\theta_{01}(ij) = \sum_{k=n_i}^{n_j} X_{k:n} \tag{2.7}$$

$$\theta_{11}(ij) = \sum_{k=n_i}^{n_j} (-X_{k:n} \log\{-\log[(k-0.5)/n]\})\ , \tag{2.8}$$

and

$$\theta_{20}(ij) = \sum_{k=n_i}^{n_j} (-\log\{-\log[(k-0.5)/n]\})^{2}\ , \tag{2.9}$$

An important property of the least squares slopes S_{ij} is that they are linear combinations of order statistics with coefficients adding up to zero. Therefore, they are location invariant. Now, our test statistic for the curvature of the upper tail of $F_n(x)$ is the ratio

$$S = S_{12}\,/\,S_{34} \tag{2.10}$$

which is both location and scale invariant. We test the hypothesis that $F(x)$ belongs to the domain of attraction of $H_{3,0}(x)$ against the alternative that $F(x)$ belongs to the domain of attraction of $H_{i,\gamma}(x)$ (i=1 or 2). The critical value of the test is determined by Monte Carlo simulation for small values of n, while for moderately large or large values of n, by an asymptotic theorem (Theorem 2 of Section 4) for the distribution of S, assuming that the null hypothesis is true.

With apparently no reason other than the discussion in the introduction, we recommend to choose

$$n_1 = n - [2\sqrt{n}]\,,\ n_2 = n_3 = n - [2\sqrt{n}]/2\ ,\ n_4 = n\ , \tag{2.11}$$

where [y] signifies the integer part of y. With this, our basic requirement of relying on the right tail only ($n_1/n \rightarrow 1$) is clearly satisfied.

3. MONTE CARLO SIMULATIONS

Because the exact distribution of S of (2.10) is very complicated, we determined this distribution by Monte Carlo simulation for varying values of n and for some parent distributions. The tables are enclosed for parent distributions whose tails were assumed to be Gumbel, uniform, Cauchy and exponential, respectively. The table for S with Gumbel tail can be utilized to select the critical values for the test described in the previous section. Although one cannot expect a good decision on the extremes from small samples, sample sizes as small as 10 are included in the tables. The tables are made on the base of 5000 replications. In order to check the goodness of the test, we tabulated in Table 5 the critical values and their associated percentages of erroneous decisions (power of the test) when testing Gumbel against Weibull type or Gumbel against Frechét type domains of attraction for different sizes of the test and parent distributions with uniform, Cauchy and exponential tail. One can see that in samples of size 200, the decisions are quite good. With further increase in the sample size, the test should improve (it seems that, with $n \geq 500$, the test becomes very reliable). Note also in the exponential column of Table 5 that the power of the test changes very little with sample size. This must be due to the relative closeness of the tails of the exponential and Gumbel distributions.

Note that an erroneous decision of rejecting a Gumbel type domain of attraction has no serious consequences on the estimation of some parameters. This is why sizes of the test as high as 0.2 or even 0.5 are included in Table 5.

The tables mentioned in this section are printed after the list of references.

4. LIMIT THEOREMS

In Section 2 we stated that if F(x) is in the domain of attraction of H_c then the upper tail of F(x) has the same shape as H_c does. By this we mean that if we fit the data with straight lines over nonoverlapping intervals in the upper tail, the ratio of the slopes of such lines will not depend on F as $n \rightarrow \infty$ but only on H_c whose domain of attraction F belongs to. Since (least squares) straight lines fitted to data are linear combinations of order statistics with coefficients adding up to zero, and since $X_{n-k:n}$, k fixed, is a good estimator of a quantile of $F^n(x)$ in the sense that $F^n(X_{n-k:n}) \rightarrow p = p(k)$ as $n \rightarrow \infty$ if F is in the domain of attraction of some H_c (see chapter 2 in Galambos (1987)), the following result contains our claim on the "approximate" shape of the tail of F.

THEOREM 1.- Let the real numbers d_j and c_j satisfy

$$\sum_{j=1}^{k} d_j = \sum_{j=1}^{k} c_j = 0 .$$

Then, if F(x) is in the domain of attraction of H_c (for the maximum), for real numbers $0 < p_j, q_j < 1$, $1 \le j \le k$, k fixed,

$$\frac{\sum_{j=1}^{k} d_j F^{-1}(p_j^{1/n})}{\sum_{j=1}^{k} c_j F^{-1}(q_j^{1/n})} \to \frac{\sum_{j=1}^{k} d_j H_c^{-1}(p_j)}{\sum_{j=1}^{k} c_j H_c^{-1}(q_j)} \quad \text{as } n \to \infty.$$

Proof. By assumption, there are constants a_n and $b_n > 0$ such that, as $n \to \infty$,

$$F^n(a_n + b_n H_c^{-1}(p_j)) \to H_c(H_c^{-1}(p_j)) = p_j . \tag{3. 1}$$

Also, if a_n^* and $b_n^* > 0$ are some constants such that

$$F^{-1}(p_j^{1/n}) = a_n^* + b_n^* H_c^{-1}(p_j) , \tag{3. 2}$$

then

$$F^n(a_n^* + b_n^* H_c^{-1}(p_j)) = F^n(F^{-1}(p_j^{1/n})) = p_j , \tag{3. 3}$$

and thus the limit of (3.3) is p_j as $n \to \infty$. We thus have from (3.2) and (3.3) (see Lemmas 2.2.2 and 2.2.3 in Galambos (1987)) that, as $n \to \infty$,

$$\frac{a_n - a_n^*}{b_n} \to 0 \quad \text{and} \quad \frac{b_n^*}{b_n} \to 1. \tag{3. 4}$$

Note that $a_n^* = a_n^*(j)$ and $b_n^* = b_n^*(j)$, and we can determine such values by considering (3.2) for two different values of p_j , which system then has a unique solution. Hence, if we write

$$b_n^{-1} \sum_{j=1}^{k} d_j F^{-1}(p_j^{1/n}) = \sum_{j=1}^{k} d_j \left[\frac{F^{-1}(p_j^{1/n}) - a_n^*}{b_n^*} \; \frac{b_n^*}{b_n} + \frac{a_n^* - a_n}{b_n} \right],$$

and a similar expression for the sum of $c_j F^{-1}(q_j^{1/n})$, in their ratio the factor b_n^{-1} cancels out, and we get the claimed limit theorem by (the exact form in) (3.2) and the limits in (3.4).

Note that the assumption that the sums of the coefficients d_i and c_i equal zero is utilized when the translation of the terms $F^{-1}(p_j^{1/n})$ by a_n is made in the last displayed formula.

If n_1 in Section 2 is chosen as n-M with large but predetermined value of M, Theorem 1 is applicable to find the limit of S of (2.10) even though the coefficients of $X_{k:n}$ in (2.5) are not constant. Since the variable k in formulas (2.6) through (2.9) has the form n-j, $1 \le j \le M$, Taylor expansions and the simplification by a power of n in the ratio (2.10) leads to a form considered in Theorem 1. However, for our test, when n is large, the actual asymptotic distribution of S can be determined under our null hypothesis, that is, that F is in the domain of attraction of $H_{3,0}$. We first note that if $\Sigma a_j=0$, then $\Sigma a_j X_{j:n}$ can be written as $\Sigma c_j(X_{j:n}-X_{j-1:n})$, where the coefficients c_j are now unrestricted. Hence, the statistics S_{ij} of (2.5) are linear combinations of the differences $D_s=X_{s:n}-X_{s-1:n}$, which are asymptotically independent exponential variables under the quoted hypothesis (Weissman (1978)). Therefore, once again assuming that n_1=n-M, M fixed, in Section 2, the asymptotic distribution of S of (2.10) is contained in the theorem below. Let us remark, that the assumption of n_1=n-M with bounded M does not make much restriction in applications, since, for most available set of data, M fixed but in the hundreds or M being the multiple of $\sqrt[3]{n}$ does not make a difference.

THEOREM 2. Let $U_1,U_2,\ldots,U_d$, $V_1,V_2,\ldots,V_{d'}$ be independent exponential variables. Then, for fixed d and d′ and for real numbers c_i and g_i, the density function of

$$Z=\frac{c_1U_1+c_2U_2+\ldots+c_dU_d}{g_1V_1+g_2V_2+\ldots+g_{d'}V_{d'}}$$

is a linear combination of functions of the form $(A_s+B_sz)^{-2}$ where A_s and B_s are constants, i.e.

$$f_Z(z)=\begin{cases}\displaystyle\sum_{\{c_i>0,g_j>0\}\cup\{c_i<0,g_j<0\}} \alpha_{1i}\alpha_{2j}\beta_{1i}\beta_{2j}(\beta_{2j}+\beta_{1i}z)^{-2} & \text{if } z>0\\[2ex] \displaystyle\sum_{\{c_i>0,g_j<0\}\cup\{c_i<0,g_j>0\}} -\alpha_{1i}\alpha_{2j}\beta_{1i}\beta_{2j}(\beta_{2j}+\beta_{1i}z)^{-2} & \text{if } z<0\end{cases}$$

where

$$\alpha_{1i}=\prod_{j\neq i}\beta_{1j}/(\beta_{1j}-\beta_{1i}) \;;\; \beta_{1i}=1/c_iE[U_i] \qquad \text{for } 1\le i\le d;$$

$$\alpha_{2j}=\prod_{k\neq j}\beta_{2k}/(\beta_{2k}-\beta_{2j}) \;;\; \beta_{2j}=1/g_jE[V_j] \qquad \text{for } 1\le j\le d';$$

Proof. Let $T_1=c_1U_1+c_2U_2+\ldots+c_dU_d$. Then the characteristic function of T_1 is the product of $e_j/(e_j-it)$, $1\le j\le d$, where $e_j=1/c_jE(U_j)$. Now, utilizing that, for all t,

$$\sum_{s=1}^{d} \prod_{j \neq s} \frac{e_j - it}{e_j - e_s} = 1,$$

the previous product becomes the sum

$$\sum_{s=1}^{d} \left(\prod_{j \neq s} \frac{e_j}{e_j - e_s} \right) \frac{e_s}{e_s - it}$$

from which the inversion formula for characteristic functions yields that T is a mixture of exponentials on the negative and positive real axes (depending on whether e_S is negative or positive). Substituting this into the well known integral formula of the density of a ratio of two independent random variables, we get the claimed closed form of density.

5. AN APPLICATION

We applied both the visual method and the proposed test of significance to the data, analyzed in Gumbel and Goldstein (1964), on the Ocmulgee river (Macon). The figure on the next page shows that the upper tail cannot be considered a straight line (the straight line in the figure is the one fitted originally to the whole set of data that we outright oppose). The value of S for this data is 4.00, which confirms the visual rejection of a Gumbel tail (at significance level 0.03 - see Table 1).

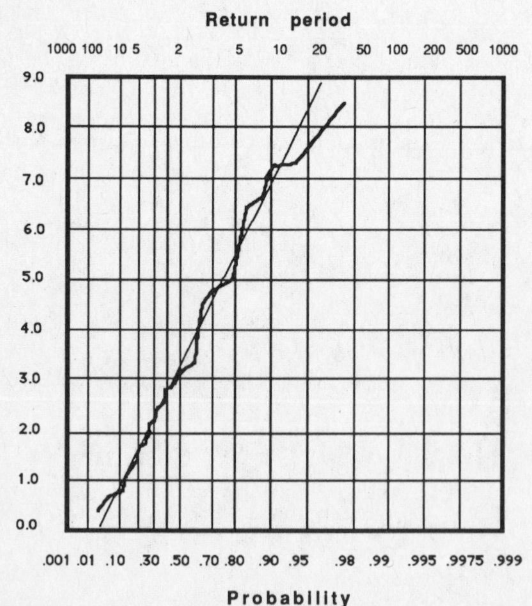

REFERENCES

Davis, R. and Resnick, S. (1984). Tail estimates motivated by extreme value theory. Ann. Statist. 12, 1467-1487.

Du-Mouchel,W.H. (1983). Estimating the stable index a in order to measure tail thickness : A critique. Ann. Statist. 11, 1019-1031.

Galambos, J. (1980). A statistical test for extreme value distributions. Colloquia Math. Soc. Janos Bolyai, Budapest, 221-229.

Galambos, J. (1987). The asymptotic theory of extreme order statistics. 2nd ed., Krieger, Melbourne, Florida.

Gomes, M.I. (1984). Extreme value theory - statistical choice. Colloquia Math. Soc. Janos Bolyai (Debrecen), Vol. 45, 195-210.

Gumbel, E. J. and Goldstein, N. (1964). Analysis of empirical bivariate extremal distributions. JASA 59, 794-816.

Janssen, A. (1988). The role of extreme order statistics for exponential families. In this proceedings.

Hill, B.M. (1975). A simple general approach to inference about the tail of a distribution. Ann. Statist. 3, 1163-1174.

Mason, D. (1982). Laws of large numbers for sums of extreme values. Ann. Probab. 10, 754-764

Otten, A. and Montfort, M.A.J. van (1978). The power of tests on the type of distributions of extremes. J. Hydrology 37, 195-199.

Pickands,J. (III) (1975). Statistical inference using extreme order statistics. Ann. Statist. 3, 119-131.

Tiago de Oliveira, J. (1981). Statistical choice of univariate extreme models. Stat. Distribution in Sci. Work 6, 367-387.

Weissman, I. (1978). Estimation of parameters and large quantiles based on the k largest observations. JASA 73, 812-815.

CDF	SAMPLE SIZE 10	20	40	60	80	100	200
0.01	0.120	0.159	0.216	0.265	0.285	0.297	0.360
0.02	0.155	0.209	0.264	0.303	0.336	0.345	0.415
0.05	0.238	0.294	0.362	0.396	0.418	0.433	0.506
0.10	0.340	0.387	0.465	0.485	0.518	0.533	0.592
0.20	0.497	0.543	0.622	0.651	0.658	0.678	0.713
0.30	0.694	0.724	0.768	0.783	0.796	0.805	0.817
0.50	1.135	1.133	1.083	1.085	1.073	1.055	1.046
0.70	1.903	1.782	1.558	1.485	1.481	1.405	1.345
0.80	2.643	2.302	1.955	1.836	1.758	1.657	1.530
0.90	4.333	3.401	2.662	2.427	2.264	2.135	1.895
0.95	6.477	4.770	3.392	3.090	2.877	2.588	2.250
0.98	11.211	6.739	4.673	4.045	3.720	3.153	2.732
0.99	15.723	9.363	5.814	4.990	4.673	3.613	2.969

Table 1.- Simulated cdf of S (Gumbel parent)

CDF	SAMPLE SIZE 10	20	40	60	80	100	200
0.01	0.252	0.469	0.801	1.063	1.219	1.340	1.766
0.02	0.368	0.588	1.008	1.235	1.370	1.486	1.979
0.05	0.554	0.873	1.319	1.613	1.756	1.857	2.343
0.10	0.809	1.176	1.650	2.005	2.172	2.283	2.720
0.20	1.245	1.686	2.210	2.548	2.720	2.830	3.247
0.30	1.691	2.131	2.680	3.086	3.234	3.324	3.715
0.50	2.778	3.281	3.765	4.098	4.288	4.288	4.655
0.70	4.587	4.854	5.319	5.605	5.708	5.593	5.734
0.80	6.361	6.614	6.545	6.720	6.720	6.510	6.614
0.90	10.42	9.881	8.711	8.803	8.651	8.197	7.911
0.95	15.62	13.44	11.11	11.06	10.82	9.843	9.191
0.98	26.32	19.08	14.97	14.12	13.15	11.73	10.96
0.99	37.88	26.32	17.61	16.78	15.62	13.66	12.44

Table 2.- Simulated cdf of S (Uniform parent)

CDF	SAMPLE SIZE 10	20	40	60	80	100	200
0.01	0.007	0.004	0.003	0.003	0.003	0.003	0.003
0.02	0.012	0.008	0.007	0.006	0.006	0.005	0.004
0.05	0.030	0.022	0.017	0.014	0.014	0.014	0.013
0.10	0.064	0.045	0.033	0.029	0.029	0.027	0.022
0.20	0.140	0.101	0.075	0.064	0.060	0.054	0.045
0.30	0.236	0.161	0.119	0.102	0.094	0.082	0.072
0.50	0.520	0.356	0.239	0.201	0.178	0.162	0.130
0.70	1.056	0.664	0.440	0.360	0.331	0.286	0.222
0.80	1.586	1.015	0.618	0.511	0.454	0.392	0.296
0.90	2.769	1.618	0.992	0.786	0.709	0.582	0.443
0.95	4.554	2.618	1.401	1.124	0.972	0.813	0.587
0.98	7.962	3.834	2.115	1.666	1.368	1.128	0.793
0.99	11.521	6.173	2.612	2.140	1.890	1.361	1.013

Table 3.- Simulated cdf of S (Cauchy parent)

CDF	SAMPLE SIZE 10	20	40	60	80	100	200
0.01	0.091	0.138	0.196	0.239	0.262	0.278	0.344
0.02	0.117	0.178	0.238	0.272	0.309	0.322	0.396
0.05	0.181	0.250	0.322	0.361	0.388	0.402	0.480
0.10	0.263	0.340	0.418	0.448	0.481	0.499	0.564
0.20	0.401	0.468	0.559	0.600	0.616	0.634	0.683
0.30	0.570	0.637	0.697	0.724	0.748	0.757	0.780
0.50	0.939	1.014	0.996	1.009	1.007	0.994	1.005
0.70	1.608	1.607	1.434	1.384	1.397	1.329	1.293
0.80	2.246	2.061	1.812	1.711	1.663	1.560	1.472
0.90	3.709	3.094	2.465	2.294	2.146	2.036	1.833
0.95	5.631	4.409	3.161	2.907	2.747	2.473	2.178
0.98	9.843	6.297	4.378	3.870	3.546	2.994	2.623
0.99	14.205	8.803	5.342	4.726	4.480	3.458	2.874

Table 4.- Simulated cdf of S (Exponential parent)

SAMPLE SIZE	SIZES OF TEST	CRITICAL VALUES WEIBULL	FRECHET	PERCENTAGES OF WRONG DECISIONS UNIFORM	CAUCHY	EXPONENTIAL
	0.05	6.477	0.238	80.5	70.0	4.0
10	0.10	4.333	0.340	68.0	61.5	8.0
	0.20	2.643	0.497	48.0	51.0	16.0
	0.50	1.135	1.135	17.5	28.5	43.5
	0.05	4.770	0.294	69.0	55.0	4.0
20	0.10	3.401	0.387	52.0	48.0	8.5
	0.20	2.302	0.543	33.0	36.0	16.5
	0.50	1.133	1.133	9.0	17.0	44.0
	0.05	3.392	0.362	44.0	36.5	4.0
40	0.10	2.662	0.465	29.5	28.0	8.0
	0.20	1.955	0.622	14.5	20.0	17.0
	0.50	1.083	1.083	2.5	8.5	45.0
	0.05	3.090	0.396	30.0	27.5	4.5
60	0.10	2.427	0.485	17.5	21.5	8.5
	0.20	1.836	0.651	7.5	14.5	17.0
	0.50	1.085	1.085	1.0	5.5	45.5
	0.05	2.588	0.433	15.5	17.0	4.5
100	0.10	2.135	0.533	8.0	12.0	8.5
	0.20	1.657	0.678	3.0	7.5	17.5
	0.50	1.055	1.055	0.4	2.5	46.0
	0.05	2.250	0.506	4.0	7.5	4.5
200	0.10	1.895	0.592	1.5	5.0	8.5
	0.20	1.530	0.713	0.5	3.0	18.0
	0.50	1.046	1.046	0.04	1.0	46.5

Table 5.- Percentages of wrong decisions for different parents and sizes of the test.

COMPARISON OF EXTREMAL MODELS THROUGH STATISTICAL CHOICE IN MULTIDIMENSIONAL BACKGROUNDS

M. Ivette Gomes

DEIOC (FCUL) and CEA (INIC), University of Lisbon

Abstract. The *multidimensional GEV model* (which generalizes the classical *Gumbel model* in Statistical Extremes) is compared with the *multivariate GEV model* introduced by Weissman, and similar in spirit to the Peaks Over Threshold approach. Assessment of usefulness of these models is here made through the comparison of power of a *Modified Locally Most Powerful* statistical choice test statistic in a multidimensional background.

1. RELEVANT MODELS IN STATISTICAL EXTREMES.

For sequences $\{X_j\}_{j\geq 1}$ of independent, identically distributed (i.i.d.) random variables (r.v.'s) the non-degenerate limiting structure, whenever it exists, of the normalized top i order statistics (o.s.), i a fixed integer, is well-known and characterized by the joint probability density function (p.d.f.)

$$h(z_1,...,z_i) = g(z_i) \prod_{j=1}^{i-1} \{g(z_j)/G(z_j)\} , z_1>...>z_i \tag{1.1}$$

where $G(z)=G_\theta(z)$ is in the class **S** of max-stable distribution functions (d.f.'s), often called *G*eneralized *E*xtreme *V*alue (*GEV*) d.f.'s, i.e.

$$G_\theta(z) = \begin{cases} \exp(-(1-\theta z)^{1/\theta}),\ 1-\theta z>0,\ z\in \mathbb{R} & \text{if } \theta\neq 0 \\ \exp(-\exp(-z)),\ z\in \mathbb{R} & \text{if } \theta=0, \end{cases} \tag{1.2}$$

$g_\theta(z)=\partial G_\theta(z)/\partial z$.

If we drop the identical distributional hypothesis — a more common set-up in applications —, and deal with sequences $\{Y_j\}_{j\geq 1}$ of r.v.'s whose associated sequence of partial maxima $\{M_j=\max(Y_1,...,Y_j)\}_{j\geq 1}$, suitably normalized, converges weakly, as $j\to\infty$, to a r.v. in Mejzler's class $\mathbf{M_1}$ [Mejzler (1956)], the limiting structure of the top i o.s. is still a multivariate extremal vector with p.d.f. given by (1.1), but where $G\in \mathbf{M_1}$ [Weissman (1975), Athayde and Gomes (1987)]. In this set-up

Keywords and phrases: Statistical Extremes; Multidimensional Extremal Models; Statistical Choice.

G(z) is such that (cf. Galambos (1978), p. 181) either (i) -log $\{G(z)\}$ convex or (ii) $z_0 = \sup\{z : G(z) < 1\}$ finite and -log $\{G(z_0 - \exp(-z))\}$ convex. If we work instead with refinements $\mathbf{M_r}$, r>1, of Mejzler's class $\mathbf{M_1}$, or with $\mathbf{M_\infty} = \bigcap_{r\geq 1} \mathbf{M_r}$ [Graça Martins and Pestana (1985)],characterized like $\mathbf{M_1}$ by (i) and (ii), with convexity replaced by complete monotonicity, analogous results are obtained. Notice that $\mathbf{M_\infty}=\{G(z)=\exp(-K(z)),\ z\in\mathbb{R}$, or $G(z)=\exp(-K(-\log(z_0-z))),\ z\leq z_0,\ K(.)$ a completely monotone function$\}$ is the smallest class containing the class of *max-stable* d.f.'s (1.2) that is closed under pointwise products and limits. This class seems thus to provide a very general framework for the study of sample maxima, justifying several models put forward by statistical users (Gomes and Pestana (1985)), the same happening to its multivariate generalizations.This assertion is strengthened by the fact that these results may be generalized to schemes of weak dependence (Hüsler (1986)).

We thus have a strong probabilistic background for the introduction of a *multivariate extremal* $\mathbf{M_1}$ *model* to analyze the set of the largest observations available, in order to infer tail properties of underlying population. This model is obviously more general and more realistic than the *multivariate GEV model,* introduced first, in a slightly different context, by Pickands (1975) and worked out by several authors [Weissman (1978); Smith (1984)]. Mainly from climatological data, where the i.d. hypothesis fails , an *extremal M_1 model*, or at least an *extremal* $\mathbf{M_\infty}$ *model*, has to be called for. Often we put ourselves in a more general situation, i.e., we consider a *multidimensional extremal model*, assuming to have a random sample $\underset{\sim}{X}_k=(X_{1,k}>X_{2,k}>...>X_{i,k})'$, $i\geq 1$, $1\leq k\leq n$, of independent *i-variate extremal vectors,* with possibly unknown location and scale parameters $\lambda\in\mathbb{R}$ and $\delta\in\mathbb{R}^+$ respectively, i.e. $\underset{\sim}{Z}_k=(\underset{\sim}{X}_k-\lambda\mathbf{1}_i)/\delta$ has a p.d.f. given by (1.1), $G\in\mathbf{M_\infty}$, $\mathbf{1}_i$ denoting, as usual, a column vector of size i with unity elements. Notice that these models being generalizations of both *multivariate extremal models* (n=1) and *univariate extremal* (or *Gumbel) models* (i=1) provide a suitable framework for comparison of these two most commonly used models in Statistical Extremes, to infer about an underlying unknown tail.

We have thus, under this context, two lines to follow in statistical analysis of extremes: either place ourselves in a general $\mathbf{M_\infty}$ context , and develop non-parametric or semi-parametric procedures for log-completely monotone d.f.'s, or to choose particular parametric $\mathbf{M_\infty}$ models, useful in applications — for instance a Generalized Pareto distribution or a Todorovich distribution — $\exp(-(1-F_\theta(x)))$, F_θ a Gamma(θ) d.f. — instead of a GEV distribution have proved to be useful for the analysis of climatological data (Gomes and Pestana (1986)) —, either univariate, multivariate or multidimensional.

This is a first paper dealing with discrimination among *multidimensional extremal models* — often referred to as *statistical choice* in a *multidimensional GEV model* — with G_0 playing a central and pro-eminent role. The statistical choice of extremes has been a stimulating statistical problem , mainly in the areas of structural reliability and of hydrology — cf. van Montfort and Gomes (1985) and Gomes and van Montfort (1986) —, and up to now has been dealt only in the *univariate GEV model* (Gumbel model) (van Montfort and Gomes (1985) and references therein) and in the *multivariate GEV model* (Gomes and Alpuim (1986), and references therein).

In section 2 of this paper we derive distributional properties of a Locally Most Powerful (LMP) test statistic to test the hypothesis H_0:θ=0 versus suitable one-sided or two-sided alternatives in a standard

(λ=0, δ=1) *multidimensional GEV(θ) model* . Under the same context, but when the parameters λ, δ and θ are all unknown, a suitable Modified Locally Most Powerful (MLMP) test statistic is studied in section 3. The ultimate aim of this paper is the comparison of the *multivariate GEV model* and the *univariate GEV model* through the power function of the MLMP test statistic. This comparison is dealt with in section 4 of this paper. For another comparison of these two models see Gomes (1985). The conclusions drawn here are identical to the ones drawn when we consider several other suitable parametric d.f.'s in $\mathbf{M}_\infty$, like the Generalized Pareto or the Generalized Gamma, or other possible statistical choice test statistics, like Gumbel test statistic. Such a statistic, balancing properly upper and lower tail of population, and being a generalization to multidimensional models of the Gumbel statistics considered before in univariate (Tiago de Oliveira and Gomes (1984)) and multivariate backgrounds (Gomes and Alpuim (1986)) will be presented in another paper.

2. DISTRIBUTIONAL PROPERTIES OF A LMP TEST STATISTIC UNDER A STANDARD i-DIMENSIONAL EXTREMAL MODEL.

Let us assume first that the parameters (λ,δ) are known, and take, without loss of generality λ=0, δ=1, i.e., we have direct access to $\underset{\sim}{Z}=(\underset{\sim}{Z}_1,\underset{\sim}{Z}_2,\ldots,\underset{\sim}{Z}_n)$. The log-likelihood of $\underset{\sim}{Z}$ is thus

$$\ln L(\theta;\underset{\sim}{Z}) = \sum_{k=1}^{n} \{-(1-\theta Z_{i,k})^{1/\theta} + (1/\theta-1) \sum_{j=1}^{i} \ln(1-\theta Z_{j,k})\}. \tag{2.1}$$

Consequentlly, the *Locally Most Powerful* (LMP) test statistic, to test H_0:θ=0, in a *multidimensional GEV(θ) model,* is asymptotically, as $n \to \infty$,

$$\mathbf{L}_{n,i} = \mathbf{L}_{n,i}(\underset{\sim}{Z}) = \left.\frac{\partial \ln L}{\partial \theta}\right|_{\theta=0}$$

i.e.,

$$\mathbf{L}_{n,i} = \mathbf{L}_{n,i}(\underset{\sim}{Z}) = \sum_{k=1}^{n} \{ Z_{i,k}^2 \exp(-Z_{i,k})/2 - \sum_{j=1}^{i} Z_{j,k}^2/2 + \sum_{j=1}^{i} Z_{j,k} \}, \tag{2.2}$$

both for one-sided and two-sided alternatives.

We shall now deal with second order properties of $\mathbf{L}_{n,i}$.

Under the validity of H_0:θ=0, $Y_{j,k}=Z_{j,k} - Z_{j+1,k}$, $1 \le j \le i-1$, $Y_{i,k}=Z_{i,k}$, $i \ge 1$, $1 \le k \le n$, are independent random variables, with p.d.f

$$\prod_{j=1}^{i-1} j\,\exp(-jy_j)\,\exp(-iy_i)\,\exp(-\exp(-y_i))\,/\,\Gamma(i),\quad y_1,y_2,\ldots,y_{i-1}>0,\ y_i\in\mathbb{R}$$

and we may write the distributional identity

$$\mathbf{L_{n,i}} = \sum_{k=1}^{n}\{\, Y_{i,k}^2\exp(-Y_{i,k})/2 - \sum_{j=1}^{i} j\,Y_{j,k}^2/2 - \sum_{j=1}^{i-1}\sum_{l=j+1}^{i} jY_{j,k}Y_{l,k} + \sum_{j=1}^{i} jY_{j,k}\,\}. \tag{2.3}$$

After a few manipulations, using either (2.2) or (2.3),we derive

$$\begin{aligned} &E[\mathbf{L_{n,i}} \mid \theta=0] = 0 \\ &Var[\mathbf{L_{n,i}} \mid \theta=0] = n\,\{i\,\Gamma^{(4)}(i) + \Gamma^{(3)}(i)+i\,\Gamma^{(2)}(i)\}\,/\,\Gamma(i) + 2\,(i-1)\,(1-\psi(i)) \end{aligned} \tag{2.4}$$

where $\Gamma^{(j)}(.)$ denotes, as usual, the j-th derivative of the Gamma function, and $\psi(.)$ the digamma function.

For a general θ, and since we are interested in values of θ close to $\theta=0$, we consider , with $Z_j=Z_{j,k},\ 1\le k\le n$, the relation

$$E_\theta[\phi(Z_1,Z_2,\ldots,Z_i)] = E_0[\phi((1-\exp(-\theta Z_1))/\theta,\ldots,(1-\exp(-\theta Z_i))/\theta)] \tag{2.5}$$

whenever such mean values do exist, and expand $\Phi(\theta) = \phi((1-\exp(-\theta Z_1))/\theta,\ldots,(1-\exp(-\theta Z_i))/\theta)$ in Taylor series around $\theta=0$. Noticing that, with $g(\theta) = (1-\exp(-\theta u))/\theta$, we have $g(0) = u$, $g'(0)= - u^2/2$ and $g''(0) = u^3/3$, we get

$$\begin{aligned} E_\theta(\phi) = E_0(\phi) - \theta \sum_{j=1}^{i} E_0(Z_j^2\,\partial\phi/\partial Z_j)/2 + \theta^2 \sum_{j=1}^{i} E_0(Z_j^3\partial\phi/\partial Z_j)/6 \\ + \theta^2 \sum_{j=1}^{i}\sum_{k=1}^{i} E_0(Z_j^2 Z_k^2\partial^2\phi/\partial Z_j\partial Z_k)/8 + o(\theta^2). \end{aligned} \tag{2.6}$$

We also have

$$\begin{aligned} E_\theta(\phi^2) = E_0(\phi^2) - \theta \sum_{j=1}^{i} E_0(\phi\, Z_j^2\,\partial\phi/\partial Z_j) + \theta^2 \sum_{j=1}^{i} E_0(\phi\, Z_j^3\partial\phi/\partial Z_j)/3 \\ + \theta^2\{\sum_{j=1}^{i}\sum_{k=1}^{i} E_0(\phi\, Z_j^2 Z_k^2\partial^2\phi/\partial Z_j\partial Z_k) + E_0((\sum_{j=1}^{i} Z_j^2\partial\phi/\partial Z_j)^2)\}/4 + o(\theta^2), \end{aligned} \tag{2.7}$$

and consequently,

$$\mathrm{Var}_\theta(\phi) = \mathrm{Var}_0(\phi) - \theta \sum_{j=1}^{i} \mathrm{Cov}(\phi, Z_j^2\, \partial\phi/\partial Z_j) + \theta^2 \sum_{j=1}^{i} \mathrm{Cov}(\phi, Z_j^3 \partial\phi/\partial Z_j)/3$$

$$+\theta^2\{ \sum_{j=1}^{i} \sum_{k=1}^{i} \mathrm{Cov}(\phi, Z_j^2 Z_k^2 \partial^2\phi/\partial Z_j \partial Z_k) + \mathrm{Var}(\sum_{j=1}^{i} Z_j^2 \partial\phi/\partial Z_j)\}/4 + o(\theta^2). \quad (2.8)$$

Since, in our particular situation $\phi(Z_1,...,Z_i) = Z_i^2 \exp(-Z_i)/2 - \sum_{j=1}^{i} Z_j^2/2 + \sum_{j=1}^{i} Z_j$, we have

$$\partial\phi/\partial Z_j = -Z_j + 1,\ 1 \le j \le i-1,\quad \partial\phi/\partial Z_i = Z_i \exp(-Z_i) - Z_i^2 \exp(-Z_i)/2 - Z_i + 1$$

$$\partial^2\phi/\partial Z_j^2 = -1,\ 1 \le j \le i-1,\quad \partial^2\phi/\partial Z_i^2 = (1 - 2Z_i + Z_i^2/2)\exp(-Z_i) - 1 \quad (2.9)$$

$$\partial^2\phi/\partial Z_j \partial Z_k = 0,\ 1 \le j \le i,\ k \ne j.$$

After a few manipulations, we get

$$E(\mathbf{L}_{n,i}/n \mid \theta) = \theta \{ \Gamma^{(4)}(i+1)/4 + \Gamma^{(2)}(i+1) - 2\Gamma^{(1)}(i+1) + 2i\Gamma(i)\} / \Gamma(i)$$

$$+\theta^2\{\Gamma^{(6)}(i+1)/16 + \Gamma^{(5)}(i+1)/3 + \Gamma^{(3)}(i+1) - 3\Gamma^{(2)}(i+1) + 6\Gamma^{(1)}(i+1) - 6i\Gamma(i)\} / \Gamma(i) + o(\theta^2), \quad (2.10)$$

the expression of Var $(\mathbf{L}_{n,i}/n \mid \theta)$ being extremely cumbersome to be presented analitically.

In table 2.1 we thus present the coefficients $a_1(i) = \mathrm{Var}(\mathbf{L}_{n,i}/n \mid \theta=0)$, $a_2(i)$, $b_1(i)$ and $b_2(i)$ in $\mu_\theta(i) = E(\mathbf{L}_{n,i}/n \mid \theta) = \theta a_1(i) + \theta^2 a_2(i) + o(\theta^2)$ and in $\sigma_\theta^2(i) = \mathrm{Var}(\mathbf{L}_{n,i}/n \mid \theta) = \mathrm{Var}(\mathbf{L}_{n,i}/n \mid \theta=0) + \theta b_1(i) + \theta^2 b_2(i) + o(\theta^2)$ respectively.

i	$a_1(i)$	$a_2(i)$	$b_1(i)$	$b_2(i)$
1	2.423606	-4.953622	-25.264701	268.205159
2	4.182242	-1.014961	-7.540957	373.130343
3	7.843915	8.940335	37.043109	670.224865
4	13.942236	27.787648	121.125272	1298.173506
5	22.807718	58.310636	256.836932	2413.607391

Table 2.1 — Coefficients $a_1(i)$, $a_2(i)$, $b_1(i)$, $b_2(i)$ of θ and θ^2 in $E(\mathbf{L}_{n,i} \mid \theta) = \theta\, a_1(i) + \theta^2 a_2(i) + o(\theta^2)$ and in $\mathrm{Var}(\mathbf{L}_{n,i} \mid \theta) = \mathrm{Var}(\mathbf{L}_{n,i} \mid \theta=0) + \theta b_1(i) + \theta^2 b_2(i) + o(\theta^2)$.

Asymptotically, we have obviously, for fixed i and $-i/4 < \theta < 1$, standard normality of $\sqrt{n}\,(\mathbf{L}_{n,i}/n - \mu_\theta(i))/\sigma_\theta(i)$, as $n \to \infty$, and asymptotic power of $\mathbf{L}_{n,i}$ is straighforwardly obtained.

3. DISTRIBUTIONAL PROPERTIES OF A MODIFIED LOCALLY MOST POWERFUL TEST STATISTIC IN AN i-DIMENSIONAL GEV(0) MODEL.

Let us assume now that we are working with a general *i-dimensional* $GEV(\theta)$ *model*, $\underset{\sim}{\mathbf{X}}=(\underset{\sim}{\mathbf{X}}_1,\ldots,\underset{\sim}{\mathbf{X}}_n)$, with unknown location $\lambda\in\mathbb{R}$ and scale $\delta\in\mathbb{R}^+$. We then consider, following Tiago de Oliveira (1982), and with $\mathbf{1}_{\mathbf{i,n}}$ denoting an ixn matrix with unity elements, the *Modified Locally Most Powerful* (MLMP) test statistic

$$\hat{\mathbf{L}}_{\mathbf{n,i}} = \mathbf{L}_{\mathbf{n,i}}\,((\underset{\sim}{\mathbf{X}} - \hat{\lambda}_0\mathbf{1}_{\mathbf{i,n}})/\hat{\delta}_0) \tag{3.1}$$

where $\hat{\lambda}_0=\hat{\lambda}_{n,i}{}^{(0)}$ and $\hat{\delta}_0=\hat{\delta}_{n,i}{}^{(0)}$ are the *Maximum Likelihood Estimates* (MLE) of (λ,δ) under $H_0{:}\theta=0$, i.e.

$$\begin{aligned}
\hat{\lambda}_0 &= \hat{\delta}_0 \ln\Big(ni \Big/ \sum_{k=1}^{n} \exp(-X_{i,k}/\hat{\delta}_0)\Big) \\
\hat{\delta}_0 &= \sum_{k=1}^{n}\sum_{j=1}^{i} X_{j,k}\,/\,(ni) - \sum_{k=1}^{n} X_{i,k}\exp(-X_{i,k}/\hat{\delta}_0) \Big/ \sum_{k=1}^{n}\exp(-X_{i,k}/\hat{\delta}_0),
\end{aligned} \tag{3.2}$$

which are asymptotically, as $n\to\infty$ and under H_0, binormal with mean value (λ,δ) and covariance matrix

$$\Sigma_i = \frac{\delta^2}{ni(1+\psi'(i+1))}\begin{bmatrix}\psi'(i+1)+\psi^2(i+1)+1 & \psi(i+1)\\ \psi(i+1) & 1\end{bmatrix} \tag{3.3}$$

(Gomes(1981).

We may write, with $\hat{\lambda}=\hat{\lambda}_0$, $\hat{\delta}=\hat{\delta}_0$,

$$\hat{\mathbf{L}}_{\mathbf{n,i}}/n = \mathbf{L}_{\mathbf{n,i}}/n + \{(\hat{\lambda}-\lambda)/n\}\,\partial\mathbf{L}_{\mathbf{n,i}}(\lambda,\delta)/\partial\lambda + \{(\hat{\delta}-\delta)/n\}\,\partial\mathbf{L}_{\mathbf{n,i}}(\lambda,\delta)/\partial\delta + R_n \tag{3.4}$$

Noticing that, with $\phi(\underset{\sim}{\mathbf{X}})=\phi(X_1,\ldots,X_i) = X_i{}^2\exp(-X_i)/2 - \sum_{j=1}^{i}(X_j{}^2/2 - X_j)$, we have

$$\begin{aligned}
&E\{\,\partial\,\phi((\underset{\sim}{\mathbf{X}}-\lambda\mathbf{1}_i)/\delta)/\partial\lambda\,\} = E_1/\delta,\ \ E_1 = \{i\,\Gamma^{(2)}(i)/2 + \Gamma^{(1)}(i)\}/\Gamma(i),\\
&E\{\,\partial\,\phi((\underset{\sim}{\mathbf{X}}-\lambda\mathbf{1}_i)/\delta)/\partial\delta\,\} = E_2/\delta,\ \ E_2 = \{-\,i\,\Gamma^{(3)}(i)/2 - 3\Gamma^{(2)}(i)/2 - i\,\Gamma^{(1)}(i) + (i-1)\Gamma(i)\}\,/\,\Gamma(i),
\end{aligned} \tag{3.5}$$

and from the fact that, under $H_0:\theta=0$, $(\hat{\lambda}-\lambda)\{\Sigma(\partial\phi((\underset{\sim}{\mathbf{X}}-\lambda\mathbf{1})/\delta)/\partial\lambda)/n-E_1/\delta\}=o_p(1/\sqrt{n})$, and $(\hat{\delta}-\delta)\{\Sigma(\partial\phi((\underset{\sim}{\mathbf{X}}-\lambda\mathbf{1})/\delta)/\partial\delta)/n - E_2/\delta\} = o_p(1/\sqrt{n})$, we have the representation

$$\hat{\mathbf{L}}_{\mathbf{n,i}}/n = \mathbf{L}_{\mathbf{n,i}}/n + E_1\{(\hat{\lambda}-\lambda))/\delta\} + E_2\{(\hat{\delta}-\delta)/\delta\} + o_p(1/\sqrt{n}). \tag{3.6}$$

But, with

$$A_i = -1,\ B_i = \psi(i+1),\ C_i = -(\psi'(i+1)+\psi^2(i+1)+1),\ D_i = i(1+\psi'(i+1)) = i(A_iC_i-B_i^2), \tag{3.7}$$

we have, under H_0, and asymptotically, as $n\to\infty$,

$$\begin{aligned}(\hat{\lambda}-\lambda)/\delta &= \delta\,[\,B_i\,\{\partial\log L/\partial\delta\}/n - C_i\,\{\partial\log L/\partial\lambda\}/n\,]\,/\,D_i + o_p(1/\sqrt{n})\\ (\hat{\delta}-\delta)/\delta &= \delta\,[\,B_i\,\{\partial\log L/\partial\lambda\}/n - A_i\,\{\partial\log L/\partial\delta\}/n\,]\,/\,D_i + o_p(1/\sqrt{n}).\end{aligned} \tag{3.8}$$

Consequently,

$$\begin{aligned}\hat{\mathbf{L}}_{\mathbf{n,i}}/n = \mathbf{L}_{\mathbf{n,i}}/n &+ \delta\,E_1[\,B_i\,\{\partial\log L/\partial\delta\}/n - C_i\,\{\partial\log L/\partial\lambda\}/n\,]\,/\,D_i\\ &+ \delta\,E_2\,[\,B_i\,\{\partial\log L/\partial\lambda\}/n - A_i\,\{\partial\log L/\partial\delta\}/n\,]\,/\,D_i + o_p(1/\sqrt{n}),\end{aligned} \tag{3.9}$$

which finally leads to the representation

$$\begin{aligned}\hat{\mathbf{L}}_{\mathbf{n,i}}/n &= \sum_{k=1}^{n}\{\,Z_{i,k}^2\exp(-Z_{i,k})/2 - \sum_{j=1}^{i}(\,Z_{j,k}^2/2 + Z_{j,k}) - \alpha_i\exp(-Z_{i,k})\\ &\qquad + \beta_i\sum_{j=1}^{i}Z_{j,k} - \beta_i Z_{i,k}\exp(-Z_{i,k}) + i(\alpha_i-\beta_i)\,\}\,/\,n\\ &= \sum_{k=1}^{n}\{\,\phi_k(\underset{\sim}{\mathbf{Z}}) + \rho_k(\underset{\sim}{\mathbf{Z}})\,\}\,/n\end{aligned} \tag{3.10}$$

where

$$\begin{aligned}\phi_k(\underset{\sim}{\mathbf{Z}}) &= Z_{i,k}^2\exp(-Z_{i,k})/2 - \sum_{j=1}^{i}(Z_{j,k}^2/2 - Z_{j,k})\\ \rho_k(\underset{\sim}{\mathbf{Z}}) &= \beta_i\sum_{j=1}^{i}Z_{j,k} - \alpha_i\exp(-Z_{i,k}) - \beta_i Z_{i,k}\exp(-Z_{i,k}) + i(\alpha_i - \beta_i),\end{aligned} \tag{3.11}$$

α_i and β_i given by

$$\alpha_i = (B_i E_2 - C_i E_1)/D_i, \ \beta_i = (B_i E_1 - A_i E_2)/D_i, \tag{3.12}$$

E_1, E_2 given by (3.5), and A_i, B_i, C_i, D_i given by (3.7).

It thus follows that, asymptotically, as $n \to \infty$,

$$E(\hat{\mathbf{L}}_{\mathbf{n,i}} \mid \theta=0) = 0$$

$$\mathrm{Var}(\hat{\mathbf{L}}_{\mathbf{n,i}} \mid \theta=0) = n\,\hat{\sigma}_i^2 \tag{3.13}$$

$$= n\,\{\, (i\,\Gamma^{(4)}(i)/4 + (1+i\beta_i)\,\Gamma^{(3)}(i) + (i+3\beta_i-i\alpha_i+i\beta_i^2)\Gamma^{(2)}(i)\,\} / \Gamma(i)$$

$$+ 2\,(\, i\beta_i - \alpha_i - (i-1) + \beta_i^2 - i\alpha_i\beta_i\,)\,\psi(i) + i\beta_i^2 + i\alpha_i^2 - 2\alpha_i\beta_i + 2(i-1)(1-\beta_i)\},$$

α_i and β_i given by (3.12).

In table 3.1, we present for i=1(1)5, the asymptotic variance of the MLMP test statistic $\hat{\mathbf{L}}_{\mathbf{n,i}}/n$, under $H_0{:}\theta=0$.

i=1	i=2	i=3	i=4	i=5
2.097970	3.342076	4.478403	5.567808	6.631130

Table 3.1 — Asymptotic variance of $\hat{\mathbf{L}}_{\mathbf{n,i}}/n$ in an i-dimensional GEV(0) model.

It thus follows that, for fixed i and $-i/4<\theta<1$,

$$\hat{\mathbf{L}}_{\mathbf{n,i}} / \{\, \hat{\sigma}_i \sqrt{n}\,\} \text{ is asymptotically, as } n\to\infty, \text{ a standard normal r.v.} \tag{3.14}$$

Just as a remark, we notice the following: if we consider n=1, $\hat{\mathbf{L}}_{\mathbf{1,i}}$ is distributionally equivalent to $i - (i^2/2) \sum_{j=1}^{i-1} \{\, V_j / \sum_{k=1}^{i-1} V_k \,\}^2$, where V_j are, under a general GEV(θ) model, i.i.d. Generalized Pareto with parameter θ. We thus have asymptotically, as $i\to\infty$ (Gomes and Alpuim(1986)), and for a general θ, standard normality of $\{\hat{\mathbf{L}}_{\mathbf{1,i}} - i\,\mu(\theta)\,\} / (\sqrt{i}\,\sigma(\theta))$,

$\mu(\theta) = \theta / (1+2\theta)$,
$\sigma(\theta) = (1+\theta)\sqrt{(6\theta^2+\theta+1)/((1+2\theta)(1+3\theta)(1+4\theta))} / (1+2\theta)$,

i.e., under $H_0{:}\theta=0$, $\hat{L}_{1,i}/\sqrt{i}$ is asymptotically, as $i\to\infty$, a standard normal r.v. We thus have that $\hat{\sigma}_i / \sqrt{i} \to 1$, as $i\to\infty$. Generally, for a fixed i, and for $-i/4<\theta<1$, we still have asymptotic normality of $\hat{L}_{n,i}$, as $n\to\infty$, with cumbersome attraction constants. We may also fix n and increase i, or increase both n and i to infinity, obtaining asymptotic normality. However, the rate of convergence is slow.

4. COMPARISON OF EXTREMAL MODELS THROUGH STATISTICAL CHOICE.

We have simulated, for different values of n and i, and alternative $H_1{:}\theta\neq0$, the power function of the test statistic $\hat{L}_{n,i}$ given in section 3. For two-sided alternatives, we present in table 4.1, for values of n=20,40, i=1,3,5 , such power function. Blank entries correspond to power greater than .995.

	n=20			n=40		
θ	i=1	i=3	i=5	i=1	i=3	i=5
-1.0	.88					
-.9	.83					
-.8	.79			.99		
-.7	.74	.99		.98		
-.6	.68	.98	.99	.96		
-.5	.62	.93	.98	.88		
-.4	.44	.85	.96	.79	.98	
-.3	.28	.64	.82	.57	.94	.99
-.2	.14	.35	.49	.28	.66	.84
-.1	.07	.12	.14	.11	.19	.29
.0	.05	.05	.05	.05	.05	.05
.1	.10	.18	.23	.16	.28	.39
.2	.22	.43	.51	.38	.68	.83
.3	.36	.66	.82	.61	.91	.98
.4	.50	.82	.93	.78	.98	
.5	.62	.92	.98	.88		
.6	.71	.97	.99	.94		
.7	.78	.98		.98		
.8	.84	.99		.99		
.9	.90					
1.0	.94					

Table 4.1.—Power function of $\hat{L}_{n,i}$, to test $H_0{:}\theta=0$ vs $H_1{:}\theta\neq0$, at a significance level $\alpha=.05$, in an *i-dimensional GEV model.*

For comparison we present in table 4.2 the values of power function of $\hat{L}_{1,k}$, for k=20,40,60,100,120,200 (the possible values of n times i in table 4.1).

θ	k=20	k=40	k=60	k=100	k=120	k=200
-1.0	.72	.99				
-.9	.66	.97				
-.8	.57	.95				
-.7	.48	.89	.98			
-.6	.38	.80	.95			
-.5	.27	.66	.87			
-.4	.18	.47	.72	.93	.98	
-.3	.13	.27	.48	.74	.84	.98
-.2	.07	.13	.22	.38	.50	.85
-.1	.06	.06	.09	.12	.15	.24
.0	.05	.05	.05	.05	.05	.05
.1	.08	.12	.14	.20	.21	.30
.2	.16	.26	.34	.47	.54	.78
.3	.24	.42	.53	.72	.79	.94
.4	.33	.56	.70	.88	.93	.99
.5	.42	.68	.82	.96	.98	
.6	.51	.79	.91	.99		
.7	.60	.85	.95			
.8	.66	.90	.98			
.9	.72	.94	.99			
1.0	.76	.97				

Table 4.2 — Power function of $\hat{L}_{1,k}$, to test $H_0:\theta=0$ vs $H_1:\theta\neq0$, at a significance level $\alpha=.05$, in a *multivariate GEV model.*

If we fix n and increase i, we obviously have an increase in power. As a measure of comparison we have here chosen, for the simulated values of θ , and taking i=1 as a pivot

$$\alpha_{n,i} = \max_{\theta\in \mathbb{R}} \{ P(\theta|\hat{L}_{n,i}) / P(\theta|\hat{L}_{n,1}) \} / i \tag{4.1}$$

which gives a measure of efficiency of $\hat{L}_{n,i}$ with respect to $\hat{L}_{n,1}$. The maximizing values of θ are always $-1/2<\theta<0$, for $n\geq10$, increasing with n.

Since there were slow variations for the different values of n chosen, we merely register the pattern for n=10,

$$\alpha_{n,2}=.90\ ;\ \alpha_{n,3}=.87\ ;\ \alpha_{n,4}=.78\ ;\ \alpha_{n,5}=.72, \tag{4.2}$$

i.e., the power almost duplicates when i goes from 1 to 2. The gain in efficiency than decreases with i. This suggests i=2 in the *multidimensional GEV model,* as a good choice for applications.

If we fix ni=k, i.e., we assume to have the same amount of data, the pattern is the following: the power function of $\hat{L}_{n,i}$ increases as $n \to k$, and $i \to 1$, i.e., the power attains its maximum for Gumbel model — see figure 4.1 for k=40.

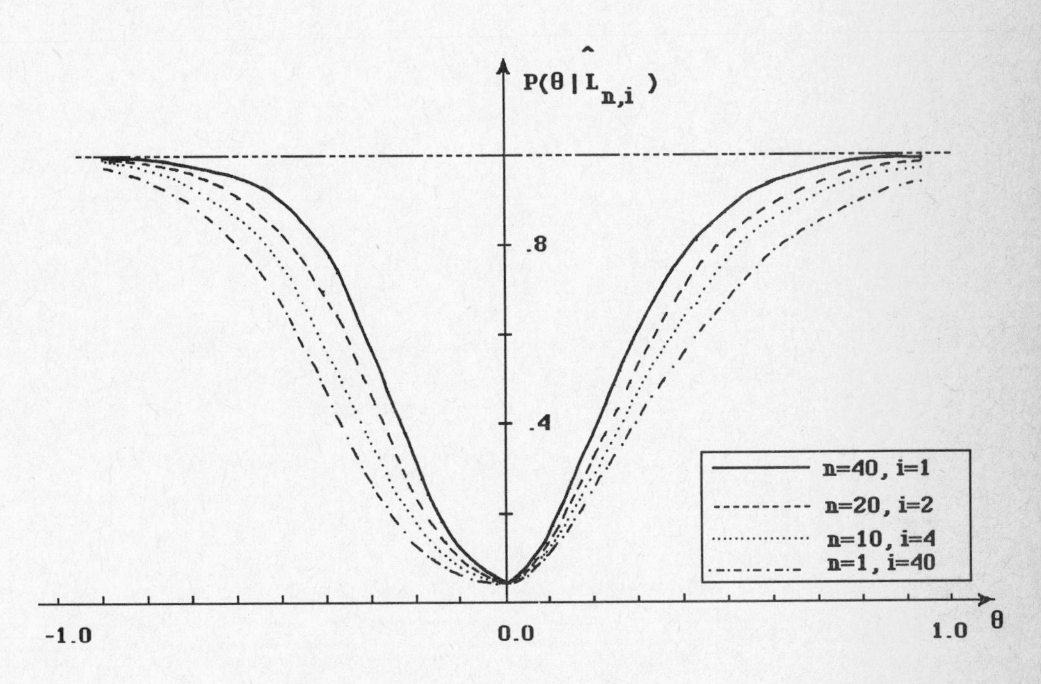

Figure 4.1 — Power function of $\hat{L}_{n,i}$, ni=40

This pattern is valid for every selection of ni=k (constant), i.e., the power function decreases as i increases to k. As k increases to infinity $\hat{L}_{k,1}$ and $\hat{L}_{1,k}$ are obviously asymptotically equivalent, once both statistics are asymptotically normal,with equivalent attraction coefficients (see Tiago de Oliveira (1982) for $\hat{L}_{k,1}$, and Gomes and Alpuim (1986) for $\hat{L}_{1,k}$). However for the values of k useful in applications $P(\theta \mid \hat{L}_{k,1})$ is always larger than $P(\theta \mid \hat{L}_{1,k})$, i.e., for such values it seems worth considering n large and i small. Notice that similar conclusions had been drawn in Gomes (1985) when comparing the extremal models through estimation of high quantiles , i.e., a generalization that includes classical Gumbel approach, i.e. the *multidimensional GEV model,* provides better results than the *multivariate GEV model*. In figure 4.2 we present the quotient $P(\theta|\hat{L}_{k,1})/P(\theta|\hat{L}_{1,k})$, for k=40 and 50.

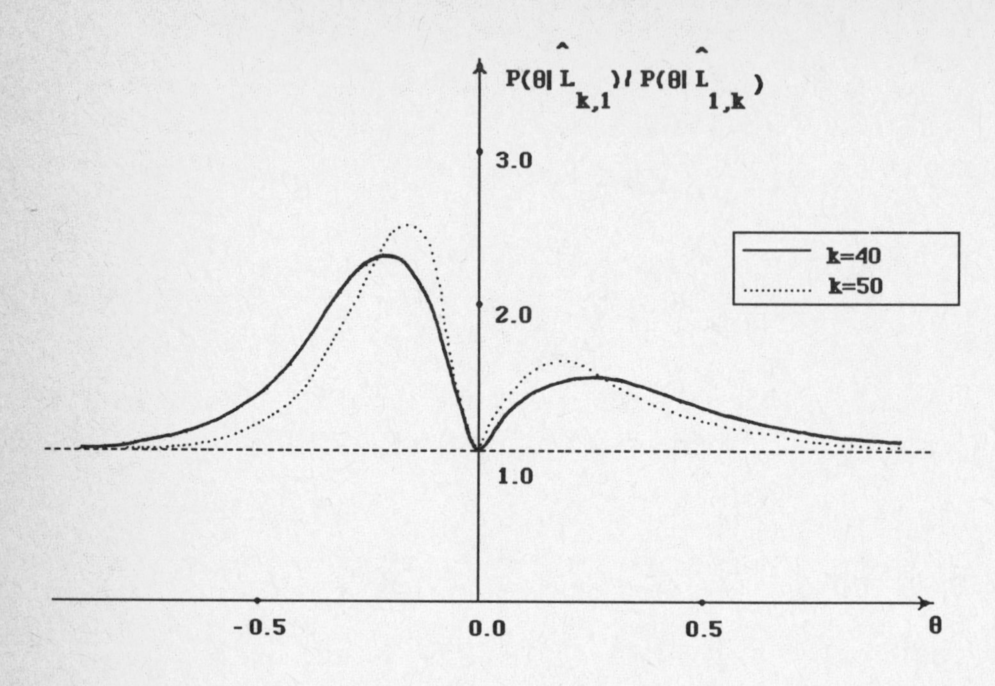

Figure 4.2 — Quotient $P(\theta|\hat{L}_{k,1})/P(\theta|\hat{L}_{1,k})$, for k=40 and 50.

From a practical point of view it is worth noticing the following: while the value of n is usually fixed, although sometimes controversial (for instance, the number of years over which we had access to observations), the value of i may easily be increased within the validation of the model. However, since large values of i easily led , in practice, to a non-validation of the multivariate structure in (1.1), we cannot be in favour of the *multivariate GEV model.* More than that: with the same amount of observations, the *multidimensional GEV model* performs better than the *multivariate GEV model,* in what regards *statistical choice* problems. As a conclusion: although with the same amount of information we have here been led to be in favour of Gumbel model (see figure 4.2), the information in the few top o.s. available is extremely relevant, and a multidimensional model is worth considering.

REFERENCES

Athayde, E. and Gomes, M.I. (1987). Multivariate extremal models under non-classical situations. In P. Bauer et al. (eds.), *Mathematical Statistics and Probability Theory,* vol. B, D. Reidel, 1-9.

Galambos, J. (1975). *The Asymptotic Theory of Extreme Order Statistics.* Wiley, New York.

Gomes, M.I. (1981). An i-dimensional limiting distribution function of largest values and its relevance to the statistical theory of extremes. In C. Taillie et al. (eds.), *Statistical Distributions in Scientific Work,* Vol. 6, 389-410, D. Reidel.

Gomes, M.I. (1985). Statistical theory of extremes — comparison of two approaches. *Statist. & Decision.* Suppl. Issue **2**, 33,37.

Gomes, M.I. and Alpuim, M.T. (1986). Inference in a multivariate Generalized Extreme Value model — asymptotic properties of two test statistics. *Scand. J. Statist.* **13**, 291-300.

Gomes, M.I. and Pestana, D.D. (1986). Non classical extreme value models. *Proc. III Internat. Conf. Statistical Climatology*, 196-200.

Gomes , M.I. and van Montfort, M.A.J. (1986). Exponentiality versus Generalized Pareto — quick tests. *Proc. III Internat.Conf. Statistical Climatology*, 185-195.

Graça Martins, M.E. and Pestana,D.D.(1985). The extremal limit problem -- extensions. In W. Grossman et al. (eds.). *Proc. 5th Pannonian Symp. Math. Statist.*, 143-153

Hüsler, J. (1986). Extreme values of non-stationary random sequences. *J. Appl. Probab.* **23**, 937-950.

Mejzler, D. (1956). On the problem of the limit distributions for the maximal term of a variational series. *Lvov. Politehn. Inst. Naucn. Zap.* (Fiz.-Mat.) **38**, 90-109.

Pickands, J. III (1975). Statistical inference using extreme order statistics. *Ann. Statist.* **3**, 119-131.

Smith, R.L. (1984). Threshold methods for sample extremes. In J. Tiago de Oliveira (ed.). *Statistical Extremes and Applications*, D. Reidel, 621-638.

Tiago de Oliveira, J.(1982). Statistical choice of univariate extreme models. In G.P.Patil et al. (eds.). *Statistical Distributions in Scientific Work,* Vol. 6, D.Reidel, 367-387.

Tiago de Oliveira , J. and Gomes, M.I.(1984). Two test statistics for choice of univariate extreme models. In J. Tiago de Oliveira (ed.). *Statistical Extremes and Applications*, D. Reidel, 651-668.

van Montfort, M.A.J. and Gomes, M.I.(1985). Statistical choice of extremal models for complete and censored data. *J. Hydrology* **77**, 77-87.

Weissman, I. (1975). Multivariate extremal processes generated by independent, non-identically distributed random variables. *J. Appl. Probab.* **12**, 477-487.

Weissman, I. (1978). Estimation of parameters and large quantiles based on the k largest observations. *J. Amer. Statist. Assoc.* **73**, 812-815.

THE ROLE OF EXTREME ORDER STATISTICS FOR EXPONENTIAL FAMILIES

Arnold Janssen

Department of Mathematics, University of Siegen, Hölderlinstr. 3
D-5900 Siegen 21

Abstract: Consider i.i.d. observations $X_1,\dots,X_n$ which arise from a natural exponential family $(P_\vartheta)_{\vartheta\geq 0}$ on $\mathbb{R}$ where P_0 lies in the domain of attraction of a stable distribution with the index $\alpha < 2$. It is pointed out that for each sequence $k(n) \to \infty$, $k(n)\leq n$, the upper order statistics $X_{n+1-k(n):n}$, $X_{n+2-k(n):n},\dots,X_{n:n}$ are asymptotically sufficient in the sense that they contain all information lying in the whole sample as $n \to \infty$. Moreover, upper bounds for the loss of information in terms of the deficiency of statistical experiments are calculated in the case when only the k largest order statistics are available. As an example we treat the exponential family of the inverse Gaussian distributions which appear in connection with the inverse sampling problem for the Wiener process with unknown non-negative drift.

1. Introduction and Preliminaries.

Consider an exponential family

$$E_{P_0} = (\mathbb{R}, \mathcal{B}, \{P_\vartheta : 0 \leq \vartheta \leq d\}) \tag{1.1}$$

with

$$\frac{dP_\vartheta}{dP_0}(x) = C(\vartheta)\exp(-\vartheta x) \tag{1.2}$$

Key words. Exponential family, order statistic, convergence of experiments, deficiency, inverse Gaussian distribution. Primary 62B15, Secondary 62B05.

where $C(\vartheta)^{-1} = \int \exp(-\vartheta x)\, dP_0(x)$ is assumed to be finite on some proper interval $[0,d]$. The following local consideration has been very fruitful in asymptotic statistics. One looks for a sequence $\delta_n \to 0$, $\delta_n > 0$, such that the rescaled family of n independent replications of E_{P_0}, denoted by

$$(1.3) \qquad E_n = (\mathbb{R}^n, \mathcal{B}^n, \{ P^n_{\delta_n \vartheta} : 0 \le \vartheta \le d\delta_n^{-1} \}),$$

can be substituted by a simpler non-degenerate limit experiment. The general idea behind this procedure can be labeled as an approximation of statistical models within the deficiency metric Δ for statistical experiments, see below. As a consequence we recall that the convergence of experiments enables us to solve all asymptotic decision problems for the limit experiment rather than for E_n, compare with LeCam (1986), Strasser (1985). The preceding papers, Janssen (1985), (1986), contain the following solution. E_n converges to an exponential family E_Q for some stable law Q with the index of stability $0 < p < 2$, $p \neq 1$, if and only if P_0 belongs to the domain of attraction of Q, i.e.: there exist normalizing constants such that

$$(1.4) \qquad \delta_n \sum_{i=1}^{n} X_i + b_n \to Q$$

in distribution as $n \to \infty$, where $X_1, X_2, \ldots$ are i.i.d. random variables with joint distribution P_0. Notably , it turns out that the normalizing constants δ_n in (1.3) and (1.4) are equal and Q is a one-sided stable distribution such that the exponential family E_Q, generated by Q as in (1.1) and (1.2), is well-defined with parameter space $[0,\infty)$.

Let $X_1, \ldots, X_n$ be i.i.d. random variables with common distribution $P_{\delta_n \vartheta}$ keeping in mind that δ_n is the correct speed of rescaling since E_n has a non-degenerate limit experiment. For continuous distributions we turn to the order statistics $X_{1:n} \le X_{2:n} \le \ldots \le X_{n:n}$ which are sufficient. The main result of the present paper shows that a finite number of the largest order statistics contain approximately all information of E_n. In addition we establish upper bounds for the loss of information in the case when only the k largest order statistics are observable. What we are doing is to compare E_n with the experiment

$$(1.5)\qquad E_{n,k} = (\mathbb{R}^k, \mathcal{B}^k, \{ \mathcal{L}((X_{n+1-k:n},\dots,X_{n:n}) \mid P^n_{\delta_n\vartheta}) : 0\le\vartheta\le d\delta_n^{-1}\})$$

within the concept of statistical experiments where $\mathcal{L}(\cdot|\cdot)$ denotes the law of a random variable. Again the idea of the approximation of statistical models plays a central role.

Before going into details let us give a few preliminary comments. In connection with absolutely continuous location families of Weibull type it is known that sparse families of order statistics are asymptotically sufficient, see Janssen and Reiss (1988) and references therein. That paper also contains rates of convergence for the loss of information when only a few extreme order statistics are observable. It is surprising that similar results can be obtained for exponential families considered here, although translation invariance and absolute continuity cannot be used and thus different arguments are required in the proofs. The present results have a formal analogous structure as recent results for sums of i.i.d. random variables. Consider the sufficient statistic (1.4) under P_0. It is well-known that the normalized sum is asymptotically mainly influenced by the extreme order statistics, see Csörgö et al. (1986) and references therein. However, this connection only seems to be of formal nature. The proofs of the present results rely on the use of conditional distributions for order statistics which are recalled from Reiss (1988).

The loss of information for $E_{n.k}$ is measured in terms of the deficiency concept for statistical experiments in the sense of LeCam (1986), see also Strasser (1985). The concept can be described as follows. Let $H_i = (\Omega_i, \mathcal{A}_i, \{ Q_{i,\vartheta} : \vartheta \in \Theta \})$ be two dominated experiments on Polish Borel spaces $(\Omega_i, \mathcal{A}_i)$ for i = 1,2. The variational distance of two probability measures is defined by

$$\| Q_1 - Q_2 \| = \sup_A | Q_1(A) - Q_2(A) |$$

where A ranges over all measurable sets. Then the deficiency is defined by

$$\delta(H_2, H_1) = \inf_K \sup_{\vartheta\in\Theta} \| Q_{1,\vartheta} - KQ_{2,\vartheta} \|$$

where K varies over all Markov kernels from $\mathcal{A}_2$ to $\mathcal{A}_1$. Here KQ is defined by

$$KQ(A) = \int K(A,\cdot)\, dQ.$$

The symmetric deficiency

$$\Delta(H_1,H_2) = \max\{\ \delta(H_1,H_2),\ \delta(H_2,H_1)\ \} \tag{1.6}$$

is used to compare the statistical experiments H_1 and H_2 whereas the kernels K are used to carry over statistical procedures from one statistical experiment to the other one. Thus we are not only interested in the deficiency but also in the construction of reasonable Markov kernels. Let

$$K_0^{(n,k)}(\cdot,x) \tag{1.7}$$

denote the conditional distribution of $(X_{1:n}, \ldots, X_{n:n})$ given $(X_{n+1-k:n}, \ldots, X_{n:n}) = (x_{n+1-k:n}, \ldots, x_{n:n}) = x$ with respect to P_0. Similar as in Janssen and Reiss (1988) we propose the kernel $K_0^{(n,k)}$ in order to compare E_n and $E_{n,k}$. Since E_n contains more information than $E_{n,k}$, i.e. $\delta(E_n,E_{n,k}) = 0$, we may restrict ourselves to the quantity

$$\rho(n,k,t) = \| \mathcal{L}((X_{1:n}, \ldots, X_{n:n})|P_t^n) - K_0^{(n,k)}\mathcal{L}((X_{n+1-k:n}, \ldots, X_{n:n})|P_t^n)\| \tag{1.8}$$

which is a measure of performance showing how good the kernel fits the two models for the parameter t. From now on we will consider all experiments relative to the parameter set $\Theta \cap [0,s]$, $s > 0$, denoting briefly the deficiency for the restricted experiments by $\Delta_s(\cdot)$. Using this notation we see that for large n

$$\Delta_s(E_n,E_{n,k}) \le \sup_{0\le t\le s} \rho(n,k,\delta_n t). \tag{1.9}$$

Before the main results are presented we give an application in advance. Observe that upper bounds for the deficiency apply at once to test problems. For instance assume that $\phi_n(X_1,\ldots,X_n)$ is a test for E_n. If only the k largest observations are available the conditional expectation

$$\psi_{n,k}(x) = E_{P_0}(\phi_n | (X_{n+1-k:n}, \ldots, X_{n:n}) = x) \tag{1.10}$$

$$= \int \phi_n(u)\, K_0^{(n,k)}(du,x)$$

suggests itself as a new test. Consequently, the following inequality holds for the corresponding power functions uniformly for $0 \le t \le s$:

$$(1.11) \qquad E_{P^n_{\delta_n t}} \phi_n \le E_{P^n_{\delta_n t}} \psi_{n,k} + \Delta_s(E_n, E_{n,k}) .$$

(1.1) Example

Consider the one-sided test problem $\{ \vartheta = 0 \}$ against $\{ \vartheta > 0 \}$ for the exponential family (1.1) and let ϕ_n denote the optimal test at sample size n and level α. Within this framework we obtain under the assumptions of section 3 that it is enough to observe the k(n) largest order statistics where k(n) may converge as slow as possible to infinity. Let $k(n) \to \infty$, $k(n) \le n$, be any sequence and $s > 0$. Then

$$(1.12) \qquad \sup_{0 \le t \le s} | E_{P^n_{\delta_n t}} \phi_n - E_{P^n_{\delta_n t}} \psi_{n,k(n)} | \le \Delta_s(E_n, E_{n,k}) \to 0$$

hold as $n \to \infty$.

Since the present paper is written in the spirit of the approximation of statistical models within the deficiency concept we first supplement the results of Janssen (1985) and (1986), where weak convergence of the underlying experiments is treated.

(1.2) Theorem

Assume that P_0 belongs to the domain of attraction of a stable distribution Q, i.e. (1.4) holds, where Q is a non trivial distribution with the index $0 < p < 2$ of stability, $p \ne 1$. Then

$$(1.13) \qquad \Delta_s(E_n, E_Q) \to 0$$

as $n \to \infty$, where E_Q is the exponential family (1.1) induced by Q.

The validity of (1.13) is an open problem for the index $p = 1$.

The paper is organized as follows. Section 2 is devoted to special exponential families on $\mathbb{R}_+$. An example shows that the results can be applied to the inverse sampling in connection with the Wiener process. In section 3 the general case is treated and the proofs are presented in section 4.

2. Exponential families on $\mathbb{R}_+$.

In this section we will consider an exponential family of the form (1.1) where P_0 is concentrated on the non-negative reals such that

$$P_0((x,\infty))/x^{-p} \to c \tag{2.1}$$

for some constants $c > 0$, $0 < p < 2$, as $x \to \infty$. Recall that then δ_n may be substituted by $n^{-1/p}$. Define

$$k_0 = k_0(p) = \begin{cases} (2-p)/(2p) & \text{if } (2-p)/(2p) \text{ is an integer} \\ [(2-p)/(2p)] + 1 & \text{otherwise} \end{cases} \tag{2.2}$$

where [ ] indicates the Gaussian bracket.

(2.1) Theorem

Let P_0 be a continuous distribution on $\mathbb{R}_+$ which satisfies (2.1). Then there exist constants K_1 and K_2 such that for all $s \geq 0$ and all $n \geq k > k_0(p)$ the following inequalities hold:

$$\sup_{0 \leq t \leq s} \rho(n,k,t/n^{1/p}) \leq K_1 s \int (X_{n+1-k:n}/n^{1/p})^{(2-p)/2}\, dP_0^n \tag{2.3}$$

$$\leq K_2 s(k - k_0)^{-(2-p)/(2p)}.$$

Utilizing (1.9) we immediately get an upper bound for the deficiency $\Delta_s(E_n, E_{n,k})$. In special situations the constants K_1 and K_2 can be evaluated.

(2.2) Example. (Inverse sampling for a Wiener process with non-negative drift). Let P_0 denote the one-sided stable law with the index $p = 1/2$ on $[0,\infty)$ having the Lebesgue density

$$g(x) = (2\pi)^{-1/2} x^{-3/2} \exp(-1/(2x))\, 1_{(0,\infty)}(x). \tag{2.4}$$

Then the corresponding exponential family

$$(2.5) \qquad \frac{dP_\vartheta}{dP_0}(x) = \exp(-\vartheta x + (2\vartheta)^{1/2}), \ \vartheta \geq 0$$

is known to be the important family of inverse Gaussian distributions. A careful study now shows that we may choose $K_1 = 2^{-1/2}$, $k_0 = 2$ and the inequality (2.3) yields

$$(2.6) \qquad \sup_{0 \leq t \leq s} \rho(n,k,t/n^2) \leq s(2/\pi)^{3/4} 2^{-1/2} (k-2)^{-3/2},$$

where K_2 equals $(2/\pi)^{3/4} 2^{-1/2} = 0.5039...$. We see that we only need the largest $k = 16$ order statistics in order to obtain a bound $s/100$ for (2.6).

Subsequently, let us briefly sketch the meaning of the inverse Gaussian distribution. Consider a standard Wiener process W_ϑ with drift $(2\vartheta)^{1/2}$, $\vartheta \geq 0$. In our model we assume that not the whole sample path of W_ϑ is observable. Only certain observations based on first hitting times are available. This concept is sometimes labeled as inverse sampling. The model can be described as follows: Let

$$T_\vartheta^j = \inf \{ t : W_\vartheta(t) \geq j \}, \ j \geq 1$$

denote the first hitting time of the line $y = j$. For certain $j \geq 1$ we observe the time $T_\vartheta^j - T_\vartheta^{j-1}$ which is needed to cross the line $y = j$ for the first time after the first entrance of the event $\{ W_\vartheta \geq j-1 \}$. We assume that our observations $X_1, \dots, X_n$ are a subset of $\{ T_\vartheta^1, T_\vartheta^2 - T_\vartheta^1, T_\vartheta^3 - T_\vartheta^2, \dots \}$. As usual in connection with inverse sampling we use the assumption that the process can be observed for a sufficiently long time. (Otherwise we may choose a conditional approach by conditioning under the number of observations.) The strong Markov property implies now that $X_1, \dots, X_n$ are i.i.d. random variables with the distribution P_ϑ of (2.5), see Feller (1971).

A concrete problem is to decide whether a positive drift exists or not. Thus we are just in the situation of the one-sided test problem described in Example (1.1). Combining (1.12) and (2.6) we see that the k largest observations with $k \geq 16$ yield one-sided tests where the loss of the power compared with the optimal tests is less or equal than $s/100$ uniformly over the interval $[0,s]$.

3. The general result.

Consider now a general exponential family as in (1.1). It is well-known that P_0 belongs to the domain of attraction of the stable law Q with index $0 < p < 2$ if and only if there exists a function L varying slowly at infinity such that

$$1 - F_0(x) := P_0((x,\infty)) = x^{-p}L(x) \tag{3.1}$$

for $x \geq x_0 > 0$, see Feller (1971), p. 577. Note that $P_0((-\infty,x))$ has at least an exponential decay as $x \to -\infty$ since $\int \exp(-dx)\, dP_0(x) < \infty$. Recall that there exists a further function R varying slowly at infinity such that for $0 < u < 1$

$$F_0^{-1}(1-u) = u^{-1/p}R(1/u), \tag{3.2}$$

where F_0^{-1} is as usual the inverse distribution function of F_0. Subsequently let us always choose

$$\delta_n = 1/F_0^{-1}(1-1/n) \tag{3.3}$$

which is known to be a proper choise of the normalizing coefficient in (1.4) whenever n is large enough, i. e. $F_0^{-1}(1-1/n) > 0$. Otherwise set $\delta_n = 1$.

Note that $\Delta_s(E_n,E_{n,k})$ is decreasing for increasing k. Thus we are mainly interested in upper bounds for small values of k. In this spirit we formulate the next result.

(3.1) Theorem

Consider an exponential family with continuous distribution P_0 such that (1.1), (1.2), (3.1) and (3.3) are satisfied for $0 < p < 2$.

(a) Let $F_0(0) < 1-\lambda < 1$ and $\varepsilon > 0$. Then there exists a constant C such that for all $k/n \leq \lambda$, $\delta_n^{-1}d \leq s$ and all n the subsequent inequality holds:

$$\sup_{0\leq t\leq s} \rho(n,k,\delta_n t) \leq sCk^{-(2-p)/(2p) + \varepsilon}. \tag{3.4}$$

(b) Assume in addition to (3.1) that $L(x) \to c > 0$ as $x \to \infty$. Then the inequality (3.4) also holds for $\varepsilon = 0$.

(c) If in the situation of (a) or (b) the probability measure P_0 is concentrated on $\mathbb{R}_+$ then the inequality (3.4) holds for all $k \le n$.

As a consequence of Theorem (3.1) we see that for each sequence $k(n) \to \infty$, $k(n) \le n$, the experiment $E_{n,k(n)}$ contains asymptotically all information, namely

$$\Delta_s(E_n, E_{n,k(n)}) \to 0 \tag{3.5}$$

as $n \to \infty$. Under the assumptions of Theorem (1.2) we obtain

$$\Delta_s(E_{n,k(n)}, E_Q) \to 0. \tag{3.6}$$

4. The proofs.

In the sequal let $d(P,Q)$ denote the Hellinger distance between probability measures P and Q defined by

$$d(P,Q) = \left(\frac{1}{2} \int \left(\left(\frac{dP}{d(P+Q)}\right)^{1/2} - \left(\frac{dQ}{d(P+Q)}\right)^{1/2}\right)^2 d(P+Q)\right)^{1/2} \tag{4.1}$$

Recall that

$$\| P - Q \| \le 2^{1/2} d(P,Q) \le 2^{1/2} \tag{4.2}$$

and

$$\| P^m - Q^m \| \le (2m)^{1/2} d(P,Q). \tag{4.3}$$

The proof of Theorem (1.2). In Janssen (1985) it has been pointed out that for some $\delta_n \to 0$ the sequence E_n is weakly convergent to E_Q in the sense that the corresponding finite dimensional marginal distributions of the log-likelihood ratio process are weakly convergent. Let now $h_n \to 1$. Then an inspection of the proof of Theorem 4 of Janssen (1985) shows that

$$F_n = (\mathbb{R}^n, \mathcal{B}^n, \{ P^n_{\delta_n h_n \vartheta} : \vartheta \ge 0 \}) \tag{4.4}$$

is also weakly convergent to E_Q. This is due to the fact that a ratio of regularly varying functions is uniformly convergent on compact sets. Here we also use the representation of the Laplace transform (10) of Theorem 4 of Janssen (1985) and Corollary 8 in Janssen (1986). Now (4.2) yields the equicontinuity of E_n, i.e.: For each $\vartheta_0 \geq 0$ and $\varepsilon > 0$ there exists $\delta > 0$ such that

$$\| P^n_{\delta_n \vartheta_0} - P^n_{\delta_n \vartheta} \| \leq \varepsilon$$

for all n whenever $| \vartheta_0 - \vartheta | < \delta$. Since E_Q is itself an exponential family we may apply the Theorem of Lindae, see LeCam (1986), p. 92 , which now yields the convergence with respect to the Δ_s - metric.□

The proof of Theorem (3.1). For fixed $k \leq n$ and $0 \leq t \leq d$ a slight modification of Lemma (2.14) of Janssen and Reiss (1988) using (4.2) and (4.3) shows

$$\rho(n,k,t) \leq (2(n-k))^{1/2} \int d(P_{t,x}, P_{0,x}) \, d\mathcal{L}(X_{n+1-k:n} | P^n_t)(x) \tag{4.5}$$

where $P_{t,x}$ denotes the probability measure

$$P_{t,x}(\cdot) = P_t(\cdot \cap (-\infty, x]) / P_t((-\infty, x]), \tag{4.6}$$

whenever (4.6) is well defined. Subsequently, we split the integral (4.5) in two parts and we will first consider $x \geq 0$ and

$$\int_0^\infty (2(n-k))^{1/2} d(P_{t,x}, P_{0,x}) \, d\mathcal{L}(X_{n+1-k:n} | P^n_t)(x). \tag{4.7}$$

Using the inequality

$$d^2(P,Q) = 1 - \int \left(\frac{dP}{dQ}\right)^{1/2} dQ \leq -\log \int \left(\frac{dP}{dQ}\right)^{1/2} dQ \tag{4.8}$$

we see that

$$d^2(P_{t,x}, P_{0,x}) \leq -\log \left[\frac{\int_{-\infty}^{x} \exp(-ty/2) dP_0(y) / P_0((-\infty, x])}{\left[\int_{-\infty}^{x} \exp(-ty) dP_0(y) / P_0((-\infty, x]) \right]^{1/2}} \right]. \tag{4.9}$$

Here we may restrict ourselves to those x which satisfy $P_0((-\infty,x]) > 0$. A Taylor expansion of the function

$$(4.10) \qquad t \longrightarrow g_x(t) = \log\left[\int_{-\infty}^{x} \exp(-ty)dP_0(y)/P_0((-\infty,x])\right]$$

at t = 0 yields for t ≥ 0

$$(4.11) \qquad g_x(t) = -tE_{P_{0,x}}(id) + \frac{t^2}{2}\operatorname{Var}_{P_{t_1,x}}(id)$$

where id is the identity on $\mathbb{R}$ and $t_1 \in (0,t)$. Thus (4.9) equals

$$(4.12) \qquad -g_x(t/2) + g_x(t)/2 \le t^2 \operatorname{Var}_{P_{t_1,x}}(id)/4$$

$$\le t^2\left[\int_{(-\infty,0)} y^2\, dP_{t_1,x}(y) + \int_{[0,x]} y^2\, dP_{t_1,x}(y)\right]/4.$$

Since $t \longrightarrow P_{t,x}$ is itself an exponential family of the form (1.2) we recall that this family is stochastically decreasing, i.e.:

$$(4.13) \qquad P_{t,x}((-\infty,y]) \ge P_{s,x}((-\infty,y])$$

for all y if t ≥ s. The inequality (4.13) can easily seen by regarding optimal tests for $P_{s,x}$ against $P_{t,x}$, cf. Witting (1985). Hence we obtain the following upper bound for (4.12):

$$(4.14) \qquad \le t^2\left[\int_{(-\infty,0)} y^2\, dP_{t,x}(y) + \int_{[0,x]} y^2\, dP_{0,x}(y)\right]/4.$$

Next we substitute t by $\delta_n t$ for $0 \le t \le s$. Thus (4.9) - (4.14) yield for the integrand of (4.7)

$$(4.15) \quad \le ((n-k)/2)^{1/2} s\delta_n\left[\left[\int_{(-\infty,0)} y^2\, dP_{\delta_n s,x}(y)\right]^{1/2} + \left[\int_{[0,x]} y^2\, dP_{0,x}(y)\right]^{1/2}\right]$$

for each $x \geq 0$. Notice now that by the dominated convergence Theorem of Lebesgue the term

$$(4.16) \qquad \int_{(-\infty,0)} y^2 \, dP_{\delta_n s,x}(y)$$

is uniformly bounded for $x \geq 0$ and all n. On the other hand we see by (3.2) and (3.3) that

$$(4.17) \qquad n^{1/2}\delta_n = n^{-(2-p)/(2p)}/R(n)$$

which is by Karamata's Theorem of the order $O(k^{-(2-p)/(2p)+\varepsilon})$ whenever $-(2-p)/(2p) + \varepsilon < 0$. Hence the first term of (4.15) has the desired rate of convergence. The second part of (4.15) will be handled as follows. Assume that $P_{0,x}$ is well-defined. Then

$$(4.18) \qquad \int_0^x y^2 \, dP_{0,x}(y) \leq x^2.$$

From Feller (1971) p. 577 we recall that

$$(4.19) \qquad \int_0^x y^2 \, dP_{0,x}(y) = O(x^2(1-F_0(x))$$

as $x \to \infty$. Combining (4.18) and (4.19) we see that

$$(4.20) \qquad \int_0^x y^2 \, dP_{0,x}(y) \leq Cx^2(1-F_0(x))$$

for each $x > 0$. Combining (4.7) and (4.15) - (4.20) it is now enough to consider

$$(4.21) \qquad n^{1/2}s\delta_n \int_0^\infty x(1-F_0(x))^{1/2} \, d\mathcal{L}(X_{n+1-k:n}|P^n_{\delta_n t})(x).$$

Let $U_1, \ldots, U_n$ be i.i.d. uniformly distributed random variables on (0,1). Let now $F_{\delta_n t}$ be the distribution function of $P_{\delta_n t}$. As in (4.13) we see that

(4.22) $$F_{\delta_n t}(x) \geq F_0(x)$$

which yields $F^{-1}_{\delta_n t}(u) \leq F_0^{-1}(u)$. Notice that $X_{n+1-k:n}$ is equal in distribution to $F^{-1}_{\delta_n t}(U_{n+1-k:n})$. Hence (4.21) is not larger than

(4.23) $$n^{1/2}s\delta_n \int \max(F_0^{-1}(u),0)(1-F_0(F_0^{-1}(u)))^{1/2}d\mathcal{L}(U_{n+1-k:n})(u)$$

if we take (4.22) into account. Since $1 - U_i$ is also uniformly distributed we arrive at

(4.24) $$\int n^{1/2}s\delta_n \max(F_0^{-1}(1-u),0)u^{1/2}d\mathcal{L}(U_{k:n})(u)$$

if we use the inequality $1 - F_0(F_0^{-1}(1-u)) \leq u$. Next we insert the definition of δ_n, see (3.3), which yields for sufficiently large n

$$\int s \max(\frac{F_0^{-1}(1-u/n)}{F_0^{-1}(1-1/n)},0)\ u^{1/2}d\mathcal{L}(nU_{k:n})(u).$$

By Karamata's representation Theorem for slowly varying functions it is possible to choose for each $\varepsilon > 0$ suitable constants $c_2 > 1$, n_0 and c_3 such that

(4.25) $$R(n/u)/R(n) \leq c_3u^{-\varepsilon}$$

for $n/u \geq c_2$ and $n \geq n_0$. Define $c_1 = F_0^{-1}(1-c_2^{-1})$. For all $0 < u < 1$ and $n \geq n_0$ we obtain

(4.26) $$\max(\frac{F_0^{-1}(1-u/n)}{F_0^{-1}(1-1/n)},0) \leq \delta_nc_1 + c_3u^{-1/p-\varepsilon}.$$

As in (4.17) we see that for $-(2-p)/(2p)+\varepsilon < 0$

$$(4.27)\qquad \int \delta_n\, u^{1/2}\, d\mathcal{L}(nU_{k:n})(u) = O(k^{-(2-p)/(2p)+\varepsilon}).$$

Thus it remains to show that for k large enough

$$(4.28)\qquad \int u^{-(2-p)/(2p)-\varepsilon}\, d\mathcal{L}(nU_{k:n})(u) \le c_3 k^{-(2-p)/(2p)-\varepsilon}.$$

For $\delta = (2-p)/(2p)+\varepsilon > 0$ choose $k_0 = \delta$ if δ is an integer and $k_0 = [\delta] + 1$ otherwise. Then by Jensen's inequality

$$(4.29)\qquad \int [u^{-k_0}]^{\delta/k_0}\, d\mathcal{L}(nU_{k:n})(u) \le \Big[\int u^{-k_0}\, d\mathcal{L}(nU_{k:n})(u) \Big]^{\delta/k_0}$$

$$\le (k-k_0)^{-\delta}$$

for $k > k_0$ since

$$(4.30)\qquad \int u^{-k_0}\, d\mathcal{L}(nU_{k:n})(u) = n^{-k_0} \frac{n\cdot\ldots\cdot(n-k_0+1)}{(k-1)\cdot\ldots\cdot(k-k_0)} \le (k-k_0)^{-k_0}.$$

If we now combine (4.15), (4.17), (4.24), (4.27) and (4.28) we see that the expression (4.7) has the desired rate of convergence uniformly for $0 \le s \le t$ if we insert $\delta_n t$ in (4.7) instead of t. Now we return to (4.5). It remains to show that

$$(4.31)\qquad \int_{-\infty}^{0} (n-k)^{1/2} d(P_{\delta_n t,x}, P_{0,x})\, d\mathcal{L}(X_{n+1-k:n} | P^n_{\delta_n t})(x)$$

$$\le (n-k)^{1/2} P^n_{\delta_n t}(\{\, X_{n+1-k:n} \le 0 \,\})$$

has uniformly for $0 \le t \le s$ the correct rate of convergence. In order to treat the right hand side of (4.31) we use the following procedure, which is well-known in the theory of large deviations. Choose $p_n = p_{n,t} = 1 - F_{\delta_n t}(0)$ and $u \geq 0$. Let $Y_1, \ldots, Y_n$ denote i.i.d. random variables with a joint Bernoulli distribution with parameter p_n. For $k \leq n\lambda$ we see that

$$(4.32) \qquad P^n_{\delta_n t}(\{ X_{n+1-k:n} \leq 0 \}) = P(\{ Y_1 + \ldots + Y_n \leq k-1 \})$$

$$\leq P(\{ Y_1 + \ldots + Y_n - n\lambda \leq 0 \}) = P(\{ \exp(-u(Y_1 + \ldots + Y_n - n\lambda) \geq 1 \})$$

$$\leq (\exp(\lambda u) \int \exp(-uY_1)\, dP)^n$$

$$(4.33) \qquad = ([1-p_n+p_n \exp(-u)]\exp(\lambda u))^n.$$

Note that (4.33) attains it's minimum for $u = \log(p_n(1-\lambda)/(\lambda(1-p_n)))$. Thus (4.32) is bounded above by

$$(4.34) \qquad \left[\frac{1-p_n}{1-\lambda} \left[\frac{p_n(1-\lambda)}{\lambda(1-p_n)} \right]^{\lambda} \right]^n .$$

Note that uniformly in $t \in [0,s]$ the value $p_{n,t}$ satisfies

$$(4.35) \qquad p_{n,t} \geq \lambda + \delta$$

for large n and some $\delta > 0$. Thus (4.34) converges with an exponential rate of convergence to zero as $n \to \infty$. Consequently, the term (4.31) has the correct rate of convergence.

(b) Under the assumption of part (b) we see that $\delta_n / n^{-1/p} = b + o(1)$ for some constant $b > 0$ and $R(n)$ tends to some positive constant as $n \to \infty$. Thus (4.17) has the order $O(k^{-(2-p)/(2p)})$. Also we may choose $\varepsilon = 0$ in the inequality (4.26). Hence the proof of part (a) carries over for $\varepsilon = 0$.

(c) For non-negative random variables the expression (4.31) is equal to zero. We see that then no restrictions concerning the range of k are necessary. Thus Theorem (3.1) is completely proved.□

The proof of Theorem (2.1). As in the proof of part (c) above the arguments become much simpler. An inspection of the proof of Theorem (3.1) shows that here

(4.36) $$\rho(n,k,t/n^{1/p}) \le$$

$$\int tn^{-1/p}((n-k)/2)^{1/2}\Big(\int_0^x y^2\, dP_{0,x}(y)\Big)^{1/2} d\mathcal{L}(X_{n+1-k:n}|P^n_{t/n^{1/p}})(x).$$

As in (4.20) we see that the integrand of (4.36) is dominated by

(4.37) $$tK_1(x/n^{1/p})^{(2-p)/2}.$$

In view of (4.22) we may then substitute $P^n_{t/n^{1/p}}$ by P^n_0. These arguments imply the first inequality of (2.3). The second one can be shown as in (4.26) - (4.30) for $\varepsilon = 0$. Thus Theorem (2.1) is proved.□

The proof of the assertion (2.6). The starting point is the inequality (4.36). In case of Example (2.2) we will prove that

(4.38) $$x^{-3/2} \int_0^x y^2\, dP_{0,x}(y) \le 1$$

for $x \ge 0$. Note that for $x \le 1$ the inequality is trivial since the integrand is dominated by x^2. Integration by parts shows that the left hand side of (4.38) equals

(4.39) $$\left[x^{-3/2} + \frac{2\exp(-1/(2x))[1-x^{-1}]}{\int_0^x y^{-3/2}\exp(-1/(2y))\, dy} \right] /3.$$

It is sufficient to show that for $x \ge 1$

(4.40) $$2\exp(-1/(2x))[1-x^{-1}] \Big/ \int_0^x y^{-3/2}\exp(-1/(2y))\, dy \le 2.$$

Let X be a standard normal distributed random variable. Then X^{-2} is distributed according to P_0. Thus

$$(4.41)\qquad \int_0^x y^{-3/2}\exp(-1/(2y))\,dy = (2\pi)^{1/2}2(1-\Phi(x^{-1/2}))$$

where Φ is the standard normal distribution function. We introduce the substitution $z^2 = 1/x$. Then it is enough to prove that for $0 < z < 1$

$$(4.42)\qquad 2\exp(-z^2/2)[1-z^2]/[(2\pi)^{1/2}2(1-\Phi(z))] \le 2.$$

Since $1-\Phi(z) \ge \Phi(1) - \Phi(z)$ we may consider

$$(4.43)\qquad \exp(-z^2/2)[1-z^2]/[(2\pi)^{1/2}(\Phi(1)-\Phi(z))]$$

which equals by the extended mean value Theorem

$$(4.44)\qquad \frac{z_1[1-z_1^2]\exp(-z_1^2/2) + \exp(-z_1^2/2)2z_1}{\exp(-z_1^2/2)} = 3z_1 - z_1^3 \le 2$$

for some $z < z_1 < 1$. Thus (4.38) is proved. If we now combine (4.36) and (4.38) we conclude that (4.37) equals

$$(4.45)\qquad t2^{-1/2}(x/n^2)^{3/4}$$

with $K_1 = 2^{-1/2}$. For the inverse Gaussian distribution we have the inequality

$$(4.46)\qquad 1 - F_0(x) \le 2(2\pi x)^{-1/2} =: 1 - H(x).$$

Note that $H^{-1}(1-u) = 2u^{-2}/\pi$ and $F_0^{-1}(1-u) \le H^{-1}(1-u)$. Inserting this in (2.3) we obtain

$$(4.47)\qquad \int (X_{n+1-k:n}/n^2)^{3/4}\,dP_0^n$$

$$= \int (F_0^{-1}(1-U_{k:n})/n^2)^{3/4}dP \le (2/\pi)^{3/4}\int((nU_{k:n})^{-2})^{3/4}dP$$

$$\le (2/\pi)^{3/4}(k-2)^{-3/2}$$

by (4.29). □

References

Csörgö, M., Csörgö, S., Horvath, L., Mason, D. M. (1986). Normal and stable convergence of integral functions of the empirical distribution function. Ann. Prob. 14, 86-118.

Janssen, A. (1985). One-sided stable distributions and the local approximation of exponential families. Selected papers, 16th EMS Meeting, Marburg 1984, Statistics and Decisions, Supplem. Issue 2, 39-45.

Janssen, A. (1986). Scale invariant exponential families and one-sided test problems. Statistics and Decisions 4, 147-174.

Janssen, A, Reiss, R.-D.(1988). Comparison of location models of Weibull type samples and extreme value processes. To appear in: Probab. Rel. Fields.

LeCam, L. (1986). Asymptotic methods in statistical decision theory. Springer series in statistics, New-York.

Reiss, R.-D. (1988). Approximate distributions of order statistics (with applications to nonparametric statistics), to appear in Springer Series in Statistics.

Strasser, H. (1985). Mathematical theory of statistics. De Gruyter Studies in Mathem. 7, Berlin, New-York.

Witting, H. (1985). Mathematische Statistik I. Teubner, Stuttgart.

MULTIVARIATE RECORDS AND SHAPE

K. Kinoshita*
Department of Statistics
Colorado State University
Fort Collins, CO 80523 USA

S.I. Resnick**
School of Operations Research and Industrial Engineering
Cornell University
Upson Hall
Ithaca, NY 14853 USA

ABSTRACT. For one dimensional data, the notion of a record is natural and universally agreed upon but this is not the case when the data is in two or more dimensions. We outline several plausible definitions of a record and settle on one induced by a partial ordering: For observations $X_1,\dots,X_n$ we say X_n is a record if each component of X_n is bigger than the corresponding components of previous observations. Some properties of such records of iid observations are reviewed and used to study the shape of the convex hull of the first n observations, as $n \to \infty$.

1. INTRODUCTION

The notion of a record in a one–dimensional series of observations is familiar, natural and unambiguous. It means of course an observation bigger (or smaller in appropriate contexts) than previous observations and is commonly used in connection with data describing sports, temperature, rainfall, stress and reliability.

When analyzing multivariate data there is no longer a commonly agreed upon definition and the simplicity resulting from the fact that $\mathbb{R}$ is totally ordered is lost. Consider $\mathbb{R}^2$–valued observations $\{X_n\} = \{(X_n^{(1)}, X_n^{(2)})'\}$. For such data, we can define the notion of record in many ways and the following are all plausible and interesting possibilities:

(a) X_n is a record if $X_n^{(p)} > \bigvee_{i=1}^{n-1} X_i^{(p)}$, $p = 1$ and 2.

(b) X_n is a record if $X_n^{(p)} > \bigvee_{i=1}^{n-1} X_i^{(p)}$, $p = 1$ or 2.

(c) X_n is a record if X_n falls outside the convex hull of $X_1,\dots,X_{n-1}$.

---

* Partially supported by NSF Grant DMS–8501673 at Colorado State University. The hospitality of the School of Operations Research and Industrial Engineering, Cornell University during the academic year 1987–1988 is gratefully acknowledged.

**Partially supported by the Mathematical Sciences Institute, Cornell University and by NSF Grant DMS–8801034 at Cornell University.

Definitions (b) and (c) are related as (c) is the infinite dimensional version of (b). One can see this by expressing the definition (c) in terms of the support function of a convex set. Mathematical analysis of (b) and (c) is more difficult than (a) because the definition (a) can be expressed in terms of a partial order, namely that $X_n > X_i$, $i = 1,\dots,n-1$, where the inequality between vectors is interpreted componentwise.

In section 2 we review some results about records defined according to (a). This material is taken from Goldie and Resnick (1989) where records on general partially ordered sets are considered. For simplicity in this paper we only consider $\mathbb{R}^2$–valued observations. In particular we need to review when an iid sequence has a finite or infinite number of records.

If X_n is a record, then previous observations must be located southwest of X_n in the plane and in this sense X_n sticks out. Another interpretation is that if a right angled template is lowered from ∞ until it contacts the sample, $X_1,\dots,X_n$, then this template contacts the sample in one point iff X_n is a record. Thus we arrive at the idea of probing the sample with various geometric objects (cf. Groeneboom, 1987) and relating the shape of the convex hull of the sample to records. This idea is explored in the last sections where the fact that a multivariate normal sequence has a finite number of records is used to prove that the boundary of the convex hull of a normal sample becomes smooth as $n \to \infty$ in the sense that the exterior angles at vertices of the convex hulls converge to π uniformly.

It is rather well known that the shape of a normal sample becomes ellipsoidal (Geoffroy, 1961; Fisher, 1969; Davis, Mulrow, Resnick, 1987) and hence the boundary becomes smooth. Using only records, it does not seem possible to obtain detailed information on the shape of the sample beyond what happens on the boundary of the convex hull.

2. MULTIVARIATE RECORDS

As in the introduction let $\{X_n = (X_n^{(1)}, X_n^{(2)})', \ n \geq 1\}$ be iid $\mathbb{R}^2$–valued random variables. X_n is a record if

$$X_n > X_j, \ j = 1,\dots,n-1$$

or equivalently

$$X_n > \bigvee_{j=1}^{n-1} X_j$$

where $(x_1, x_2)' \vee (y_1, y_2)' = (x_1 \vee y_1, x_2 \vee y_2)'$. As a convention, make X_1 a record. The first order of business is to decide when the sequence has an infinite number of records so define the counting variables

$$N(A) = \sum_{n=1}^{\infty} 1_{[X_n \in A, \ X_n \text{ is a record}]}$$

for $A \in \mathscr{B}(\mathbb{R}^2)$. Let $N = N(\mathbb{R}^2)$ be the total number of records in the sequence. From the Hewitt–Savage 0–1 Law, N is finite or infinite with probability 1 and either case can occur. For example, if the bivariate distribution of X_1 is a product of two continuous distributions so that for each n we have $X_n^{(1)}$ and $X_n^{(2)}$ are independent random variables then

$$EN = \sum_{n=1}^{\infty} P[X_n \text{ is a record}]$$

$$= \sum_{n=1}^{\infty} P[X_n^{(1)} > \bigvee_{j=1}^{n-1} X_j^{(1)}] P[X_n^{(2)} > \bigvee_{j=1}^{n-1} X_j^{(2)}]$$

$$= \sum_{n=1}^{\infty} (1/n)(1/n) = \sum_{n=1}^{\infty} 1/n^2 < \infty$$

so that $P[N < \infty] = 1$. On the other hand if the distribution of X_1 concentrates on the diagonal so that $P[X_n^{(1)} = X_n^{(2)}] = 1$ for $n \geq 1$, and the distribution of $X_n^{(1)}$ is continuous then $EN = \sum_{n=1}^{\infty} P[X_n^{(1)} > \bigvee_{j=1}^{n-1} X_j^{(1)}] = \sum_{n=1}^{\infty} (1/n) = \infty$ and from one dimensional considerations it is clear $P[N = \infty] = 1$.

In general we have for $A \in \mathscr{B}(\mathbb{R}^2)$

$$EN(A) = \sum_{n=1}^{\infty} P[X_n \in A,\ X_n \text{ is a record}]$$

and if we denote the distribution of X_1 by F we obtain

$$EN(A) = \sum_{n=1}^{\infty} \int_A F(dx) P\left\{ \bigcap_{j=1}^{n-1} [X_j < x] \right\}$$

$$= \sum_{n=1}^{\infty} \int_A F(dx)(F(-\infty, x))^{n-1}$$

$$= \sum_{n=1}^{\infty} \int_A \frac{F(dx)}{F((-\infty, x)^c)} =: H(A) .$$

The criterion for when the number of records is finite is in terms of H.

Theorem 2.1. (Goldie and Resnick, 1989). We have $P[N(A) < \infty] = 1$ or $P[N(A) = \infty] = 1$ according as $H(A) < \infty$ or $H(A) = \infty$.

The converse half of this result is proved using a standard converse to the Borel–Cantelli lemma.

It is not easy to calculate the measure H explicitly. Cases where this can be done are when (i) components are independent or (ii) the components are totally dependent so that effectively the data are one dimensional. In general, it is a helpful heuristic to think of

$$H(dx) \approx P[\text{some record occurs at } dx] .$$

Because H is difficult to calculate, examples in $\mathbb{R}^2$ are best checked by not computing H but rather using a notion of asymptotic independence. Earlier we saw that independent components resulted in a finite number of records but in fact asymptotic independence in the following sense is all that is necessary. Let the common distribution of $\{X_n\}$ be $F(x, y)$. Then we say F is in the domain of attraction of a multivariate extreme value distribution G if there exist $a_n^{(i)} > 0$, $b_n^{(i)} \in \mathbb{R}$, $n \geq 1$, $i = 1,2$ such that

$$F[(\bigvee_{j=1}^{n} X_j^{(i)} - b_n^{(i)})/a_n^{(i)} \leq x_i,\ i = 1,2] = F^n(a_n^{(1)}x_1 + b_n^{(1)}, a_n^{(2)}x_2 + b_n^{(2)}) \to G(x_1,x_2)$$

(cf. Resnick, 1987). Then

THEOREM 2.2. (Goldie and Resnick, 1989). N is finite or infinite with probability 1 according as G is or is not a product measure.

Criteria based on F which guarantee G is a product measure are well known (Resnick, 1987). Using these, there is a surprising result of Sibuya (1960) (cf. Resnick, 1987) that states that the bivariate normal distribution is in a domain of attraction of a bivariate extreme value distribution which is a product measure, provided the correlation ρ of the bivariate normal density is not equal to 1. Thus we get the next result.

COROLLARY 2.3. (Goldie and Resnick, 1989). If $\{X_n\}$ is iid from a bivariate normal whose correlation ρ satisfies $\rho \neq 1$ then $\{X_n\}$ has a finite number of records:

$$P[N < \infty] = 1 .$$

Further properties of multivariate records are given in Goldie and Resnick (1989) including expressions for the moments and and Laplace functional of $N(A)$, $A \in \mathcal{B}(\mathbb{R}^2)$ and an expression for the fundamental quantity $Q(A) = P[N(A) = 0]$. In terms of Q, the Markovian structure of the records falling in the set A can be discussed and the transition probabilities characterized. A partial independent increment property for the point process $N(\cdot)$ is also given.

3. GEOMETRICAL INTERPRETATION

Let H_n be the convex hull of $X_1,\ldots,X_n$. Whether or not X_n is a record (i.e., whether or not $X_n > \bigvee_{i=1}^{n-1} X_i$) can be examined graphically in the following way.

Consider a right angled template which has edges of infinite length and whose bisecting line has inclination $\pi/4$ with the positive x–axis. Suppose the template is moving down from $\boldsymbol{\infty} = (\infty, \infty)$ to $-\boldsymbol{\infty} = (-\infty, -\infty)$ without changing the inclination and that it cannot pass any point of H_n. The template stops when it hits H_n. If the sample is from a continuous distribution, there are only two possible outcomes: Either the template hits H_n at one point or two points. These possibilities are shown in Figure 3.1.

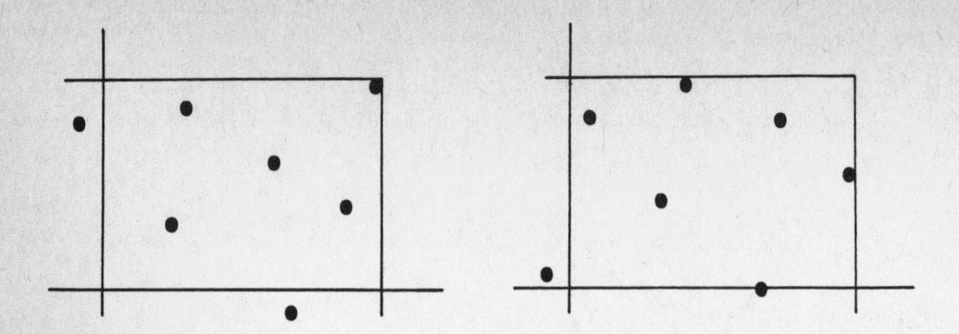

Template hits H_n at one point

Template hits H_n at two points

FIGURE 3.1

Notice that no matter how the template approaches H_n, it hits H_n at the same point (or points). Consider the situations in Figure 3.2 and 3.3. In Figure 3.2, the template cannot move down, but still can move to the left until it hits X_2. Figure 3.3 shows the opposite case and the template eventually hits H_n at X_1 and X_2 in both cases. Remember that the only restriction when we move the template is to keep the inclination $\pi/4$. Alternatively, this procedure may be viewed as moving a horizontal line down until it contacts the sample and then moving a vertical line to the left until the sample is encountered. Do the lines meet the sample at different points or at the same point?

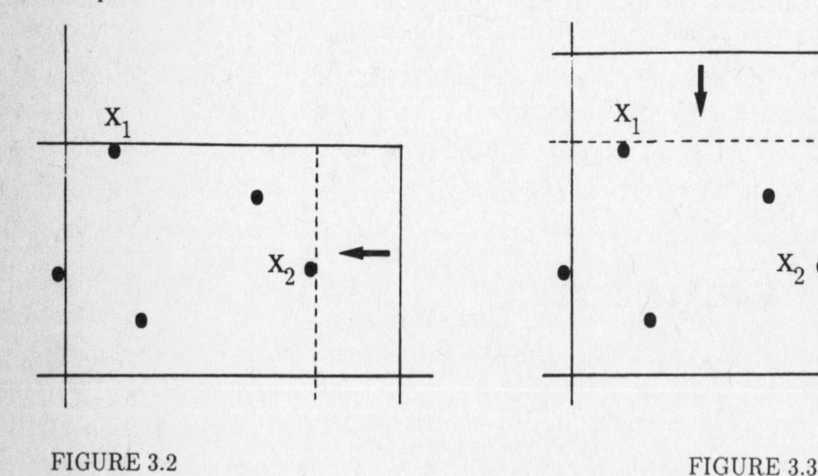

FIGURE 3.2

FIGURE 3.3

It is then clear that

$$X_n \text{ is a record iff the template hits } H_n \text{ only at } X_n \tag{3.1}$$

and

$$[\text{Template hits } H_n \text{ at two points}] = \bigcap_{j=1}^{n} [X_j > \bigvee_{\substack{i=1 \\ i \neq j}}^{n} X_i]^c . \tag{3.2}$$

Now we consider more general templates. We denote by Temp (θ, φ) the template with the exterior angle θ $(\pi < \theta \le 3\pi/2)$ and whose bisecting line makes angle φ $(0 \le \varphi < 2\pi)$ with the positive x–axis. (See Figure 3.4). The template used in the previous discussion is now denoted Temp $(3\pi/2, \pi/4)$. Then bringing down (or up depending on φ) Temp (θ, φ) from the point at infinity in the direction φ, we again see that Temp (θ, φ) hits H_n at either one point or two points. We can use this fact to define another record property as follows.

Definition 3.1. X_n is a (θ, φ)–record iff Temp (θ, φ) hits H_n only at X_n.

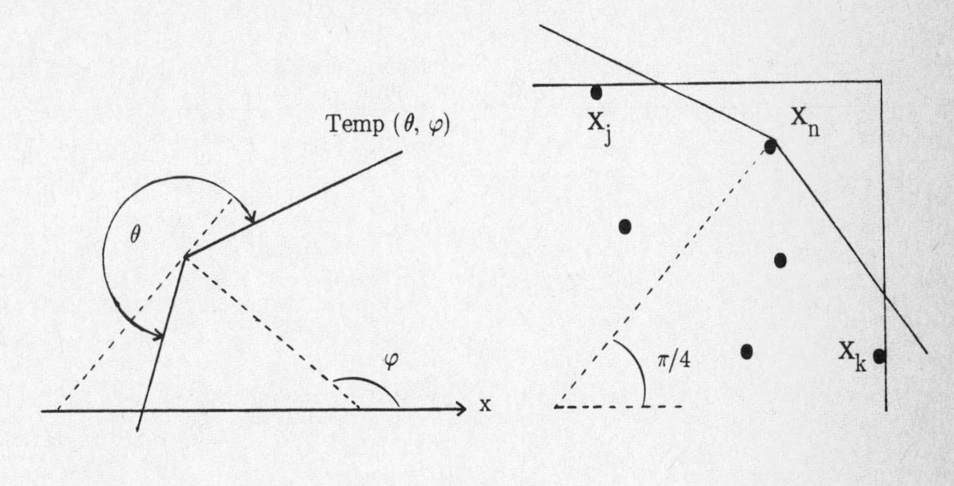

FIGURE 3.4

FIGURE 3.5

Let $\pi < \theta_1 < \theta_2 \le 3\pi/2$, then it follows from the definition that if X_n is a (θ_2, φ)–record, then it is always a (θ_1, φ)–record and that X_n can possibly be a (θ_1, φ)–record even though it is not a (θ_2, φ)–record. For example, in Figure 3.5, X_n is not a $(3\pi/2, \pi/4)$–record (or simply a record) since Temp $(3\pi/2, \pi/4)$ hits H_n at X_j and X_k, but it is a $(\theta, \pi/4)$–record since Temp $(\theta, \pi/4)$ hits H_n only at X_n.

The following proposition explains the relation between these two records.

Proposition 3.2. X_n is a (θ, φ)–record $(\pi < \theta \le 3\pi/2,\ 0 \le \varphi < 2\pi)$ iff Y_n is a record among $Y_1,\dots,Y_n$ where $Y_j = S_\theta \cdot T_\varphi \cdot X_j$, $j = 1,2,\dots,n$,

$$T_\varphi = \begin{bmatrix} \cos(\frac{\pi}{4} - \varphi) & -\sin(\frac{\pi}{4} - \varphi) \\ \sin(\frac{\pi}{4} - \varphi) & \cos(\frac{\pi}{4} - \varphi) \end{bmatrix},$$

$$S_\theta = \begin{bmatrix} \alpha & 1 \\ 1 & \alpha \end{bmatrix},$$

$$\alpha = \tan(\tfrac{3}{4}\pi - \tfrac{\theta}{2}).$$

Proof. Observe by trigonometry that

$$T_\varphi \begin{bmatrix} r\cos t \\ r\sin t \end{bmatrix} = \begin{bmatrix} r\cos(t + \pi/4 - \varphi) \\ r\sin(t + \pi/4 - \varphi) \end{bmatrix}$$

so that the mapping T_φ rotates points in the plane through an angle $\pi/4 - \varphi$. Let $Z_j = T_\varphi X_j$.

It follows that $\mathbf{X}_n$ is a (θ, φ)–record iff $\mathbf{Z}_n$ is a $(\theta, \pi/4)$–record. Thus it suffices to show that

$$\mathbf{Z}_n \text{ is a } (\theta, \pi/4)\text{–record iff } \mathbf{Y}_n \text{ is a record.} \tag{3.3}$$

Note (3.3) holds when $\theta = 3\pi/2$ since in this case $Y_j = (Z_j^{(2)}, Z_j^{(1)})'$, $j = 1,2,\ldots,n$.

So assume $\pi < \theta < 3\pi/2$, which implies $0 < \alpha < 1$. The two edges of Temp $(\theta, \pi/4)$ are given by the lines (see Figure 3.6)

$$y - Z_n^{(2)} = -\alpha(x - Z_n^{(1)}) \qquad \text{for } x \geq Z_n^{(1)}, \text{ and}$$

$$y - Z_n^{(2)} = -\frac{1}{\alpha}(x - Z_n^{(1)}) \qquad \text{for } x < Z_n^{(1)}.$$

Thus, we have

$$\{\mathbf{Z}_n \text{ is a } (\theta, \pi/4)\text{–record}\}$$

$$= \bigcap_{j=1}^{n-1} \{Z_j^{(2)} - Z_n^{(2)} < -\alpha(Z_j^{(1)} - Z_n^{(1)}),\ Z_j^{(2)} - Z_n^{(2)} < -\frac{1}{\alpha}(Z_j^{(1)} - Z_n^{(1)})\}$$

$$= \bigcap_{j=1}^{n-1} \{\alpha Z_j^{(1)} + Z_j^{(2)} < \alpha Z_n^{(1)} + Z_n^{(2)},\ Z_j^{(1)} + \alpha Z_j^{(2)} < Z_n^{(1)} + \alpha Z_n^{(2)}\}$$

$$= \bigcap_{j=1}^{n} \left\{ S_\theta \begin{bmatrix} Z_j^{(1)} \\ Z_j^{(2)} \end{bmatrix} < S_\theta \begin{bmatrix} Z_n^{(1)} \\ Z_n^{(2)} \end{bmatrix} \right\}$$

$$= \bigcap_{j=1}^{n-1} \{Y_j^{(1)} < Y_n^{(1)},\ Y_j^{(2)} < Y_n^{(2)}\}$$

$$= \quad \{\mathbf{Y}_n \text{ is a record}\}$$

so that (3.3) holds. □

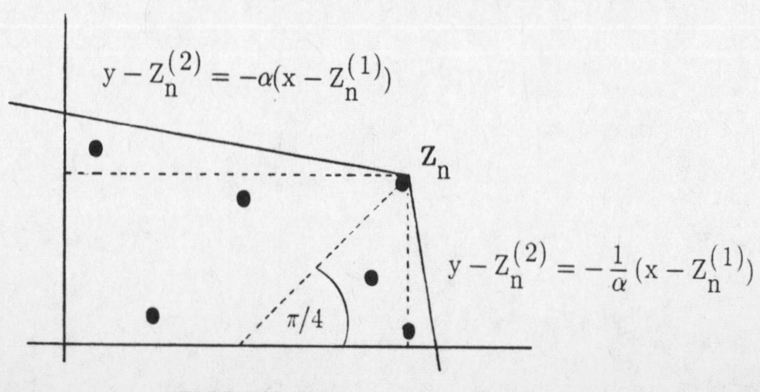

FIGURE 3.6

4. APPLICATION TO A RANDOM SAMPLE FROM A BIVARIATE NORMAL DISTRIBUTION

Let $\mathbf{X}_n = (X_n^{(1)}, X_n^{(2)})'$ be an iid sequence of random vectors in $\mathbb{R}^2$ from the bivariate normal distribution. We assume $E\mathbf{X}_n = 0$ and that the correlation ρ of $\mathbf{X}_1$ satisfies $|\rho| < 1$. Define events as follows:

$$A_n = \{\mathbf{X}_n \text{ is a record}\}$$

$$= \{X_n^{(i)} > \bigvee_{j=1}^{n-1} X_j^{(i)},\ i = 1,2\}$$

$$A_n^{(i)} = \{X_n^{(i)} \text{ is a record}\}$$

$$= \{X_n^{(i)} > \bigvee_{j=1}^{n-1} X_j^{(i)}\},\ i = 1,2\ .$$

$$B_n = \{\text{Temp}\ (3\pi/2,\ \pi/4) \text{ hits } H_n \text{ at one point}\}$$

$$B_n(\theta, \varphi) = \{\text{Temp}\ (\theta, \varphi) \text{ hits } H_n \text{ at one point}\}.$$

Notice that $A_n = A_n^{(1)} \cap A_n^{(2)}$ and $A_n \subset B_n$. We also define

$$N = \sum_{n=1}^{\infty} 1_{A_n}, \text{ and}$$

$$N_i = \sum_{n=1}^{\infty} 1_{A_n^{(i)}},\ i = 1,2\ .$$

It follows from Corollary 2.3 and one dimensional theory (Resnick, 1987, p. 169) that

$$P(N < \infty) = 1 \tag{4.1}$$

$$P(N_1 = \infty) = P(N_2 = \infty) = 1\ . \tag{4.2}$$

Lemma 4.1. Temp $(3\pi/2, \pi/4)$ hits H_n at two points for all large n, i.e., $P(B_n \text{ i.o.}) = 0$.

Proof. We have (4.1) and (4.2) are respectively equivalent to

$$P(A_n \text{ i.o.}) = 0, \text{ and} \tag{4.3}$$

$$P(A_n^{(1)} \text{ i.o.}) = P(A_n^{(2)} \text{ i.o.}) = 1\ . \tag{4.4}$$

These imply that

$$1 = P(\liminf_{n\to\infty} A_n^c \cap \limsup_{n\to\infty} A_n^{(1)})\ .$$

Now pick $\omega \in \liminf_{n\to\infty} A_n^c \cap \limsup_{n\to\infty} A_n^{(1)}$ and there exists some m such that $\omega \in A_m^{(1)}(A_m^{(2)})^c$ and $\omega \in A_n^c$ for any $n \geq m$. If there exists some $n \geq m$ such that for some $\ell \leq n$

$$X_\ell > \bigvee_{\substack{i=1\\ i\neq \ell}}^{n} X_i(\omega)$$

then if $\ell < m$, $X_m^{(1)}(\omega)$ would not be a record (a contradiction to $\omega \in A_m^{(1)}$) and if $n \geq \ell \geq m$, $X_\ell(\omega)$ is a record (a contradiction to $\omega \in A_\ell^c$). From (3.2) we conclude $\omega \in B_n^c$, $n \geq m$, whence

$$1 = P[\liminf_{n\to\infty} A_n^c \cap \limsup_{n\to\infty} A_n^{(1)}] \leq P[\liminf_{n\to\infty} B_n^c]$$

which gives the desired result. □

Lemma 4.2. For any θ $(\pi < \theta \leq 3\pi/2)$ and φ $(0 \leq \varphi < 2\pi)$, Temp (θ, φ) hits H_n at two points for all large n a.s.; i.e.,

$$P(B_n(\theta, \varphi) \text{ i.o.}) = 0 .$$

Proof. Let $\mathbf{X}_n$, $\mathbf{Y}_n$, $\mathbf{Z}_n$ be as in Proposition 3.2. Let $H(\mathbf{Y}_1,\dots,\mathbf{Y}_n)$ be the convex hull of the indicated points. As in Proposition 3.2

$$[\text{Temp } (\theta, \varphi) \text{ hits } H_n \text{ at two points}]$$

$$= [\text{Temp } (3\pi/2, \pi/4) \text{ hits } H(\mathbf{Y}_1,\dots,\mathbf{Y}_n) \text{ at two points}].$$

Since $\{\mathbf{Y}_n\}$ are iid normal, the desired result is a direct application of Lemma 4.1 provided $\rho_{\mathbf{Y}} = \text{Corr}\,(Y_1^{(1)}, Y_1^{(2)}) < 1$. This is readily checked as follows: Since $\alpha < 1$, $\det S_\theta = \alpha^2 - 1 \neq 0$ so S_θ is invertible. The rotation T_φ is also invertible. Since $|\rho| = |\text{corr}\,(X_1^{(1)}, X_1^{(2)})| < 1$ the same is true for $\rho_{\mathbf{Y}}$, else there is an a.s. linear relationship between $Y_1^{(1)}$ and $Y_1^{(2)}$ (see Brockwell and Davis, 1987, p. 43) which translates into an a.s. linear relationship between $X_1^{(1)}$ and $X_1^{(2)}$ which violates $|\rho| < 1$. □

For any sequences of events $\{C_n\}$, $\{D_n\}$ we have

$$\limsup_{n\to\infty} C_n \cup D_n = \limsup_{n\to\infty} C_n \cup \limsup_{n\to\infty} D_n$$

and we therefore obtain from Lemma 4.2 the following result.

Lemma 4.3. For θ_j $(\pi < \theta_j \leq 3\pi/2)$ and φ_j $(0 \leq \varphi_j < 2\pi)$, $j = 1,2,\dots,m$, all the m templates Temp (θ_j, φ_j), $j = 1,2,\dots,m$ eventually hit H_n at two points for all large n a.s.; i.e.,

$$P(\bigcup_{j=1}^{m} B_n(\theta_j, \varphi_j) \text{ i.o.}) = 0 .$$

Next consider a vertex v of H_n and Temp (θ, φ). Let ψ be the exterior angle of v and ϕ be the inclination between the bisecting line of v and the x–axis. (See Figure 4.1). Then we see that

$$\text{Temp}(\theta, \varphi) \text{ hits } H_n \text{ only at v iff } \psi > \theta > \pi \text{ and } |\phi - \varphi| < \frac{\psi - \theta}{2}. \tag{4.5}$$

We denote by V_n the number of vertices of H_n and further define $v_{n,j}$, $\psi_{n,j}$, and $\phi_{n,j}$, $j = 1,2,...,V_n$ as the vertices of H_n, the exterior angles, and the inclinations, respectively. (See Figure 4.2). The next proposition shows that the exterior angles converge to π uniformly.

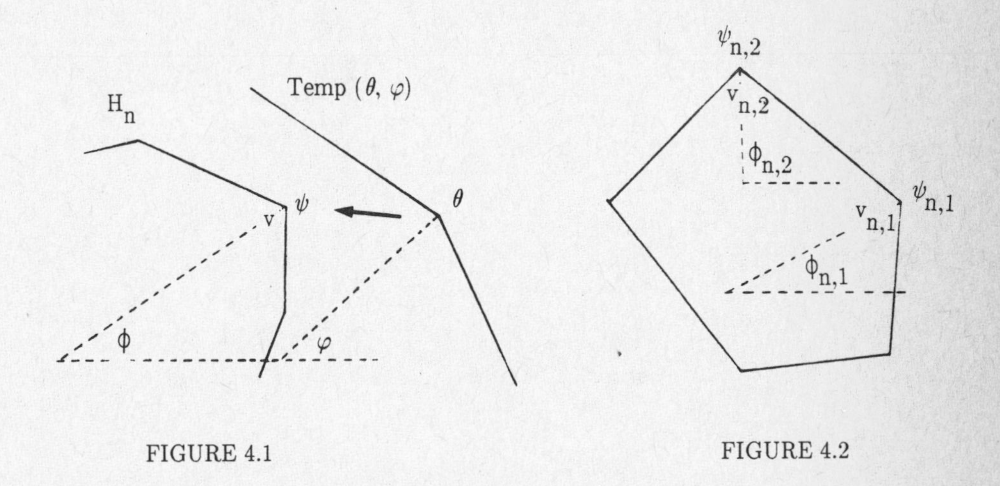

FIGURE 4.1

FIGURE 4.2

Proposition 4.4. We have for a bivariate normal sample whose correlation ρ satisfies $|\rho| < 1$:

$$\lim_{n\to\infty} \bigvee_{j=1}^{V_n} \psi_{n,j} = \pi \text{ a.s.}$$

Proof. Pick any $\epsilon > 0$. Then since $\psi_{n,j} > \pi$ for any j, it suffices to show

$$\bigvee_{j=1}^{V_n} \psi_{n,j}(\omega) \le \pi + \epsilon, \quad n \ge n_0(\omega). \tag{4.6}$$

Choose m large enough so that $m > 4\pi/\epsilon$. Then for any $\phi_{n,j}$, there exists some $\ell(j) \in \{1,2,...,m\}$ such that

$$|\phi_{n,j} - \frac{2\pi \cdot \ell(j)}{m}| < \frac{\epsilon}{4}. \tag{4.7}$$

For m so chosen, we take m templates

$$\{\text{Temp}(\pi + \tfrac{\epsilon}{2}, \tfrac{2\pi}{m} j),\ 1 \le j \le m\}$$

and it follows from Lemma 4.3 that

$$P(\bigcup_N \bigcap_{n \ge N} \bigcap_{j=1}^{m} \{\text{Temp}(\pi + \tfrac{\epsilon}{2}, \tfrac{2\pi}{m} j) \text{ hits } H_n \text{ at two points}\}) = 1. \tag{4.8}$$

Suppose for $n \geq N(\omega)$

$$\omega \in \bigcap_{j=1}^{m} [\text{temp}\ (\pi + \tfrac{\epsilon}{2}, \tfrac{2\pi}{m} j) \text{ hits } H_n \text{ at two points}] .$$

For any $j \leq V_n$ there exists $\ell(j, \omega)$ as in (4.7), whence from (4.5) either

$$\begin{cases} \psi_{n,j} \leq \pi + \frac{\epsilon}{2} \quad \text{or} \\ |\phi_{n,j}(\omega) - \frac{2\pi\, \ell(j,\omega)}{m}| > \frac{\psi_{n,j} - (\pi + \frac{\epsilon}{2})}{2} \end{cases}$$

and therefore from (4.7)

$$\frac{\psi_{n,j} - (\pi + \frac{\epsilon}{2})}{2} < \frac{\epsilon}{4} ,$$

whence

$$\psi_{n,j} < \pi + \frac{\epsilon}{2} + \frac{\epsilon}{2} = \pi + \epsilon$$

giving (4.6) as desired. □

COROLLARY 4.5. For a bivariate normal sample whose correlation ρ satisfies $|\rho| < 1$ we have

$$V_n \to \infty \text{ a.s.}$$

PROOF. We have by Proposition 4.4 that for any $\epsilon > 0$, there exists n_0 such that for any $n > n_0$,

$$\max_{1 \leq j \leq V_n} \psi_{n,j} < \pi + \epsilon \text{ a.s.} \tag{4.9}$$

We also have

$$\sum_{j=1}^{V_n} \psi_{n,j} = (V_n + 2)\pi \text{ a.s.} \tag{4.10}$$

It follows from (4.9) and (4.10) that

$$(V_n + 2)\pi \leq V_n(\pi + \epsilon) \text{ a.s.}$$

This implies

$$V_n > \frac{2\pi}{\epsilon} \text{ for any } n \geq n_0 \text{ a.s.}$$

□

REFERENCES

Brockwell, P. and Davis, R. (1987). Time Series: Theory and Methods. Springer–Verlag, New York.

Davis, R., Mulrow, E. and Resnick, S. (1988). Almost sure limit sets of random samples in $\mathbb{R}^d$. Adv. Appl. Prob.

Fisher, L. (1969). Limiting sets and convex hulls of samples from product measures. Ann. Math. Statist. 40, 1824–1832.

Geffroy, J. (1961). Localisation asymptotique du polyedre d'appui d'un echantillon laplacien a k dimensions. Publ. Inst. Statist. Univ. Paris. X, 212–228.

Goldie, C.M. and Resnick, S. (1989). Records in a partially ordered set. Forthcoming: Ann. Prob.

Groeneboom, P. (1987). Limit theorems for convex hulls. Report 87–07, Mathematics Institute, University of Amsterdam. Forthcoming: Probability Theory and Related Fields.

Resnick, S. (1987). Extreme Values, Regular Variation, and Point Processes. Springer–Verlag, New York.

Sibuya, M. (1960). Bivariate extreme statistics. Ann. Inst. Stat. Math. 11, 195–210.

LIMIT DISTRIBUTIONS OF MULTIVARIATE EXTREME VALUES IN NONSTATIONARY SEQUENCES OF RANDOM VECTORS

J.Hüsler

Department of Math. Statistics, University of Bern

Sidlerstr. 5, CH-3012 Bern, Switzerland

Abstract: We discuss conditions which characterize the limit distributions of multivariate extreme values in general nonstationary sequences of random vectors. We deal mainly with results such that the limit distributions can be found easily with the use of certain asymptotic independence properties. This paper extends known results for multivariate Gaussian sequences to non-Gaussian ones.

1. Introduction

We denote by $\{\mathbf{X}_i, i \geq 1\}$ a sequence of d-dimensional random vectors, neither necessarily stationary nor independent, and by F_i the (multivariate) distribution of $\mathbf{X}_i$. We discuss the limit behaviour of the multivariate extreme values $\mathbf{M}_n = (M_{n1}, M_{n2}, \ldots, M_{nd})'$, where M_{nj} denotes the maximum value of the j-th components of $\mathbf{X}_i$, up to time n:

$$M_{nj} = \max(X_{1j}, \ldots, X_{nj}),\ j \leq d.$$

It is known in the iid. case that if the limit distribution of $\mathbf{M}_n$ exists, with suitable normalization, i.e.

$$P\{(\mathbf{M}_n - \mathbf{b}_n)/\mathbf{a}_n \leq \mathbf{z}\} \xrightarrow{d} G(\mathbf{z}) \ \text{ as } n \to \infty, \tag{1}$$

the limit distribution G is a so-called (multivariate) extreme value distribution. Note that all algebraic operations are meant componentwise. The extreme value distributions are characterized by the max-stability property (cf. de Haan and Resnick (1977), Galambos (1978, 1986), Marshall and Olkin (1983), Deheuvels (1984)).

The convergence in (1) is treated for the case of general stationary random sequences by Hsing (1987) and Hüsler (1987), as well as for particular sequences e.g. for Gaussian sequences by Haiman (1984) and Hüsler and Schüpbach (1988), or bivariate exchangeable sequences Villasenor (1976).

The limit distributions for multivariate extreme values are discussed in more general cases only in a few papers. In particular, we showed that in the independent, but nonidentical case any multivariate distribution can occur as the limit of (1) with suitable sequences $\{F_i, i \geq 1\}, \{\mathbf{a}_n, n \geq 1\}$, and $\{\mathbf{b}_n, n \geq 1\}$ (cf. Hüsler (1988)). This generalizes an old result in the univariate case of Juncosa (1949) and Meijzler (1950). Hence, it is important to characterize classes of limits G under reasonable restrictions, as e.g. the uniform asymptotic negligibility (UAN) condition (see condition $\mathbf{A}_d$ below). A thorough discussion of such a class is given by Gerritse (1986); he dealt with the restriction, that $\mathbf{a}_n \equiv \mathbf{1} = (1, \ldots, 1)' \in R^d$, which implies the class of so-called sup self-decomposable random vectors $\mathbf{Z} \sim G$. It means that there exists for every $t > 0$ a random vector $\mathbf{Z}_t$ such that

$$\mathbf{Z} \stackrel{d}{=} (\mathbf{Z} - t\mathbf{1}) \vee \mathbf{Z}_t$$

where $\mathbf{Z}$ and $\mathbf{Z}_t$ are independent. Equivalently, this means for G, that for every $t > 0$ there exists a distribution G_t^* such that

$$G(\mathbf{z}) = G(\mathbf{z} + t\mathbf{1})G_t^*(\mathbf{z}) \quad \text{for every } \mathbf{z} \in R^d.$$

We see that $G_t^*(\cdot) = G(\cdot)/G(\cdot + t\mathbf{1})$ is in this case a multivariate distribution, for every positive t.

A similar characterization is possible in the general independent case with a general linear normalization, if the UAN condition holds. This is dicussed in our paper (Hüsler (1988)), giving also some results on the dependence structure of these distributions.

The purpose of this paper is to generalize some of these results to the general case of a nonstationary sequence which satisfies certain mixing conditions on the dependence of $\{\mathbf{X}_i\}$. The approach is based on the same ideas as in related situations of univariate stationary or nonstationary sequences (cf. Leadbetter et al. (1983), Hsing (1987), Hüsler (1983), (1987)). The particular case of nonstationary Gaussian sequences was treated by Hüsler and Schüpbach (1988). This class of sequences can be considered as the classical starting point for extending the results to non-Gaussian sequences. Note also that the Gaussian case does not fit into the subclass of sup self-decomposable distributions of Gerritse (1986).

2. General convergence results

Instead of the linear normalization, we deal in this section more generally with any normalization $\{\mathbf{u}_{in}, i \leq n, n \geq 1\}$ with $\mathbf{u}_{in} \in R^d$ and components u_{inj} such that

$$P\{\mathbf{X}_i \leq \mathbf{u}_{in}, i \leq n\} \tag{2}$$

converges. Our aim is to state conditions such that the limit can be evaluated rather easily. For instance, a simple but rather important case to discuss in the next section is the special case with linear normalization and limit distributions with independent components.

In order to discuss the possible limit distributions, the following condition $\mathbf{A}_d$ is generally used. It is the general form of the uniform asymptotic negligibility of the random vectors with respect to the (extreme) boundary values $\mathbf{u}$. We define for every i and j: $\bar{F}_{ij}(x) = 1 - F_{ij}(x)$ with $X_{ij} \sim F_{ij}$, the j-th marginal distribution of F_i.

Condition $\mathbf{A}_d$: Let $F_n^* = \sum_{i=1}^n \sum_{j=1}^d \bar{F}_{ij}(u_{inj})$. Assume that

$$\limsup_{n\to\infty} F_n^* < \infty \text{ and } \lim_{n\to\infty} \max_{i\leq n, j\leq d} \bar{F}_{ij}(u_{inj}) =: \lim_{n\to\infty} F_{max,n}^* = 0.$$

That $F_{max,n}^*$ converges to 0, means that the random vectors $\mathbf{X}_i$ are uniform asymptotically negligible. The case of a linear normalization $\mathbf{a}_n\mathbf{z} + \mathbf{b}_n$ with a convergent sum F_n^*, defines an important class of distributions and it is possible to characterize the limits G belonging to this class (Hüsler (1988)).

To control the dependence of the random sequence, mixing conditions are usually introduced. The type of condition we are using, originates in a paper by Leadbetter (1974) and was extended to a more general form for the univariate non-stationary case (Hüsler (1983)). We denote by $B_n(I) = \{\mathbf{X}_i \leq \mathbf{u}_{in}, i \in I\}$ where $I \subset \{1, \ldots, n\}$, which usually depends also on n.

Condition $\mathbf{D}_d$ $= \mathbf{D}_d(\{\mathbf{u}_{in}, i \leq n, n \geq 1\})$. We assume that there exists an array $\{\alpha_{nm}, m \leq n, n \geq 1,\}$ such that

i) $|P(B_n(I \cup J)) - P(B_n(I))P(B_n(J))| \leq \alpha_{nm}$
for every pair of subsets I and J of $\{1, \ldots, n\}$ which are m-separated
(i.e. $\min_{i\in J}(i) - \max_{i\in I}(i) \geq m$ or $\min_{i\in I}(i) - \max_{i\in J}(i) \geq m$)
and

ii) $\lim_{n\to\infty} \alpha_{n,m_n^*} = 0$ for some sequence $\{m_n^*, n \geq 1\}$ with $m_n^* \to \infty$ and $m_n^* F_{max,n}^* \to 0$ as $n \to \infty$.

This condition implies the asymptotic independence of extreme values which are separated by at least m_n^*; it restricts the long-range dependence. It implies the following result.

Lemma 2.1: Let I_k, $k \leq r$, r some integer, be r pairwisely m-separated subsets of $\{1, \ldots, n\}$. Assume that $\mathbf{D}_d$ holds with respect to $\mathbf{u}_{in}$. Then

$$|P(B_n(\cup_{k=1}^r I_k)) - \prod_{k=1}^r P(B_n(I_k))| \leq r\alpha_{nm}.$$

The proof follows as in Leadbetter (1974) or Hüsler (1983). To control the local behaviour of extreme values a second mixing type condition is introduced. Because of the multivariate setup we distinguish between the local behaviour of a single component and of the local cross behaviour between two and more components. The conditions are such that a clustering of extreme values is excluded in a small time interval with respect to a single component or with respect to two and more components, respectively (see the simple examples below). Let $d_n(I, \delta)$ denote for any $I \subset \{1, \ldots, n\}$ and $\delta > 0$

$$d_n(I, \delta) = \min_{I^* \subset I} \sum_{i<h \in I^*} P\{\mathbf{X}_i \not\leq \mathbf{u}_{in}, \mathbf{X}_h \not\leq \mathbf{u}_{hn}\}$$

where $\sum_{i \in I \setminus I^*} P\{\mathbf{X}_i \not\leq \mathbf{u}_{in}\} < \delta$. Let $F_n^*(I) = \sum_{i \in I} P\{\mathbf{X}_i \not\leq \mathbf{u}_{in}\}$.

Condition $\mathbf{D}_d'$= $\mathbf{D}_d'(\{\mathbf{u}_{in}, i \leq n, n \geq 1\})$. We assume that there exist an array $\{\alpha_{nr}^*, n \geq 1, r \geq 1\}$ and a sequence $\{g_r, r \geq 1\}$ such that $\lim_{r \to \infty} r g_r = 0$, $\lim_{r \to \infty} \limsup_{n \to \infty} r\alpha_{nr}^* = 0$ and for every $r \geq 1$ and for all $n \geq n_0(r)$: $d_n(I, g_r) \leq \alpha_{nr}^*$ for all $I \subset \{1, \ldots, n\}$ such that $F_n^*(I) \leq F_n^*/r$.

The introduction of I^* in $d_n(I, \delta)$ is rather useful in the particular Gaussian case, to exclude in a simple way some random vectors with too heavy weight in the $d_n(I, \delta)$-term (cf. Hüsler (1983)). But this is certainly true in other non-Gaussian cases also. Without taking the minimum on subsets I^*, the condition $\mathbf{D}_d'$ would be more restrictive.

Our first result shows that under these three conditions $\mathbf{A}_d, \mathbf{D}_d$ and $\mathbf{D}_d'$, the possible convergence is in the non-stationary case equal to the convergence in the independent, non-identical case.

Theorem 2.2: *Let $\{\mathbf{X}_i, i \geq 1\}$ be a sequence of random vectors and $\{\mathbf{u}_{in}, i \leq n, n \geq 1\}$ an array of extreme boundary values. Assume that the conditions $\mathbf{A}_d, \mathbf{D}_d$ and $\mathbf{D}_d'$ hold, then*

$$P\{\mathbf{X}_i \leq \mathbf{u}_{in}, i \leq n\} - \prod_{i=1}^n P\{\mathbf{X}_i \leq \mathbf{u}_{in}\} \to 0 \text{ as } n \to \infty.$$

The proof uses the technique given in Hüsler(1983) which is based on the ideas of Leadbetter (1974). We use the following blocking technique. For any integer r, divide the set $\{1, \ldots, n\}$, for given $n \geq r$, into r subinterval $I_1 = \{1, \ldots, i_1\}$, $I_2 = \{i_1 + 1, \ldots, i_2\}$, ..., $I_r = \{i_{r-1} + 1, \ldots, i_r\}$ such that $F_n^*(I_k) \leq F_n^*/r$ with successively maximally chosen i_k (i.e. $F_n^*(I_k) + P\{\mathbf{X}_{i_k+1} \not\leq \mathbf{u}_{i_k+1,n}\} > F_n^*/r$), $k \leq r$. Obviously, we have $i_r \leq n$. In addition, subdivide for a given $\varepsilon > 0$, I_k into I_{k1} and I_{k2} with $I_{k1} = \{i_{k-1} + 1, \ldots, i_k - m_k\}$, $I_{k2} = I_k \setminus I_{k1}$ and $F_n^*(I_{k2}) \leq \varepsilon F_n^*/r$ with m_k maximally chosen, for every k. The maximality of m_k implies that $m_k + 1 \geq \varepsilon F_n^*/rdF_{max,n}^*$, for every k; it follows as in Hüsler (1983). In the following we choose a particular ε:

$$\varepsilon^*(n) := (m_n^* + 1)F_{max,n}^*/rdF_n^*.$$

Note that by this choice $\varepsilon^*(n) \to 0$ as $n \to \infty$ for each r and that also for all $k \leq r$: $m_k > m_n^*$. For these intervals we state first a result which is used in the following proofs.

Lemma 2.3: *Assume that* $\mathbf{A}_d$ *and* $\mathbf{D}_d$ *hold. Then for every* r

$$\lim_{n\to\infty} |P\{\mathbf{X}_i \le \mathbf{u}_{in}, i \le n\} - \prod_{k=1}^{r} P(B_n(I_k))| = 0$$

and for every $j \le d$ *with* $B_{nj}(I_k) = \{X_{ij} \le u_{inj}, i \in I_k\}$:

$$\lim_{n\to\infty} |P\{X_{ij} \le u_{inj}, i \le n\} - \prod_{k=1}^{r} P(B_{nj}(I_k))| = 0.$$

The statements hold also with I_{k1} *instead of* I_k.

Proof: It follows by the construction of the subintervals that for every r fixed and $\varepsilon \to 0$ as $n \to \infty$:

$$0 \le P\{\mathbf{X}_i \le \mathbf{u}_{in}, i \in \cup_{k=1}^{r} I_{k1}\} - P\{\mathbf{X}_i \le \mathbf{u}_{in}, i \le n\} \to 0 \text{ as } n \to \infty. \tag{3}$$

For, the difference of the two probabilities in (3) is obviously positive and bounded above by

$$\sum_{k=1}^{r} \sum_{i \in I_{k2}} P\{\mathbf{X}_i \not\le \mathbf{u}_{in}\} + \sum_{i=i_r+1}^{n} P\{\mathbf{X}_i \not\le \mathbf{u}_{in}\}$$

$$= \sum_{k=1}^{r} F_n^*(I_{k2}) + (F_n^* - \sum_{k=1}^{r} F_n^*(I_k)) \le \varepsilon F_n^* + o(1) = o(1)$$

as $n \to \infty$, since $\limsup F_n^* < \infty$ and by the maximality of the division into I_k : $F_n^* - \sum_{k=1}^{r} F_n^*(I_k) \le dr F_{max,n}^* \to 0$.

By Lemma 2.1 it follows that

$$|P(B_n(\cup_{k=1}^{r} I_{k1})) - \prod_{k=1}^{r} P(B_n(I_{k1}))| \le r\alpha_{nm}$$

which tends to 0 as $n \to \infty$, by choosing $m = m_n^*$, and $\varepsilon = \varepsilon^*(n)$.

Furthermore, $0 \le P(B_n(I_{k1})) - P(B_n(I_k)) \le \varepsilon F_n^*/r$, which implies that for every r

$$0 \le \prod_{k=1}^{r} P(B_n(I_{k1})) - \prod_{k=1}^{r} P(B_n(I_k)) = O(\varepsilon F_n^*) \to 0 \text{ as } n \to \infty.$$

This implies the first statement. In the same way the componentwise and the other statements follow.

Proof of Theorem 2.2: Let r be fixed. For every $k \le r$ we have

$$\sum_{i \in I_{k1}} P\{\mathbf{X}_i \not\le \mathbf{u}_{in}\} \le F_n^*(I_k) \le F_n^*/r.$$

By $\mathbf{D}'_d$, there exists a $I_{k1}^* \subset I_{k1}$ such that

$$d_n(I_{k1}, g_r) \le \alpha_{nr}^* \text{ for all } n \ge n_0(r).$$

Hence by Bonferroni's inequalities, for all $k \le r$

$$1 - P\{\mathbf{X}_i \le \mathbf{u}_{in}, i \in I_{k1}^*\} \le 1 - P\{\mathbf{X}_i \le \mathbf{u}_{in}, i \in I_{k1}\} \le F_n^*/r$$

and

$$1 - P\{\mathbf{X}_i \le \mathbf{u}_{in}, i \in I_{k1}^*\} \ge \sum_{i \in I_{k1}^*} P\{\mathbf{X}_i \not\le \mathbf{u}_{in}\} - d_n(I_{k1}^*, g_r)$$

$$\geq \sum_{i \in I_{k1}} P\{\mathbf{X}_i \not\leq \mathbf{u}_{in}\} - g_r - \alpha^*_{nr}.$$

Since F^*_n/r is arbitrarily small for r sufficiently large, we get with $d_r > 0$ suitably chosen ($d_r \to 0$ as $r \to \infty$),

$$\exp\{-(1+d_r)S_{nr}\} \leq \prod_{k=1}^{r} P\{\mathbf{X}_i \leq \mathbf{u}_{in}, i \in I_{k1}\} \leq \exp\{-S_{nr} + rg_r + r\alpha^*_{nr}\}$$

and in the same way also

$$\exp\{-(1+c_n)S_{nr}\} \leq \prod_{k=1}^{r} \prod_{i \in I_{k1}} P\{\mathbf{X}_i \leq \mathbf{u}_{in}\} \leq \exp\{-S_{nr}\}$$

with $c_n = O(F^*_{max,n})$ and $S_{nr} = \sum_{k=1}^{r} \sum_{i \in I_{k1}} P\{\mathbf{X}_i \not\leq \mathbf{u}_{in}\}$.

Therefore it follows by assumption $\mathbf{A}_d$ and $\mathbf{D}'_d$

$$\lim_{r\to\infty} \limsup_{n\to\infty} | \prod_{k=1}^{r} P\{\mathbf{X}_i \leq \mathbf{u}_{in}, i \in I_{k1}\} - \prod_{k=1}^{r} \prod_{i \in I_{k1}} P\{\mathbf{X}_i \leq \mathbf{u}_{in}\}| = 0.$$

By combining this statement with Lemma 2.3, the proof is complete, since also

$$\limsup_{n\to\infty} | \prod_{i=1}^{n} P\{\mathbf{X}_i \leq \mathbf{u}_{in}\} - \prod_{k=1}^{r} \prod_{i \in I_{k1}} P\{\mathbf{X}_i \leq \mathbf{u}_{in}\}| = 0.$$

A consequence of the conditions $\mathbf{A}_d$ and $\mathbf{D}_d$ is the following basic lemma, which will be used to deduce the positive dependence of the possible limit distributions.

Lemma 2.4: *Assume that* $\mathbf{A}_d$ *and* $\mathbf{D}_d$ *hold, then*

$$\lim_{n\to\infty} \Big(P\{\mathbf{X}_i \leq \mathbf{u}_{in}, i \leq n\} - \prod_{j=1}^{d} P\{X_{ij} \leq u_{inj}, i \leq n\} \Big) \geq 0$$

Proof: We use the construction of I_{k1} and I_{k2} of Theorem 2.2. Hence we get for every r, with $\varepsilon = \varepsilon^*(n)$, $m_k > m^*_n$ by Lemma 2.3

$$\lim_{n\to\infty} |P\{\mathbf{X}_i \leq \mathbf{u}_{in}, i \leq n\} - \prod_{k=1}^{r} P(B_n(I_k))| = 0$$

as well as for every component. We bound each factor of the product:

$$P(B_n(I_k)) = P\{\mathbf{X}_i \leq \mathbf{u}_{in}, i \in I_k\} = P(\cap_{j=1}^{d} \{X_{ij} \leq u_{inj}, i \in I_k\}) =$$

$$= P(\cap_{j=1}^{d} B_{nj}(I_k)) \geq 1 - \sum_{j=1}^{d} P(B^c_{nj}(I_k)).$$

By construction $\sum_{j=1}^{d} P(B^c_{nj}(I_k)) \leq dF^*_n/r$ for every $k \leq r$, which implies

$$\prod_{j=1}^{d} P(B_{nj}(I_k)) \leq \exp(-\sum_{j=1}^{d} P(B^c_{nj}(I_k))) \leq 1 - \frac{1}{1+\delta} \sum_{j=1}^{d} P(B^c_{nj}(I_k))$$

and hence

$$\Big(\prod_{j=1}^{d} P(B_{nj}(I_k))\Big)^{1+\delta} \leq \exp(-(1+\delta)\sum_{j=1}^{d} P(B^c_{nj}(I_k))) \leq 1 - \sum_{j=1}^{d} P(B^c_{nj}(I_k))$$

with a suitable $\delta > 0$. Note that δ depends on r and can be chosen arbitrarily small if r is sufficiently large. Together we get for every $k \leq r$

$$\Big(\prod_{j=1}^{d} P(B_{nj}(I_k))\Big)^{1+\delta} \leq P(B_n(I_k))$$

and finally

$$\Big(\prod_{k=1}^{r}\prod_{j=1}^{d} P(B_{nj}(I_k))\Big)^{1+\delta} \leq \prod_{k=1}^{r} P(B_n(I_k)),$$

which implies the statement by letting $r \to \infty$.

Theorem 2.2 deals with the case where the random vectors are asymptotically independent with respect to the extreme behaviour. This implies that the joint exceedances $\mathbf{X}_i \not\leq \mathbf{u}_{in}$ occur not in time clusters; but exceedances in more than one component at the same timepoint are not excluded. E.g. the simple example with independent bivariate r. vectors with $X_{i1} \equiv X_{i2}$ is covered by Theorem 2.2. Another possible and important case arises if also the components of each random vector are asymptotically independent with respect to the extreme boundary values. This excludes a possible joint clustering of extreme values in two or more components at the same timepoint, as in the above example.

Condition $\mathbf{D}_d^* = \mathbf{D}_d^*(\{\mathbf{u}_{in}, i \leq n, n \geq 1\})$ holds if

$$\sum_{i=1}^{n} \sum_{1\leq j<l\leq d} P\{X_{ij} > u_{inj}, X_{il} > u_{inl}\} =: \sum_{i=1}^{n} f_{in}^* \to 0 \text{ as } n \to \infty.$$

Lemma 2.5: *If* $\mathbf{A}_d$ *and* $\mathbf{D}_d^*$ *hold, then*

$$\prod_{i=1}^{n} P\{\mathbf{X}_i \leq \mathbf{u}_{in}\} - \prod_{i=1}^{n}\prod_{j=1}^{d} F_{ij}(u_{inj}) \to 0 \text{ as } n \to \infty. \tag{4}$$

Proof: The left hand side of (4) can be bounded by

$$\sum_{i=1}^{n} |P\{\mathbf{X}_i \leq \mathbf{u}_{in}\} - \prod_{j=1}^{d} F_{ij}(u_{inj})|. \tag{5}$$

We have

$$\sum_{i=1}^{d} \bar{F}_{ij}(u_{inj}) - f_{in}^* \leq P\{\mathbf{X}_i \not\leq \mathbf{u}_{in}\} \leq \sum_{i=1}^{d} \bar{F}_{ij}(u_{inj}) \leq dF_{max,n}^*.$$

Since $F_{max,n}^* \to 0$ as $n \to \infty$, it follows that

$$P\{\mathbf{X}_i \leq \mathbf{u}_{in}\} - \prod_{j=1}^{d} F_{ij}(u_{inj}) \leq f_{in}^* + C_n \sum_{i=1}^{d} \bar{F}_{ij}(u_{inj}) \tag{6}$$

with $C_n = o(1)$ as $n \to \infty$. In the same way the left hand side of (6) is bounded below by

$$\tilde{C}_n \sum_{i=1}^{d} \bar{F}_{ij}(u_{inj})$$

with $\tilde{C}_n = o(1)$ as $n \to \infty$. Hence the absolute value of the difference in (5) is bounded by

$$(C_n + \tilde{C}_n)F_n^* + \sum_{i=1}^{n} f_{in}^* \to 0 \text{ as } n \to \infty,$$

since $\sum_i \sum_j \bar{F}_{ij}(u_{inj}) = F_n^*$ is bounded. This finishes the proof.

Hence we get immediately the following result, by combining Theorem 2.2 with Lemma 2.5.

Theorem 2.6: *Let $\{\mathbf{X}_i, i \geq 1\}$ be a sequence of random vectors and $\{\mathbf{u}_{in}, i \leq n, n \geq 1\}$ an array of extreme boundary values. Assume that the conditions $\mathbf{A}_d, \mathbf{D}_d, \mathbf{D}'_d$ and $\mathbf{D}^*_d$ hold, then*

$$P\{\mathbf{X}_i \leq \mathbf{u}_{in}, i \leq n\} - \prod_{i=1}^{n} \prod_{j=1}^{d} P\{X_{ij} \leq u_{inj}\} \to 0 \text{ as } n \to \infty. \tag{7}$$

We proved that the assumption of negative dependent random vectors $\mathbf{X}_i$ in the stationary case implies already the independence of the components Z_j, assuming the conditions $\mathbf{A}_d$ and $\mathbf{D}_d$. This is true also in the nonstationary case based on the following Proposition 2.7. Therefore we recall this weak negative dependence assumption: a multivariate distribution H is *pairwise negative quadrant dependent* (PNQD), if for every bivariate marginal H_{jl} of H

$$H_{jl}(x, y) \leq H_j(x) H_l(y)$$

where H_j and H_l are the corresponding univariate marginal distributions of H.

Proposition 2.7: *If $\mathbf{A}_d$ holds, and if every $\mathbf{X}_i \sim F_i$ is PNQD, $i \geq 1$, then $\mathbf{D}^*_d$ holds. Hence* (7) *holds by assuming also $\mathbf{D}_d$ and $\mathbf{D}'_d$.*

Proof: Since F_i is PNQD, the sum f^*_{in} is bounded by

$$\sum_{1 \leq j < l \leq d} \bar{F}_{ij}(u_{inj}) \bar{F}_{il}(u_{inl}) \leq d F^*_{max,n} \sum_{j=1}^{d} \bar{F}_{ij}(u_{inj}).$$

Thus

$$\sum_{i=1}^{n} f^*_{in} \leq d F^*_{max,n} F^*_n = o(1) \quad \text{as} \quad n \to \infty.$$

Note that this proves that $\mathbf{D}^*_d$ is a rather weak assumption, much weaker than PNQD. Theorem 2.6 deals with the case that neither a single component nor two or more components show a cluster behaviour of extremes. A clustering possibility is already discussed in Theorem 2.2, where in the dependent components of $\mathbf{M}_n$, a joint clustering is only possible at the same timepoint. But there is still another interesting case where the components of $\mathbf{M}_n$ are asymptotically independent with the possibility for a clustering of extremes in single components. This is best illustrated by a simple example again. Let $X_{ij} = Y_{[i/c_j]+1,j}$ for every $i \geq 1$ and $j \leq d$, with Y_{ij} being iid. r.v.'s, and with integers $c_j \geq 1$. Then obviously, the M_{nj}'s are independent ($j \leq d$), but in every component there are always exceedances of clustersize c_j. This idea is treated generally by assuming instead of $\mathbf{D}'_d$ a similar condition, say $\mathbf{D}''_d$, where we redefine in $\mathbf{D}'_d$ the expression $d_n(I, \delta)$ by

$$d_n(I, \delta) = \min_{I^* \subset I} \sum_{i,h \in I^*} \sum_{j < l} P\{X_{ij} > u_{inj}, X_{hl} > u_{hnl}\}$$

with everthing else unchanged. Then in a similar way we find

Theorem 2.8: *Let $\{\mathbf{X}_i, i \geq 1\}$ be a sequence of random vectors and $\{\mathbf{u}_{in}, i \leq n, n \geq 1\}$ an array of extreme boundary values. Assume that the conditions $\mathbf{A}_d, \mathbf{D}_d$ and $\mathbf{D}''_d$ hold, then*

$$P\{\mathbf{X}_i \leq \mathbf{u}_{in}, i \leq n\} - \prod_{j=1}^{d} P\{X_{ij} \leq u_{inj}, i \leq n\} \to 0 \text{ as } n \to \infty.$$

3. Limit distributions

In this section we deal with the characterization of the nondegenerate limit distributions for the extreme values M_n. We apply the results of Section 2 by using $\mathbf{u}_n = \mathbf{u}_n(\mathbf{z}) = \mathbf{a}_n\mathbf{z} + \mathbf{b}_n$ with $\mathbf{z}$ such that $G(\mathbf{z}) > 0$.

As mentioned in the introduction any multivariate distribution can occur as limit of (1) in the general case (Hüsler (1988)). By using the UAN condition $\mathbf{A}_d$ in the case of independent r. vectors $\mathbf{X}_i$, the class of limits G is rather restricted to be a subclass of the max-infinitely divisible (max-id.) distributions, introduced by Balkema and Resnick (1977). Under the conditions $\mathbf{D}_d$ and $\mathbf{D}'_d$ we get also in the general case the same class of limits G as in the independent case. Hence Theorem 1 and 2 of Hüsler (1988) and the results of Section 2 imply immediately the following statements.

We call a univariate distribution function G *sup-log-concave* if either i) $\log G$ is concave, or ii) $x^*(G) < \infty$ and $\log G(x^*(G) - \exp(-\cdot))$ is concave, or iii) $x_*(G) > -\infty$ and $\log G(x_*(G) + \exp(\cdot))$ is concave, with $x^*(G)$ and $x_*(G)$ denoting upper and lower endpoints of G, respectively.

We denote by $F_n^*(t, \mathbf{z}) = \sum_{i \le nt}(1 - F_i(\mathbf{a}_n\mathbf{z} + \mathbf{b}_n))$ with respect to some normalization $\mathbf{a}_n$ and $\mathbf{b}_n$. We use in the following, that also these partial sums $F_n^*(t, \mathbf{z})$ converge, for all $t \in (0, 1]$ and z,

$$F_n^*(t, \mathbf{z}) \to w(t, \mathbf{z}) \tag{8}$$

Theorem 3.1: *Let $\{\mathbf{X}_i, i \ge 1\}$ be a sequence of random vectors in R^d. Assume that the conditions $\mathbf{A}_d, \mathbf{D}_d$ and $\mathbf{D}'_d$ hold for every $\mathbf{u}_n(\mathbf{z}) = \mathbf{a}_n\mathbf{z} + \mathbf{b}_n$ with $\mathbf{z}$ such that $G(\mathbf{z}) > 0$, and suitable normalizations $\mathbf{a}_n(> 0), \mathbf{b}_n$.*
i) *Then as $n \to \infty$*

$$P\{\mathbf{M}_n \le \mathbf{a}_n\mathbf{z} + \mathbf{b}_n\} \xrightarrow{d} G(\mathbf{z}) \tag{9}$$

is equivalent to

$$\sum_{i \le n}(1 - F_i(\mathbf{a}_n\mathbf{z} + \mathbf{b}_n)) \to -\log G(\mathbf{z}). \tag{10}$$

ii) *If also* (8) *holds, then the class of nondegenerate limits G in i) is precisely given by distributions G satisfying for every $t \in (0, 1]$:*

$$G_t^*(\mathbf{z}) := G(\mathbf{z})/G(\mathbf{A}(t)\mathbf{z} + \mathbf{B}(t))$$

is a distribution in R^d, with suitable functions $\mathbf{A}(t) : (0, 1] \to R_+^d$ and $\mathbf{B}(t) : (0, 1] \to R^d$, where $\mathbf{A}(s)\mathbf{A}(t) = \mathbf{A}(st)$ and $\mathbf{B}(st) = \mathbf{A}(s)\mathbf{B}(t) + \mathbf{B}(s)$ for any $s, t \in (0, 1]$.
iii) *Assume in addition that* (8) *holds. For every nondegenerate limit G of* (9) *and every $\mathbf{v} \in [0, \infty)^d$*

$$\tilde{G}_{\mathbf{v}}(z) := P\{\max_{j \le d: v_j > 0} v_j Z_j \le z\} = G(\mathbf{z})$$

is sup-log-concave with $z \in R, \mathbf{Z} \sim G$, and $z_j = z/v_j$, if $v_j > 0$, and $= +\infty$, else.

Remark: i) Note that (10) is obviously equivalent to $\prod_{i \le n} F_i(\mathbf{a}_n\mathbf{z} + \mathbf{b}_n) \xrightarrow{d} G(\mathbf{z})$ as $n \to \infty$.

ii) We mention also that the limits are max id. (see Resnick (1987) for a definition). Hence another converse statement can be formulated being equivalent to ii) (cf. Hüsler (1988)).

iii) Since the limits G in (9) are max id. , they are associated (see Resnick (1987)). Hence they are PLOD (positive lower orthant dependent); see Joag-Dev (1983). $\mathbf{Z} \sim G$ is PLOD if

$$G(\mathbf{z}) \ge \prod_{j=1}^{d} G_j(z_j), \quad \text{for every } \mathbf{z} \in R^d.$$

The PLOD property, being a rather useful inequality in applications, holds already under less restrictive assumptions on the dependence structure of the random sequence, with respect to the extreme boundary values.

Theorem 3.2: *Let* $\{\mathbf{X}_i, i \geq 1\}$ *be a sequence of random vectors in* R^d. *Assume that the conditions* $\mathbf{A}_d$ *and* $\mathbf{D}_d$ *hold for every* $\mathbf{u}_n(\mathbf{z}) = \mathbf{a}_n\mathbf{z} + \mathbf{b}_n$ *with* $\mathbf{z}$ *such that* $G(\mathbf{z}) > 0$, *and suitable normalizations* $\mathbf{a}_n(> 0), \mathbf{b}_n$. *If*

$$P\{\mathbf{M}_n \leq \mathbf{a}_n\mathbf{z} + \mathbf{b}_n\} \xrightarrow{d} G(\mathbf{z})$$

as $n \to \infty$, *then* G *is* PLOD.

This is an immediate consequence of Lemma 2.4. By improving slightly the technique of Section 2, by introducing $r = r(n) \to \infty$ in a suitable way, it is even possible to show that under the assumptions of Theorem 3.2, the existing G is max id. , hence associated. This shows again the reasonability of Condition $\mathbf{D}_d$. We state the improved result.

Theorem 3.3: *Let* $\{\mathbf{X}_i, i \geq 1\}$ *be a sequence of random vectors in* R^d. *Assume that the conditions* $\mathbf{A}_d$ *and* $\mathbf{D}_d$ *hold for every* $\mathbf{u}_n(\mathbf{z}) = \mathbf{a}_n\mathbf{z} + \mathbf{b}_n$ *with* $\mathbf{z}$ *such that* $G(\mathbf{z}) > 0$, *and suitable normalizations* $\mathbf{a}_n(> 0), \mathbf{b}_n$. *If* $P\{\mathbf{M}_n \leq \mathbf{a}_n\mathbf{z} + \mathbf{b}_n\} \xrightarrow{d} G(\mathbf{z})$ *as* $n \to \infty$, *then* G *is max id., hence associated.*

Proof: Use the technique of Section 2 with

$$r = r(n) = \min((\alpha_{n,m_n^*})^{-1/2}, (F^*_{max,n})^{-1/2}).$$

Then Lemma 2.3 holds with I_k depending on $r(n)$. Define

$$\mathbf{M}_k^* = \max\{\mathbf{X}_i, i \in I_k\}, \ k \leq r(n).$$

By using again Resnick (1987) p. 262, the existing limit distribution of

$$\prod_{k=1}^{r(n)} P\{\mathbf{M}_k^* \leq \mathbf{a}_n\mathbf{z} + \mathbf{b}_n\}$$

is max id.; thus the Lemma 2.3, improved as mentioned above, implies that the existing G is max id..

The next result considers the more restrictive asymptotic independence of the components M_{nj}, $j \leq d$. The statements are implied by Theorem 2.6, Proposition 2.7 and Theorem 2.8.

Theorem 3.4: *Let* $\{\mathbf{X}_i, i \geq 1\}$ *be a sequence of random vectors in* R^d. *Assume that the conditions* $\mathbf{A}_d$ *and* $\mathbf{D}_d$ *hold for every* $\mathbf{u}_n(\mathbf{z}) = \mathbf{a}_n\mathbf{z} + \mathbf{b}_n$ *with* $\mathbf{z}$ *such that* $G(\mathbf{z}) > 0$, *and suitable normalizations* $\mathbf{a}_n(> 0), \mathbf{b}_n$. *Then* $G(\mathbf{z}) = \prod_{j=1}^d G_j(z_j)$ *if either for every* $\mathbf{u}_n(\mathbf{z})$
i) *also* $\mathbf{D}'_d$ *and* $\mathbf{D}^*_d$ *hold, or*
ii) $\mathbf{D}'_d$ *holds and every* $\mathbf{X}_i$ *is* PNQD, *or*
iii) *also* $\mathbf{D}''_d$ *holds and* $G_j(z) \stackrel{d}{=} \lim_{n\to\infty} P\{(M_{nj} - b_{nj})/a_{nj} \leq z\}$ *exists for every* $j \leq d$.

The condition $\mathbf{D}_d$ cannot be simply replaced by any of the dependence concepts as e.g. positive or negative association, or positve or negative lower orthant dependence, since we get only one bound for the difference in $\mathbf{D}_d$. More precisely, if a positive dependence is assumed, we have e.g.

$$P(B_n(I \cup J)) \geq P(B_n(I))P(B_n(J));$$

an additional restriction is necessary since also totally dependent random vectors are positive dependent. Such strongly dependent random sequences may have rather different limit distributions for the maxima. But the condition $\mathbf{D}'_d$ can be replaced by such an argument, similar to Proposition 2.7; note again that the following dependence condition is more restrictive than $\mathbf{D}'_d$.

Proposition 3.5: *Assume that* $\{\mathbf{X}_i, i \geq 1\}$ *is negative dependent, in the sense that for every pair* $i, k \geq 1$*:*

$$P\{\mathbf{X}_i \leq \mathbf{x}, \mathbf{X}_k \leq \mathbf{y}\} \leq F_i(\mathbf{x})F_k(\mathbf{y}) \text{ for all } x, y \in R^d.$$

Then Condition $\mathbf{D}'_d$ *holds with respect to every boundary* $\{\mathbf{u}_{in}, i \leq n, n \geq 1\}$ *such that the condition* $\limsup_{n\to\infty} F_n^* < \infty$ *holds. Hence, if in addition* $\mathbf{A}_d$ *and* $\mathbf{D}_d$ *hold, then* G *has independent components, if existing with repect to a suitable* $\mathbf{u}_n(\mathbf{z})$.

Notice that the negative dependence assumption in Proposition 3.5 as well as in Theorem 3.4 are rather strongly restrictive in relation to the used information on extreme values. Unfortunately, we cannot replace in these results the negative dependence by the positive analogon, since the same fact mentioned above with totally dependent random vectors, is crucial again.

Our last result deals with the opposite case of totally dependent random vectors, which means that

$$P\{Z_1 = Z_2 = \ldots = Z_d\} = 1$$

by assuming $G_j(\cdot) = G_1(\cdot)$ for every marginal distribution G_j of G. Here we use only the univariate convergence condition

$$P\{M_{nj} \leq a_{nj}z + b_{nj}\} \xrightarrow{d} G_1(z) \text{ as } n \to \infty \tag{11}$$

for every $j \leq d$ with suitable a_{nj}, b_{nj}. Sufficient conditions for the convergence in (11) are discussed in Hüsler (1986).

Theorem 3.6: *Let* $\{\mathbf{X}_i, i \geq 1\}$ *be a sequence of random vectors in* R^d. *Assume that* (11) *holds for every* $j \leq d$. *If*

$$\sum_{i=1}^{n} P\{X_{ij} > u_{nj}(z), X_{il} > u_{nl}(z)\} = \sum_{i=1}^{n} P\{X_{ij} > u_{nj}(z)\} + o(1) \tag{12}$$

as $n \to \infty$, *for all* $1 \leq j < l \leq d$ *and every* z *with* $G_1(z) > 0$ *and* $u_{nj}(z) = a_{nj}z + b_{nj}$, *the limit distribution* G *in* (1) *exists with*

$$G(\mathbf{z}) = G_1(\min_j(z_j)).$$

Proof: It is easily seen that (12) is equivalent to

$$\sum_{i=1}^{n} P\{X_{ij} > u_{nj}(x), X_{il} > u_{nl}(y)\} = \sum_{i=1}^{n} P\{X_{il} > u_{nl}(y)\} + o(1) \tag{13}$$

for every $x \leq y$ and that

$$\sum_{i=1}^{n} P\{X_{ij} > u_{nj}(x)\} = \sum_{i=1}^{n} P\{X_{il} > u_{nl}(x)\} + o(1).$$

We use the inequality

$$P(B_{n1}) \geq P(B_n) = P(B_{n1}) - P(\cup_{j=2}^{d}(B_{n1} \cap B_{nj}^c))$$

$$\geq P(B_{n1}) - \sum_{i=1}^{n}\sum_{j=2}^{d} P\{X_{ij} > u_{nj}(z_j), X_{i1} \leq u_{n1}(z_1)\}$$

with $B_n = B_{n1} \cap \cdots \cap B_{nd}$ and $B_{nj} = \{X_{ij} \leq u_{nj}(z_j), i \leq n\}$, by assuming without loss of generality that $z_1 = \min_j(z_j)$. Now (12) implies that for every $2 \leq j \leq d$

$$\sum_{i=1}^{n} P\{X_{ij} > u_{nj}(z_j), X_{i1} \le u_{n1}(z_1)\} = \sum_{i=1}^{n} P\{X_{ij} > u_{nj}(z_j)\} -$$

$$-\sum_{i=1}^{n} P\{X_{ij} > u_{nj}(z_j), X_{i1} > u_{n1}(z_1)\}$$

$$= o(1) \text{ for } z_j > z_1,$$

which shows that $P(B_n) \to G_1(z_1) = G_1(\min_j(z_j))$.

Necessary and sufficient conditions for the above result are given by Sibuya (1960) for iid. sequences, that are equal to condition (12) in this particular case. The following simple example demonstrates that (12) cannot be the necessary condition in the general situation. Take X_{i1} an iid. random sequence and let $X_{i2} = X_{i+l,1}$ for any $i \ge 1$ and any integer $l \ge 1$. Then G is totally dependent if it exists (i.e if F_{11} belongs to the domain of attraction of an extreme value distribution). But (12) does not hold, the l.h.s. sum converges to 0 for any z. Without assuming an additional dependence restriction on the random sequence, we find easily from the above proof that the convergence (1) with G totally dependent implies that

$$P(B_{nj}^c \cap B_{nl}^c) = P(B_{nj}^c) + o(1)$$

as $n \to \infty$ for any $1 \le j < l \le d$. This is in the case of independent random variables naturally equivalent to (12). Therefore, (12) is the necessary condition only if the random sequence is asymptotically equal to an independent random sequence.

It is also interesting to know a criteria such that G is a multivariate extreme value distribution. Besides of the stationary case, this holds in fact for a class of random sequences which are in some sense asymptotically stationary. Such approaches were discussed in Turkman and Walker (1983) and Hüsler (1986), (1988). In particular, the statement of Theorem 8 of Hüsler (1988), dealing with this question for the independent sequences, holds also for general random sequences by assuming in addition the conditions $\mathbf{D}_d$ and $\mathbf{D}'_d$, because of Theorem 2.2. In the case of Theorem 3.6, note that G is obviously a multivariate extreme value distribution, if G_1 is an extreme value distribution.

In the same way results could be formulated with the help of the point processes approach, giving with similar proofs more detailed information on the extremes and the exceedances (cf. Leadbetter et al. (1983), Resnick (1987)).

As mentioned in the introduction, our results extend the particular statements in the Gaussian case. In order that it is really an extension, one has to verify the conditions $\mathbf{D}_d, \mathbf{D}'_d, \mathbf{D}''_d$ and $\mathbf{D}^*_d$ for Gaussian sequences, which satisfy certain conditions on the autocorrelation functions, as e.g. Berman's condition. This follows by the method introduced in Hüsler (1983), generalized in an obvious way for the higher dimensions. Therefore we omit the verification (for a related verification see Schüpbach (1986)). The details and the results are given in Hüsler and Schüpbach (1988) as mentioned.

References

Amram, F. (1985) Multivariate extreme value distributions for stationary Gaussian sequences. J. Multiv. Analysis 16, 237-240.

Deheuvels, P. (1984) Probabilistic aspects of multivariate extremes. In *Statistical Extremes and Applications*, ed. J. Tiago de Oliveira, Reidel, Dordrecht, 117-130.

Galambos, J. (1978) *The asymptotic theory of extreme order statistics.* Wiley, New York. (2nd. ed. (1987) Krieger, Florida).

Gerritse, G. (1986) Supremum self-decomposable random vectors. Probab. Theory and Rel. Fields 72, 17-34.

Haan, L. de, and Resnick, S.I. (1977) Limit theory for multivariate sample extremes. Z. Wahrscheinlichkeitstheorie verw. Geb. 40, 317-337.

Hsing, T. (1987) Extreme value theory for multivariate stationary sequences. Techn. Report, Texas A&M University.

Hüsler, J. (1983) Asymptotic approximation of crossing probabilities of random sequences. Z. Wahrscheinlichkeitstheorie verw. Geb. 63, 257-270.

Hüsler, J. (1986) Extreme values and rare events of non-stationary random sequences. In *Dependence in Probability and Statistics*, eds. E. Eberlein, M. S. Taqqu, Birkhäuser, Boston, 438-456.

Hüsler, J. (1987) Multivariate extreme values in stationary random sequences. Techn. Report, University of Bern.

Hüsler, J. (1988) Limit properties for multivariate extremes in sequences of independent non-identically distributed random vectors. Techn. Report, University of Bern.

Hüsler, J., and Schüpbach, M. (1988) Limit results for maxima in non-stationary multivariate Gaussian sequences. Stoch. Proc. Appl. 28, 91-99.

Joag-Dev, K. (1983) Independence via uncorrelatedness under certain dependence structures. Ann. Probab. 11, 1037-1041.

Juncosa, M. L. (1949) On the distribution of the minimum in a sequence of mutually independent random variables. Duke Math. J. 16, 609-618.

Leadbetter, M. R. (1974) On extreme values in stationary sequences. Z. Wahrscheinlichkeitstheorie verw. Geb. 28, 289-303.

Leadbetter, M. R., Lindgren, G., and Rootzén, H. (1983) *Extremes and related properties of random sequences and processes.* Springer, Series in Statistics, Berlin.

Marshall, A. W., and Olkin, I. (1983) Domains of attraction of multivariate extreme value distributions. Ann. Probab. 11, 168-177.

Meijzler, D. G. (1950) On the limit distribution of the maximal term of a variational series. Dopovidi Akad. Nauk Ukrain. SSR 1, 3-10.

Resnick, S. I. (1987) *Extreme values, regular variation, and point processes.* Springer, Berlin.

Schüpbach, M. (1986) Grenzwertsätze und Punktprozesse für Minima und Maxima nichtstationärer Zufallsfolgen. Ph.d.thesis, University of Bern.

Sibuya, M. (1960) Bivariate extreme statistics. Ann. Inst. Stat. Math. 11, 195-210.

Turkman, K. F., and Walker, A. M. (1983) Limit laws for the maxima of a class of quasi-stationary sequences. J. Appl. Probab. 20, 814-821.

Villasenor, J. (1976) On univariate and bivariate extreme value theory. Ph.d.thesis, Iowa State University.

STATISTICAL DECISION FOR BIVARIATE EXTREMES

J. Tiago de Oliveira

Academy of Sciences of Lisbon - 19, R. Academia das Ciências, 1200, Lisboa, Portugal

Abstract. Bivariate extreme random pairs, with extreme margins, have for the distribution function, in the case of maxima, or for the survival function, in the case of minima, a dependence function. For the cases of Gumbel margins for maxima or of the exponential margins for minima an index of dependence as well the correlation coefficient are obtained; analogous results could be obtained for other margins and also study the non-parametric correlation coefficients.

Parametric models either for the cases where a density in $|R^2$ does exist (differentiable models) or does not (non-differentiable models) are considered and some statistical procedures are proposed.

Finally a reference to the intrinsic estimation of the dependence function (i.e., verifying some conditions and, thus, with restrictions on its coefficients) is made and some difficulties of the usual technique are referred.

1. Introduction

Bivariate asymptotic distributions of maxima are useful for the analysis of many concrete problems as largest ages of death for men and women each year, whose distribution naturally splits in the product of the margins by independence; floods at two different places of the same river, each year; bivariate extreme meteorological data (pressures, temperature, wind velocity, etc.), each week; largest waves, each week; etc. The same can be said about minima. Extensions to multivariate distributions, although more complex, can be made.

Evidently the target of a study of asymptotic distributions of bivariate extremes is to obtain their asymptotic probabilistic behaviour and, also, to provide bivariate models of (asymptotic) extremes that fit observed data.

2. Asymptotic Behaviour of Bivariate Maxima

Consider a sequence of i.i.d. random pairs $\{(X_n, Y_n)\}$ with distribution function $F(x,y)$. Evidently, $\text{Prob}\{\max(X_1, ..., X_n) \leq x \, , \, \max(Y_1, ..., Y_n) \leq y\} = F^n(x,y)$. We may search a pair of sequences $\{(\lambda_n, \delta_n) \, , (\lambda'_n, \delta'_n) \, , \delta_n > 0 \, , \delta'_n > 0\}$ such that

$$\text{Prob}\left\{ \frac{\max(X_1,..., X_n)-\lambda_n}{\delta_n} \leq x, \; \frac{\max(Y_1,..., Y_n)-\lambda'_n}{\delta'_n} \leq y \right\} =$$

$$= F^n(\lambda_n + \delta_n x, \lambda'_n + \delta'_n y)$$

Keywords: Bivariate extremes, dependence functions, statistical decision. AMS Subject classifications Primary 62H12, Seconday 62H15, 62J02, 60E99.

does have a (weak) limiting non-degenerate distribution function $L(x,y)$. If this happens, the margins also have (weak) limiting distributions (the ones of the marginal maxima) and thus are of the three well known forms. We will begin by supposing that the limiting distributions of the margins are of Gumbel standard form $\Lambda(z) = \exp(-e^{-z})$, i.e.

$$L(x, +\infty) = \Lambda(x), \quad L(+\infty, y) = \Lambda(y).$$

Using Khintchine's theorem, as done for the univariate case, and supposing Gumbel standard margins, we can show that $L(x,y)$ must satisfy the (stability) relation

$$L^k(x,y) = L(x-\log k, y-\log k)$$

for any positive integer k. Passing from the positive integer k to rationals $r(>0)$ and finally to reals $t(>0)$ we get

$$L^t(x,y) = L(x-\log t, y-\log t).$$

Taking now $x = \log t$ we have $L(x,y) = L^{e^x}(0, y-x)$. Puttting now $L(0,w) = \exp(-(1+e^{-w})\,k(w))$ we have shown, finally, that the limiting (and stable) distribution of maxima with Gumbel margins are of the form

$$L(x,y) = \Lambda(x,y) = \exp\{-(e^{-x} + e^{-y})\,k(y-x)\} = \{\Lambda(x)\Lambda(y)\}^{k(y-x)}.$$

We have to study the <u>dependence</u> function $k(w)$, continuous and non-negative, for $\Lambda(x,y)$ to be a distribution function. Those results are well known. They can be found, with different forms for the margins, in Finkelshteyn (1953), Tiago de Oliveira (1958), Geffroy (1958-59) and Sibuya (1960) with a synthesis of the results in Tiago de Oliveira (1962-63). Subsequent results are in Tiago de Oliveira (1975), (1980) and (1984). Galambos (1968) contains a recent account.

A direct characterization of the distribution function $\Lambda(x,y)$ can be made in the following way. It is immediate that a random pair (X,Y) with distribution $\Lambda(x,y)$ is such that $V = \max(X+a, Y+b)$ has a Gumbel distribution function with a location parameter. But, also, the converse is true, which characterizes $\Lambda(x,y)$ as the only distribution function $L(x,y)$ such that $\mathrm{Prob}\,\{\max(X+a, Y+b) \le z\} = L(z-a, z-b) = \Lambda(z-\lambda(a,b))$; see Tiago de Oliveira (1984).

Let us now consider the dependence function $k(w)$. Although a continuous function, it does not have necessarily derivatives and consequently we cannot expect all bivariate maxima random pairs with distribution function $\Lambda(x,y) = \{\Lambda(x)\,\Lambda(y)\}^{k(y-x)}$ to have a density in $\mathbb{R}^2$. From the Boole-Fréchet inequality

$$\max(0, \Lambda(x) + \Lambda(y) - 1) \le \Lambda(x,y) \le \min(\Lambda(x), \Lambda(y))$$

we have, replacing x and y by $x + \log n$ and $y + \log n$, raising to the power n and letting $n \to \infty$, the limit inequality

$$\Lambda(x)\,\Lambda(y) \le \Lambda(x,y) \le \min(\Lambda(x), \Lambda(y)).$$

The upper limit corresponds to the <u>diagonal</u> case where the pair with reduced margins (X,Y) is concentrated, with probability 1, in the first diagonal; it has a singular distribution. The lower limit corresponds to <u>independence.</u> For the dependence functions we have then

$$\left(\tfrac{1}{2} \le\right) \text{(diagonal) } k_D(w) = \frac{\max(1, e^{-w})}{1+e^{-w}} \le k(w) \le 1 = k_I(w) \text{ (independence).}$$

Note that $k(-\infty) = k(+\infty) = 1$. If there is a density in $\mathbb{R}^2$, i.e. if $k''(w)$ exists, taking derivatives we see that $k(w)$ must satisfy the relations:

$$k(-\infty) = k(+\infty) = 1,$$

$$[(1+e^{w})\ k(w)]' \geq 0,$$

$$[(1+e^{-w})\ k(w)]' \leq 0,$$

and

$$(1+e^{-w})\ k''(w) + (1-e^{-w})\ k'(w) \geq 0;$$

for the general case the corresponding conditions, because $\Delta^2_{x,y}\ \Lambda(x,y) \geq 0$, are:

$$k(-\infty) = k(+\infty) = 1,$$

$(1+e^{w})\ k(w)$ a non-decreasing function,

$(1+e^{-w})\ k(w)$ a non-increasing function,

and

$$\Delta^2_{x,y}[(e^{-x} + e^{-y})\ k(y-x)] \leq 0.$$

Some other properties can be ascribed to the set of k(w). The first one is symmetry property, i.e., if k(w) is a dependence function, then k(-w) is also a dependence function. The proof is immediate if we consider the conditions in the differentiable case (where a density in $\mathbb{R}^2$ does exist) and slightly longer in the general case. If k(w) = k(-w) then (X,Y) is an exchangeable pair and $\Lambda(x,y) = \Lambda(y,x)$.

It is immediate that if $k_1(w)$ and $k_2(w)$ are dependence functions, any mixture $\theta k_1(w) + (1-\theta)\ k_2(w)$, $0 \leq \theta \leq 1$ is also a dependence function showing that the set of dependence functions is, then, convex. This convexity property, written in terms of distribution functions as

$$\Lambda(x,y) = \Lambda_1^{\theta}(x,y)\ .\ \Lambda_2^{1-\theta}(x,y)\ ,$$

is very useful in obtaining models: the mixed model, as well as the Gumbel model, are such examples of the use of mix-technique.

Another method of generating models is the following. Let (X,Y) be an extreme random pair, with dependence function k(w) and standard Gumbel margins, and consider the new random pair $(\bar{X},\bar{Y})$ with $\bar{X} = \max(X+a, Y+b)$, $\bar{Y} = \max(X+c, Y+d)$. To have standard Gumbel margins we must have $(e^a + e^b)\ k(a-b) = 1$ and $(e^c + e^d)\ k(c-d) = 1$. Then we have

$$\bar{k}(w) = \frac{[e^{\max(a+w,c)} + e^{\max(b+w,d)}]\ k[\max(a+w,c)-\max(b+w,d)]}{1 + e^{w}}$$

with (a,b) and (c,d) satisfying the conditions given before. This max-technique can be used to generate the biextremal and natural models; see also Marshall and Olkin (1983) about those techniques.

Let us stress that independence has a very important position as a limiting situation. If we denote by P(a,b) the structural survival function defined by Prob $\{X > x, Y > y\} = P(F(x,+\infty), F(+\infty, y))$, Sibuya (1960) has shown that the necessary and sufficient condition to have limiting independence is that $P(1-s, 1-s)/s \to 0$ as $s \to 0$. He also showed that the necessary and sufficient condition for having the diagonal case as a limit situation is that $P(1-s,1-s)/s \to 1$ as $s \to 0$. With the first result we can show, easily, that maxima of the binormal distribution have independence as a limiting distribution if $|\rho| < 1$.

Sibuya conditions are easy to interpret: we have limiting independence if Prob $\{X > x, Y > y\}$ is a vanishing summand of Prob $\{X > x \text{ or } Y > y\}$ and the diagonal case as limit if Prob $\{X > x, Y > y\}$ is the leading summand of Prob $\{X > x \text{ or } Y > y\}$. Asymptotic independence of maxima pairs is very common.

A random pair (X,Y) with distribution function $F(x,y)$ is said to have positive association (as in Sibuya, 1960) or positively quadrant dependent (as in Marshall and Olkin, 1983 or Joag-Dev, 1983) if

$$\text{Prob}\{X \le x, Y \le y\} + \text{Prob}\{X > x, Y > y\}$$

is larger than or equal to be corresponding probabilities in the case of independence; intuitively this means that large (small) values of one variable are associated with large (small) values of the other. Positive association or positive quadrant dependence is equivalent to

$$F(x,y) \ge F(x, +\infty)\ F(+\infty, y).$$

The inequality $\Lambda(x)\ \Lambda(y) \le \Lambda(x,y) \le \min(\Lambda(x), \Lambda(y))$ shows positive association for bivariate maxima pairs, a result due to Sibuya (1960). As a consequence the correlation coefficients are non-negative.

A very important function is $D(w) = \text{Prob}\{Y - X \le w\}$.

We have, immediately,

$$D(w) = \frac{k'(w)}{k(w)} + \frac{1}{1+e^{-w}} .$$

We have $D(w) = 1/(1 + e^{-w})$ in the independence case and $D(w) = H(w)$ where $H(w)$ is the Heavside jump function at 0 - $H(w) = 0$ if $w < 0$ and $H(w) = 1$ if $w \ge 0$ - for the diagonal case.

In the same way, we could suppose that we would be dealing with bivariate minima with exponential standard margins $(\tilde{x},\tilde{y})$ which is equivalent to the use of the transformation $\tilde{x} = e^{-x}$, $\tilde{y} = e^{-y}$. Then the survival function is

$$S(\tilde{x},\tilde{y}) = \exp\{-(\tilde{x}+\tilde{y})\ A(\frac{\tilde{y}}{\tilde{x}+\tilde{y}})\}\ (\tilde{x},\tilde{y} \in \mathbb{R}_+),$$

proposed by Pickands (1981); the new dependence function $A(u) = k(\log((1-u)/u))$ $(k(w) = A(1/(1+e^{w})))$, defined for $u \in [0,1]$, must satisfy, in the general case, the simpler conditions

$A(0) = A(1) = 1,$

$A(u)/u$ non-increasing,

$A(u)/(1-u)$ non-decreasing,

$A(u)$ convex.

But with the last property the conditions on $A(u)/u$ and $A(u)/(1-u)$ can be substituted by $-1 \le A'(0) \le 0$ and $0 \le A'(1) \le 1$.

In the case of existence of $A''(u)$ - equivalent to the existence of a density in $\mathbb{R}^2_+$ for $(\tilde{x},\tilde{y})$ - the last relation takes the form

$$A''(u) \ge 0.$$

We also get for $A(u)$ the double inequality

$$\max(u, 1-u) \le A(u) \le 1.$$

All the other results obtained for the dependence function $k(w)$ can be translated to the corresponding dependence function $A(u)$ by using the decreasing transformation $w = \log((1-u)/u)$; for instance, the set of $\{A(u)\}$ is also convex and a min-technique, for the generation of new models, can be translated from the max-technique: for $X' = \min(a' \tilde{X}, b' \tilde{Y})$, $Y' = \min(c' \tilde{X}, d' \tilde{Y})$ ($a' = e^{-a}$, $b' = e^{-b}$, $c' = e^{-c}$, $d' = e^{-d}$), we get the conditions, $A(a'/(a'+b')).(a'+b') = a' b'$, $A(c'/(c'+d')).(c'+d') = c'd'$ and the new dependence function, noted (temporarily) $A'(u)$ for analogy, is $A'(u) = (\max((1-u)/a', u/c') + \max((1-u)/b', u/d')) \times A(\max((1-u)/b', u/d')/(\max((1-u)/a' + u/c') + \max((1-u)/b', u/d'))$.

Other results can be obtained; for example, corresponding to $D(w) = \text{Prob}\{Y - X \leq w\}$ we get $B(u) = \text{Prob}\{\tilde{Y}/(\tilde{X} + \tilde{Y}) \leq u\} = \text{Prob}\{Y - X \geq \log((1-u)/u)\} = 1 - D(\log((1-u)/u)) = u + u(1-u) A'(u)/A(u)$ with the corresponding results for independence and the diagonal case.

3. Dependence Results

The results on correlation that follow and the regression results given later, in some way, do illuminate the situation.

At it is well known, the covariance between X and Y can be written

$$\operatorname{cov}(X,Y) = \iint_{-\infty}^{+\infty} [F(x,y) - F(x, +\infty)\, F(+\infty, y)]\, dx\, dy.$$

In our case we obtain

$$\operatorname{cov}(X,Y) = - \int_{-\infty}^{+\infty} \log k(w)\, dw$$

and the correlation coefficient is

$$\rho = - \frac{6}{\pi^2} \int_{-\infty}^{+\infty} \log k(w)\, dw$$

or equivalently

$$\rho = 1 - \frac{3}{\pi^2} \int_{-\infty}^{+\infty} w^2\, d\, D(w)\, .$$

Since $k(w) \leq 1$ we have $0 \leq \rho$, as could be expected. It is very easy to show that for the diagonal case we have $\rho = 1$. The value of ρ does not identify the dependence function (or distribution): ρ is the same for $k(w)$ and $k(-w)$. But $\rho = 0$, as $k(w) \leq 1$ and $\log k(w) \leq 0$, implies $k(w) = 1$, or independence. Writing now ρ under the form

$$\rho = 1 - \frac{6}{\pi^2} \int_{-\infty}^{-\infty} \log \frac{k(w)}{k_D(w)}\, dw$$

we see, analogously, that $\rho = 1$ as $k(w) \geq k_D(w)$ implies $k(w) = k_D(w)$, or the diagonal case; this is evidently trivial as $\rho = 1$.

If we use the formulation for minima with standard exponential margins it is immediate that

$$\mathrm{cov}(\tilde{X},\tilde{Y}) = \iint_0^{+\infty} (S(\tilde{x},\tilde{y}) - e^{-(\tilde{x}+\tilde{y})})\, d\tilde{x}d\tilde{y} = \int_0^1 \frac{du}{A^2(u)} - 1$$

and the correlation coefficient, as the standard exponential margins have unit variance, has the same value.

Linear regression lines are easily drawn once the correlation coefficients are known.

The index of dependence $\sup_{x,y} | \Lambda(x,y) - \Lambda(x)\,\Lambda(y)) | = (1-k_0)\, k_0^{\,k_0/(1-k_0)} = (1-k_0)\,.$ $\exp\{k_0 \log k_0/(1-k_0)\}$ where $k_0 = \min k(w)$ for the case of bivariate maxima with Gumbel standard margins and $\sup_{\tilde{x},\tilde{y}>0} |(S(\tilde{x},\tilde{y}) - e^{-(\tilde{x}+\tilde{y})})| = (1- A_0)\exp\{A_0 . \log A_0/(1- A_0)\}$ for the case of bivariate minima with exponential margins where $A_0 (= k_0) = \max A(u)$. This index of dependence is the uniform distance between a bivariate model and independence.

4. Some General Remarks

In what regards statistical decision for bivariate extremes we will use an approach similar to the used for univariate extremes: we will replace, for large samples, $F^n(x,y)$ by an asymptotic approximation

$$L\left(\frac{x-\lambda}{\delta}, \frac{y-\lambda'}{\delta'}\right)$$

where $L(x,y)$ will have the expressions given before, thus reducing the difficulty of not knowing $F^n(n,y)$ to the simpler (!) difficulties of not knowing the dependence function ($k(.)$ or $A(.)$ for instance) and 4 independent margin parameters. Recall that we have the results similar to the ones used to justify the substitution of $F^n(x)$ by $L((x-\lambda)/\delta)$ in the univariate case: stability and uniformity of convergence are valid and results similar to the ones for random sequences show that, even in the case of some dependence waning out between the sucessive observations, the limiting distributions are yet the same.

In the asymptotic approximation of $L((x-\lambda)/\delta), (y-\lambda')/\delta')$ to the actual and unknown $F^n(x,y)$ we will use, in principle, a dependence function. The existence of a strong correlation between the maxima (or minima), correlation not completely dissolved in some applications by the relatively large number of observations from which maxima (or minima) are taken, suggests the use of an asymptotic bivariate extremes distribution function as a better approximation to the actual (unknown) distribution function of the bivariate extremes. This stand is justified also by the fact that the discrepancy between independence and the dependence cases may be very large in a probabilistic sense but not expressed by the usual indicators. As an example we can note that the uniform distance between the biextremal model with standard Gumbel margins where the distribution function is $\Lambda(x,y|\theta) = \exp\{-(e^{-x} + e^{-y} - \min(e^{-x}, \theta e^{-y}))\}$ (notice that $\Lambda(x,y|0) = \Lambda(x)\,.\,\Lambda(y)$ - independence - and that $\Lambda(x,y|1) = \Lambda(\min(x,y))$, the diagonal case) and the independence case is, with easy algebra,

$$\sup_{x,y} |\Lambda(x,y|\theta) - \Lambda(x,y|0)| = \theta\,(1+\theta)^{-1-1/\theta}\,;$$

consequently if we take the independence model as an approximation to a biextremal model with θ close to 1, i.e. close to the diagonal case, we could obtain probability evaluations for quadrants with an error around 1/4, the value of the distance for $\theta = 1$, but that can differ of almost 1 if the set used is a strip containing the 1st diagonal !

Experience will, finally, decide of the fitness and accuracy of those approximations. But, for the moment, as the (2,N) natural model shows, we can have any finite number of parameters, forgetting the 4 of the margins. And, also, it was shown in Tiago de Oliveira (1962-63) that for no one-parameter models (with standard Gumbel margins) exist sufficient statistics. An alternative is the intrinsic estimation (see the last section) of the dependence function (k or A).

For some models we do not have the best test, the best estimation procedure, etc. - although we have one - but in other situations nothing at all is known. An example: statistical choice of bivariate extreme models - so important for applications - is now beginning to be considered ! In general, the few papers up to now choose one model from the beginning or compare two of them by the use of Kolmogorov--Smirnov or another lest; see for instance, Gumbel and Goldstein (1964).

As a general rule in statistical decision for bivariate maxima with Gumbel margins the papers Tiago de Oliveira (1970), (1971), (1975), (1975'), (1980) and (1984) contain also results other to ones given here.

In what regards modelling we have to deal both with cases where sometimes exists a density in $|R^2$ but with others, which appear naturally, have a singular component.

Finally, as it can be sometimes used, let us recall that as the margins in both cases (Gumbel or exponential) have all moments, we have also crossed moments, etc..

Note that for models that are going to be considered we only gave the correlation coefficient and index of dependence when they have a compact form. The correlation coefficient can then easily be used for statistical decision. Non-parametric correlation coefficients and the index of dependence are the same for corresponding $\Lambda(x,y|\theta)$ and $S(\tilde{x},\tilde{y}|\theta)$; the correlation coefficients are evidently different.

5. The Differentiable Models: Statistical Decision

Until recently only two models of bivariate extremes with a density in $|R^2$ (with a point exception in one case) have been considered and studied in the literature: the logistic model and the mixed model, appearing in Gumbel (1961'). Both are symmetrical but nothing precludes the existence of asymmetric models, which is the case in the majority of non-differentiable models too be considered in the next section. Here, as well as for non-differentiable models, we will always study them under the form of bivariate maxima with standard Gumbel margins, i.e. with the distribution function $\Lambda(x,y) = (\Lambda(x)\,\Lambda(y))^{k(y-x)}$ with the dependence function $k(w)$; at the end of each case we will consider the corresponding forms for bivariate minima with standard exponential margins, i.e., with the survival function $S(\tilde{x},\tilde{y}) = \exp\{-(\tilde{x}+\tilde{y})\, A(\tilde{y}/(\tilde{x}+\tilde{y}))\}$ $(\tilde{x},\tilde{y} \geq 0)$ obtained by the transformation $\tilde{X} = e^{-X}$, $\tilde{Y} = e^{-Y}$ and with $A(u) = k((\log(1-u)/u)$, as said.

The logistic distribution function is $\Lambda(x,y|\theta) = \{\exp -(e^{-x/(1-\theta)} + e^{-y/(1-\theta)})^{1-\theta}\}$, $0 \leq \theta < 1$ (for $\theta = 0$ we have independence and for $\theta = 1^-$ we get the diagonal case, which is the only situation without density). It is so called because

$$\text{Prob}\{Y-X \leq x|\theta\} = D(w|\theta) = (1 + e^{-w/(1-\theta)})^{-1}$$

corresponding to the dependence function

$$k(w|\theta) = \frac{(1 + e^{-w/(1-\theta)})^{1-\theta}}{1 + e^{-w}};$$

note that we must have $0 \leq \theta \leq 1$, as comes from the inequality $k_D(w) = \max(1,e^w)/(1+e^w) \leq k(w) \leq 1 = k_I(w)$.

The margins are exchangeable as shown by the expression of $\Lambda(x,y|\theta)$, by the fact that $k(-w|\theta) = k(w|\theta)$ or yet because $D(w|\theta) + D(-w|\theta) = 1$.

The correlation coefficient $\rho(\theta) = \theta(2-\theta)$ increases from 0 to 1 as θ increases from 0 to 1. Linear regression of Y in X is, evidently, $L_y(x|\theta) = \gamma + \theta\,(2-\theta)\,(x-\gamma)$. It can be shown that the index of dependence is

$$\sup_{x,y} |\Lambda(x,y|\theta) - \Lambda(x,y|0)| = (1-2^{-\theta})\, 2^{-\theta/(2^{\theta}-1)}$$

increasing from 0 (at $0 = 0$) to $1/4 = .25$ (at $\theta = 1$). It is, thus, intuitive that the distance between the independence ($\theta = 0$) and the assumed model for $\theta > 0$ is small in general, and for small samples it will naturally be difficult to separate between models.

Owing to the theoretical importance of independence, as a reference pattern because in general we can expect association, we can formulate the confirmation of dependence by a L.M.P. (locally most powerful) test of $\theta = 0$ vs $\theta > 0$, as a first step in the statistical analysis of extremes.

The likelihood of the sample $(x_1, y_1), \ldots, (x_n, y_n)$ is

$$L(x_i, y_i|\theta) = \prod_1^n \frac{\partial^2 \Lambda(x_i, y_i|\theta)}{\partial x_i \partial y_i}$$

and so the rejection region for the L.M.P. test has the form,

$$\frac{\theta}{\partial \theta}\Big(\sum_1^n \log \frac{\partial^2 \Lambda(x_i, y_i|\theta)}{\partial x_i \partial y_i}\Big)\Big|_{\theta = 0} \ge c_n \quad \text{or equivalently} \quad \sum_1^n v(x_i, y_i) \ge c_n$$

where

$$v(x,y) = x+y+x\, e^{-x} + y\, e^{-y} + (e^{-x} + e^{-y}-2)\, \log(e^{-x} + e^{-y}) + (e^{-x} + e^{-y})^{-1}.$$

It can be shown that, in case of independence, $v(X,Y)$ has zero mean value but infinite variance; evidently $1/n \,.\, \Sigma_1^n\, v(x_i,y_i) \overset{a.s.}{\to} 0$, in case of indepencence, by Khinchine's theorem. Although the usual central limit theorem is not applicable it was shown by Tawn (1987) that $\Sigma_1^n\, v(x_i,y_i)/ \sqrt{(n \log n)/2}$ is asymptotically standard normal. Estimation by the maximum likelihood is possible and Cramer - regular for $\theta \neq 0$, having for $\theta = 0$ the difficulties arising from the infinite variance of $v(X,Y)$. Thus we have $c_n \sim \sqrt{n \log n/2}\; N^{-1}(1-\alpha)$, where $N(.)$ denotes the standard normal distribution function and α the significance level.

As θ is a dependence parameter, it is natural to use the correlation coefficient, whose estimator is independent of the margin parameters, to test $\theta = 0$ vs $\theta > 0$. It was shown, in Tiago de Oliveira (1965) and (1975), that the most efficient one is the correlation coefficient in comparison with the non-parametric ones. Denoting by r its usual estimator, in the case of independence, $\sqrt{n}.r$ is asymptotically standard normal. As $\rho(\theta) = \theta(2-\theta) \ge 0$, a point estimator of θ is given by $\max(0,r) = \theta^*(2-\theta^*)$ or $\theta^* = 1 - \sqrt{1 - \max(0,r)}$.

The logistic survival function for bivariate minima with standard exponential margins is immediately obtained by the transformation $\tilde{X} = e^{-X}$, $\tilde{Y} = e^{-Y}$($A(u) = k(\log((1-u)/u))$ is $S(\tilde{x},\tilde{y}|\theta) = P(\tilde{X} > \tilde{x}\ ,\ \tilde{Y} > \tilde{y}) = \exp\{-[\tilde{x}^{1/(1-\theta)} + \tilde{y}^{1/(1-\theta)}]^{1-\theta}]$ for $\tilde{x},\tilde{y} \ge 0$; also we get $A(u|\theta) = (u^{1/(1-\theta)} + (1-u)^{1/(1-\theta)})^{1-\theta}$.

The function $v(\tilde{x},\tilde{y})$ is given by $v(\tilde{x},\tilde{y}) = -(\log \tilde{x} + \tilde{x} \log \tilde{x} + \log \tilde{y} + \tilde{y} \log \tilde{y}) + (\tilde{x} + \tilde{y} - 2).\log(\tilde{x} + \tilde{y}) + 1/(\tilde{x} + \tilde{y})$.

The correlation $\tilde{\rho} = \int_0^1 A^{-2}(u|\theta)du-1$, by the transformation $u = (1+((1-t)/t)^{1-\theta})^{-1}$, takes the form

$$\rho(\theta) = (1-\theta) \int_0^1 t^{-\theta}(1-t)^{-\theta}\, dt-1 = (1-\theta)\ B(1-\theta,\ 1-\theta) - 1 = (1-\theta)\frac{\Gamma^2(1-\theta)}{\Gamma(2(1-\theta))} -1$$

which increases from $\tilde{\rho}\ (0) = 0$ to $\tilde{\rho}\ (1^-) = 1$, because $(1-\theta)\ \Gamma(1-\theta) \to 1$ for $\theta \to 1^-$; θ^* can be obtained in an way analogous to what was done above.

Let us now consider the mixed distribution function $\Lambda(x,y|\theta) = \exp(-e^{-x} - e^{y} + \theta/(e^{x} + e^{y}))$, $0 \le \theta \le 1$, where for $\theta = 0$ we have independence but for $\theta = 1$ it does not degenerate in $Y = X$ with probability 1 as it happens with the logistic model.

The model has the dependence function $k(w|\theta)=1- \theta\, e^{w}/(1+e^{w})^2 = k(-w|\theta)$ and is exchangeable. The dependence function $k(w|\theta)$ is the $(1-\theta,\theta)$ mixture of the dependence functions 1 and $1 - e^{w}/(1+e^{w})^2$; the inequality $\max(1,e^{w})/(1+e^{w}) \le k(w) \le 1$ shows that the domain [0,1] of θ can not be enlarged.

The correlation coefficient has the expression $\rho(\theta) = 6/\pi^2.\ (\text{arc cos}\ (1-\theta/2))^2$ which increases from $\rho(0) = 0$ to $\rho(1) = 2/3$. The model is, thus, not fit to express a strong dependence as confirmed below by the index of dependence. Linear regression of Y is X is evidently $L_y(x|\theta) = \gamma + 6/\pi^2\ (\text{arc cos}\ (1-\theta/2))^2\ (x-\gamma)$. We have also

$$\sup_{x,y} |\Lambda(x,y|\theta) - \Lambda(x,y|0)| = \frac{\theta}{4-\theta}\ (1-\theta/4)^{4/\theta}$$

increasing from 0 (at $\theta = 0$) to $3^3/4^4 = .105$ at $\theta = 1$. The smaller variation of the correlation coefficient and of the index of depedence shows that the deviation from independence is smaller and more difficult to detect. For other remarks see Gumbel (1961).

The L.M.P. test of $\theta = 0$ vs $\theta > 0$ is given by the rejection region $\Sigma_1^n\ v(x_i,y_i) \ge c_n$ with $v(x,y) = 2\ e^{2x}.\ e^{2y}/(e^{x} + e^{y})^3 - (e^{2x} + e^{2y})/(e^{x} + e^{y})^2 + (e^{x} + e^{y})^{-1}$.

As before, the situation is marred by the fact that for independence although $v(X,Y)$ has mean value zero the variance is infinite. The usual central limit theorem can not be applied but, as before, $1/n\ .\ \Sigma_1^n\ v(x_i,y_i) \overset{a.s.}{\to} 0$, and, also as shown by Tawn (1987), $\Sigma_1^n\ v(x_i,y_i)/\ \sqrt{(n\ \log\ n)/15}$ is also asymptotically normal; then $c_n\ \ \sqrt{(n\ \log\ n)/15}\ N^{-1}(1-\alpha)$.

Other correlation coefficients, independent of the margin parameters, could be used to test $\theta = 0$ vs $\theta > 0$. Once more the correlation coefficient is the best in comparison with the non-parametric ones , see Tiago de Oliveira (1965) and (1975). The point estimator of θ, if $0 \le r \le 2/3$, is given by $\theta^* = 2(1-\cos(\sqrt{(\pi^2/6)}\ r))$, with the truncation $\theta^* = 0$ if $r < 0$ and $\theta^* = 1$ if $r > 2/3$.

The mixed survival function with standard exponential margins is $S(\tilde{x},\tilde{y}|\theta)=\exp\{-(\tilde{x} + \tilde{y}) + \theta\, x\, \tilde{y}/(\tilde{x} + \tilde{y})\}$, $\tilde{x}$, $\tilde{y} \ge 0$; we have $A(u|\theta) = 1- \theta\, u(1-u)$. The function $v(\tilde{x},\tilde{y})$ is $v(\tilde{x},\tilde{y}) = 2\ \tilde{x}\ .\ \tilde{y}/ /(\tilde{x} + \tilde{y})^3 - (\tilde{x}^2 + \tilde{y}^2)/(\tilde{x} + \tilde{y})^2 + \tilde{x}\ \tilde{y}/(\tilde{x} + \tilde{y})$. The correlation coefficient is simply $\tilde{\rho}(\theta)=8/((4-\theta).\ \sqrt{\theta(4-\theta)})\ tg^{-1}\ \sqrt{\theta/(4-\theta)}-(2-\theta)/(4-\theta)$ increasing from $\tilde{\rho}(0)=0$ to $\tilde{\rho}(1) = (4\pi/3\sqrt{3}-1)/3 = .4727997$.

Finally, we discuss statistical choice between the non-separated models (logistic and mixed), non-separated because they have in common the independence case ($\theta = 0$). As developed in Tiago de Oliveira (1984) and generalized in Tiago de Oliveira (1985), the decision rule (with v_L and v_M denoting the v functions of the logistic and mixed models, given before) is:

decide for the independence if $\sum_1^n v_L(x_i, y_i) \leq a_n$, $\sum_1^n v_M(x_i, y_i) \leq b_n$;

decide for the logistic model if $\sum_1^n v_L(x_i, y_i) > a_n$, $\sum_1^n \frac{v_L(x_i, y_i)}{a_n} \geq \sum_1^n \frac{v_M(x_i, y_i)}{b_n}$

decide to the mixed model if $\sum_1^n v_M(x_i, y_i) > b_n$, $\sum_1^n \frac{v_M(x_i, y_i)}{b_n} \geq \sum_1^n \frac{v_L(x_i, y_i)}{a_n}$.

Tawn (1987) has shown that the asymptotic distribution of the pair $(\Sigma_1^n v_L(x_i,y_i)/\sqrt{(n \log n)/2},\ \Sigma_1^n v_M(x_i,y_i)/\sqrt{(n \log n)/15})$ is asymptotically binormal independent with standard margins and so we should take, with the usual meanings, $a_n \sim \sqrt{(n \log n)/2}\ N^{-1}(1-\alpha/2)$ and $b_n \sim \sqrt{(n \log n)/15}\ N^{-1}(1-\alpha/2)$.

It must be strongly emphasized that the tests and estimators based on the correlation coefficients are independent of the values of the margin parameters. But in the other cases we have to "estimate" reduced values by using $\hat{x}_i = (x_i - \hat{\lambda}_x)/\hat{\delta}_x$ and $\hat{y}_i = (y_i - \hat{\lambda}_y)/\hat{\delta}_y$ in Gumbel margins case, whose effect is not disturbing in general.

6. The Non-differentiable Models; Statistical Decision

We know essentially three forms of the non-differentiable models - i.e. with a singular component, thus not having a density in $\mathbb{R}^2$ - the Gumbel model, the biextremal model and the natural model, the last one being recently generalized by the author as sketched below; no statistical result have appeared until now except when it degenerates in the natural model.

Gumbel model has the distribution function $\Lambda(x,y|\theta)=\exp-(e^{-x} + e^{-y} - \theta \min(e^{-x}, e^{-y}))$, $0 \leq \theta \leq 1$; its dependence function is $k(w|\theta)=1-\theta \min(1,e^{-w})/(1+e^{-w})$ which is the $(1-\theta, \theta)$ mixture of $k_I(w) = 1$ (independence) and $k_D(w) = \max(1,e^{-w})/(1+e^{-w})$ (diagonal case). The model is exchangeable as $k(w|\theta) = k(-w|\theta)$ and $D(w|\theta)$ has a jump of $\theta/(2-\theta)$ at $w = 0$, which is the probability $P(Y=X)$; if $w \neq 0$ we have $D(-w|\theta) + D(-w|\theta) = 1$. This model was introduced in Tiago de Oliveira (1971) and is a conversion, to Gumbel margins, of a bivariate model with exponential margins, due to Marshall and Olkin (1967); see also Marshall and Olkin (1983).

The correlation coefficient takes the form $\rho(\theta) = 12/\pi^2 . \int_0^\theta \log(2-t)/(1-t)\, dt$ and increases from 0 to 1 as θ varies from 0 to 1. The index of dependence is $\sup_{x,y}|\Lambda(x,y|\theta)-\Lambda(x,y|0)| = \theta/(2-\theta)$. $(1-\theta/2)^{2/\theta}$ increasing from 0 to 1/4 as θ increases from 0 to 1.

The non-linear regression is

$$\bar{y}(x|\theta) = \gamma + \log(1-\theta) + \int_{(1-\theta)e^{-x}}^{+\infty} e^{-t}/t.dt - (1-\theta)\ \exp(\theta\ e^{-x}) .$$

$$\int_{e^{-x}}^{+\infty} e^{-t}/t.dt = x + \int_0^{(1-\theta)e^{-x}} \frac{1-e^{-t}}{t}\ dt - (1-\theta)\exp(\theta\ e^{-x}) \int_{e^{-x}}^{+\infty} \frac{e^{-t}}{t}\ dt.$$

The form of $y(x|\theta)$ is very different from the linear regression $\gamma+\rho(\theta)\ (x-\gamma)$ but the non-linearity index NL(*) oscilates between 0.000 (for $\theta = 0$ and $\theta = 1$) and .006; improvement is very small in the passage from linear regression to non-linear regression: in fact, a large difference between the two regressions only happens for very small or very large values of X or Y, whose probability is very small; see Tiago de Oliveira (1974).

Let us consider now statistical decision for this model.

The estimation of the margin parameters is avoided using that correlation coefficient. The test for independence can be done, as before, based on the fact that if $\theta = 0$ (independence) then $\sqrt{n}\,.\,r$ is asymptotically standard normal. θ can be estimated from the correlation coefficient, taking the value 0 if the sample correlation coefficient is negative.

Gumbel model for bivariate minima with standard exponential margins, was presented, in a different approach by Marshall and Olkin (1967). We have $S(\tilde{x},\tilde{y}|\theta) = e^{-\tilde{x}-\tilde{y}+\theta\min(\tilde{x},\tilde{y})}, 0 \le \theta \le 1$; we have $A(u|\theta) = 1 - \theta\min(u,1-u)$, a mixture $(\theta, 1-\theta)$ of independence $(A(u) = 1)$ and diagonal case $(A(u) = \max(u,1-u))$. $B(u|\theta)$ has a jump of $\theta/(2-\theta)$ at $u = 1/2$ as could be expected from the result for $D(w|\theta)$.

The correlation coefficient is $\tilde{\rho}(\theta) = \theta/(2-\theta)$ increasing from $\tilde{\rho}(0)=0$ to $\tilde{\rho}(1) = 1$.

Let us consider now the biextremal model. It appears naturally in the study of extremal processes; see Tiago de Oliveira (1968).

It can defined directly as follows: consider (X,Z) independent reduced Gumbel random variables and form the new pair (X,Y) with $Y = \max(X-a, Z-b)$, a and b such that $\mathrm{Prob}\{Y \le y\} = \mathrm{Prob}\{\max(X-a, Z-b) \le y\} = \Lambda(y)$ which implies $e^{-a} + e^{-b} = 1$. Putting $e^{-a} = \theta$, $e^{-b} = 1-\theta\ (0 \le \theta \le 1)$ we have $Y = \max(X + \log\theta, Z + \log(1-\theta))$. This was the initial use of the max-technique to generate new models and is a particular case of a natural model with $a = 0$, $b = +\infty$, $c = -\log\theta$ and $d = -\log(1-\theta)$. The distribution function is $\Lambda(x,y|\theta) = \exp\{-\max(e^{-x} + (1-\theta)e^{-y}, e^{-y})\}$, $0 \le \theta \le 1$ and the dependence function is immediately $k(w|\theta) = 1-\min(\theta,e^{w})/(1+e^{w}) = (1-\theta+\max(\theta,e^{w})/(1+e^{w}))$. $D(w|\theta)$ jumps at $w = \log\theta$ from 0 to $\theta = \mathrm{Prob}\{Y = X + \log\theta\}$ and, so, the singular part is concentrated at the line $y = x + \log\theta$, with probability θ. Evidently $Y \ge X + \log\theta$ with probability one. For $\theta = 0$ and $\theta = 1$ we obtain independence and the diagonal case. Remark that $k(w|\theta) = k(-w|\theta)$ and so the model is not exchangeable.

The correlation coefficient of the biextremal model is

$$\rho(\theta) = -\frac{6}{\pi^2}\int_0^\theta \frac{\log t}{1-t}\,dt = 1 + \frac{6}{\pi^2}L(\theta),\ L(\theta) = \int_1^\theta \frac{\log w}{w-1}\,dw$$

being the Spence integral; $\rho(\theta)$ increases from $\rho(0) = 0$ to $\rho(1) = 1$. Linear regression is thus $L_y(x) = \gamma + \rho(\theta)(x-\gamma)$.

The index of dependence is

$$\sup_{x,y} |\Lambda(x,y|\theta) - \Lambda(x,y|0)| = \theta/(1+\theta)^{(1+\theta)/\theta},$$

(*) - The non-linearity index $NL = (R^2 - \rho^2)/(1-\rho^2)$ was introduced in Tiago de Oliveira (1974) as a measure of the percentage reduction of the variance when, for prediction (or estimation) of one component based in the other, when we pass from linear to non-linear regression.

increasing from 0 to 1/4 as θ increases from 0 to 1.

The non-linear regressions - here we do not have exchangeability - are

$$\bar{y}(x|\theta) = x + \log\theta + \int_0^{(1-\theta)/\theta . e^{-x}} \frac{1-e^{-t}}{t}\,dt = x + \log(1-\theta) + \int_{(1-\theta)/\theta . e^{-x}}^{+\infty} \frac{e^{-t}}{t}\,dt$$

and

$$\bar{x}\,(y|\theta) = y - \log\theta - (1-\theta)\exp(\theta\, e^{-y}) \int_{\theta e^{-y}}^{+\infty} \frac{e^{-t}}{t}\,dt.$$

The comparison of non-linear with linear regressions $L_y(x|\theta) = \gamma + \rho(\theta)(x-\gamma)$ and $L_x(y|\theta) = \gamma + \rho(\theta)(y-\gamma)$ shows that the non-linearity index varies from 0 to 0.068 in the first case and from 0 to 0.007 in the second regression; see Tiago de Oliveira (1974). We do not have, also, a substantial improvement in using non-linear regression, although the comparison between the curves $\bar{y}(x|\theta)$ and $L_y(x|\theta)$ and between $\bar{x}(y|\theta)$ and $L_x(y|\theta)$ shows a great difference; the explanation is the one before: the small probability mass attached to the regions where the regressions (linear and non-linear) differ strongly.

The test of independence ($\theta = 0$ vs $\theta > 0$) can be made by using the correlation coefficient as usual.

In what regards the estimation of θ, as $y_i - x_i \geq \log\theta$ and $0 \leq \theta \leq 1$, the natural estimator of the left-end point $\log\theta$ is $\log\tilde{\theta}$ where $\tilde{\theta} = \min(e^{y_i - x_i}, 1)$. We have, for $z < 1$, $\text{Prob}\{\tilde{\theta} \leq z|\theta\} = \text{Prob}\{\min(e^{y_i - x_i}) \leq z|\theta\} = 1 - \text{Prob}\{\min(y_i - x_i) > \log z|\theta\} = 1 - (1 - D(\log z\,|\theta))^n$ and so we have $\text{Prob}\{\theta \leq z|\theta\} = 0$ if $z < \theta$, $= 1 - (1-\theta+z)^{-n}$ if $\theta \leq z < 1$ and $= 1$ if $z > 1$; the jumps have the values $1 - (1-\theta)^n$ at $z = 0$ and $((1-\theta)/(2-\theta))^n$ at $z = 1$; if $\theta = 1$ we see that $\theta \to \theta$; see also Tiago de Oliveira (1970) and (1975') for other details.

The biextremal model for minima has $S(\tilde{x},\tilde{y}|\theta) = e^{-\max(\tilde{x}+(1-\theta)\tilde{y},\tilde{y})}$ $0 \leq \theta \leq 1, \tilde{x},\tilde{y} \geq 0$; we obtain $A(u|\theta) = \max(u, 1-\theta u)$ and $B(u|\theta)$ has a jump of θ at $u = 1/(1+\theta)$. The correlation coefficient takes the value $\tilde{\rho}\,(\theta) = \theta$.

Let us consider now the natural model. Using the max-technique, from a pair (X,Y) of independent reduced Gumbel random variables we get a new (dependent) random pair $\bar{X} = \max(X-a, Y-b)$, $\bar{Y} = \max(X-c, Y-d)$ with Gumbel margins if and only if $e^{-a} + e^{-b} = 1$ and $e^{-c} + e^{-d} = 1$. The dependence function is $\bar{k}(w|a,b,c,d) = (e^{-\min(a,c+w)} + e^{-\min(b,d+w)})/(1+e^{-w})$, whose tails coincide with the ones of the diagonal case $k_D(w) = \max(1,e^w)/(1+e^w)$. It is easy to see that $\bar{W} = \bar{Y} - \bar{X} = \min(a+W, b) - \min(c+W, d)$ is such that $a-c \leq \bar{W} \leq b-d$ if $a+d < b+c$ and $b-d \leq \bar{W} \leq a-c$ if $b+c < a+d$. The case $a+d = b+c$ is irrevelant because $\bar{W}$ is then an almost sure random variable with $\bar{W} = a-c = b-d$ and as $\bar{W}$ must be zero (its mean value is zero) we have $a = c$, $b = d$, $\bar{X} = \bar{Y}$.

From now on we will suppose $a+d < b+c$, the case $b+c < a+d$ being dealt with by exchange of $\bar{X}$ and $\bar{Y}$. Let us put $\alpha = a-c$, $b-d = \beta$; by the fact that $\bar{W}$ has a mean value zero we have $\alpha < 0 < \beta$. Using the relations $e^{-a} + e^{-b} = 1$ and $e^{-c} + e^{-d} = 1$, with $a-c = \alpha$ and $b-d = \beta$ we get the final expression

$$\bar{k}(w|\,\alpha,\beta) = \frac{(e^{\beta}-1)\max(1,e^{\alpha-w}) + (1-e^{\alpha})\max(1,e^{\beta-w})}{(e^{\beta}-e^{\alpha})\,(1+e^{-w})}$$

as said before for $w < \alpha$ and for $w = \beta$ we have $\bar{k}(w|\alpha, \beta) = \bar{k}_D(w)$. Note that $\alpha = 0$ implies $\beta = 0$ and vice-versa.

We have that $\bar{k}(w|-\beta,-\alpha) = \bar{k}(-w|\alpha, \beta)$; thus we get an exchangeable model if $\alpha = -\beta$. For $\alpha = \log\theta$, $\beta = +\infty$ we get the biextremal model and for $\alpha = -\infty$, $\beta = -\log\theta$ its dual.

Using now the parameters (α, β) $(\alpha < 0 < \beta)$ we see that $\Lambda(x,y|\alpha, \beta) = \exp\{-[(e^{\beta}-1)\max(e^{-x}, e^{\alpha-y}) + (1-e^{\alpha})\max(e^{-x}, e^{\beta-y})]/(e^{\beta}-e^{\alpha})\}$, $-\infty < \alpha < 0 < \beta < +\infty$ and $\alpha \le \bar{W} \le \beta$. $D(w|\alpha, \beta)$ has jumps of $(e^{\beta}-1)e^{\alpha}/(e^{\beta}-e^{\alpha})$ at $w = \alpha$ and of $(1-e^{\alpha})/(e^{\beta}-e^{\alpha})$ at $w = \beta$; $[\alpha, \beta]$ is, thus, the support of $\bar{W} = \bar{Y}-\bar{X}$, $\bar{R} = \beta - \alpha$ the range and so when we have $-\infty < \alpha < 0 < \beta < +\infty$ there exists, as $\alpha \le \bar{W} = \bar{Y}-\bar{X} \le \beta$, a strong stochastic connection between $\bar{X}$ and $\bar{Y}$; the model has, thus, flexibility to express such situations.

The correlation coefficient can be written as

$$\rho(\alpha,\beta) = 1 - \frac{6}{\pi^2}\int_{-\infty}^{+\infty} \log\frac{\bar{k}(w|\alpha,\beta)}{k_D(w)}\,dw =$$

$$= 1 - \frac{6}{\pi^2}\int_{\alpha}^{\beta} \log\frac{e^{\beta} - 1 + (1-e^{\alpha})\,e^{\beta-w}}{(e^{\beta}-e^{\alpha})\max(1,e^{-w})}\,dw;$$

it is immediate that $\rho(\alpha,\beta) \to 1$ as $\alpha, \beta \to 0$ and that $\rho(\alpha,\beta) \to 0$ as $-\alpha, \beta \to +\infty$. The index of dependence is

$$(1-k_0)\exp\{\frac{k_0}{1-k_0}\log k_0\} \text{ with } k_0 = \min(\frac{1}{e^{\alpha}+1}, \frac{e^{\beta}}{e^{\beta}+1}) = (1 + \exp(-\min(-\alpha,\beta)))^{-1}.$$

The linear regressions with reduced margins are given by the straight lines $L_y(x) = \gamma + \rho(\alpha,\beta)(x-\gamma)$ and $L_x(y) = \gamma + \rho(\alpha,\beta)(y-\gamma)$. For the non-linear regressions, as the interchange of X and Y is equivalent to the interchange of $-\alpha$ and β, it is sufficient to compute the regression curve

$$\bar{y}(x|\alpha,\beta) = x+\beta - \frac{e^{\beta}-1}{e^{\beta}-e^{\alpha}}\exp(\frac{1-e^{\alpha}}{e^{\beta}-e^{\alpha}}e^{-x})\int_{u(x)}^{v(x)}\frac{e^{-t}}{t}\,dt,$$

$$\text{with } u(x) = \frac{1-e^{\alpha}}{e^{\beta}-e^{\alpha}}e^{-x} \text{ and } v(x) = e^{\beta-\alpha}\,u(x).$$

Analogously to the situation for the biextremal model we may expect that linear regression is a good approximation, also in the same sense, to non-linear regression.

As $\alpha \le \bar{w}_i \le \beta$, with $\alpha < 0 < \beta$, we can take as estimators of α and β the statistics $\alpha^* = \min(-\varepsilon_n, \bar{y}_i - \bar{x}_i)$ and $\beta^* = \max(\varepsilon_n, \bar{y}_i - \bar{x}_i)$, with $\varepsilon_n \downarrow 0$ to garantee $\alpha^* < 0 < \beta^*$. If we have exchangeability $(\bar{k}(w|\alpha,\beta) = \bar{k}(-w|\alpha,\beta))$ the formulae simplify strongly and we have $(-\alpha)^* = \beta^* = \max(\varepsilon_n, \max|\bar{w}_i|)$. For details see Tiago de Oliveira (1982) and (1985).

Passing to standard exponential margins and introducing the new parameters $\alpha = e^{-\alpha} > 1 > \beta = e^{-\beta}$ we get

$$S(\tilde{x},\tilde{y}|\tilde{\alpha},\tilde{\beta}) = \exp\{-\frac{(1-\tilde{\beta})\max(\tilde{\alpha}\,\tilde{x},\tilde{y}) + (\tilde{\alpha}-1)\max(\tilde{\beta}\,\tilde{x},\tilde{y})}{\tilde{\alpha}-\tilde{\beta}}\}.$$

the dependence function being

$$A(u|\tilde{\alpha},\tilde{\beta}) = \frac{(1-\tilde{\beta})\max(\tilde{\alpha}(1-u),u) + (\tilde{\alpha}-1)\max(\tilde{\beta}(1-u),u)}{\tilde{\alpha}-\tilde{\beta}}.$$

$B(u|\tilde\alpha,\tilde\beta)$ has jumps at $u = \tilde\beta/(1+\tilde\beta)$ and $u = \tilde\alpha/(1+\tilde\alpha)$ of respectively $(\tilde\alpha-1)\ \tilde\beta/(\tilde\alpha-\tilde\beta)$ and $(1-\tilde\beta)/(\tilde\alpha-\tilde\beta)$. The support for $\tilde Y/(\tilde X+\tilde Y)$ is thus $[\tilde\beta/(1+\tilde\beta), \tilde\alpha/(1+\tilde\alpha)]$ or equivalently the angle $\tilde\beta\ \tilde X < \tilde Y < \tilde\alpha\ \tilde X$ and not all the 1st quadrant as in the other examples.

Just a quick reference to a generalization of the natural model, the (2,N) natural model, a particular case of the (m,N) natural model (m = 2) presented in Tiago de Oliveira (1987).

Let $Z_1, \ldots, Z_N$ be N independent standard Gumbel random variables and define $X = \max(Z_j-p_j)$, $Y = \max(Z_j-q_j)$.

The random pair (X,Y) has standard Gumbel margins iff $\Sigma_1^N e^{-p_j} = 1$ and $\Sigma_1^N e^{-q_j} = 1$; then we have the distribution function $\Lambda(x,y|p,q)=\{\text{Prob } X \le x, Y \le y\}=\exp\{-\Sigma^N{}_{j=1} \max(e^{-x-p_j}, e^{-y-q_j})\}$, with the dependence function

$$k(w) = \frac{\sum_1^N \max(e^{-p_j}, e^{-q_j-w})}{1+e^{-w}}.$$

The model has, thus, 2(N-1) parameters. The corresponding model for bivariate minima with standard exponential margins has the survival function $S(\tilde x,\tilde y|p,q) = \exp\{-\Sigma_1^N \max(x\ \tilde p_j, y\ \tilde q_j)\}$, where $\tilde p_j = e^{-p_j}$, $\tilde q_j = e^{-q_j}$, $\Sigma_1^N\ \tilde p_j = 1$, $\Sigma_1^N\ \tilde q_j = 1$ with the dependence function $A(u) = \Sigma_1^N \max(\tilde p_j(1-u),\tilde q_j u)$.

7. Intrinsic Estimation of the Dependence Function

We gave before the necessary and sufficient conditions on $k(w)$ and $A(u)$ for $\Lambda(x,y)=\exp\{-(e^{-x} + e^{-y})\ k(y-x)\}$ to be a distribution function of bivariate maxima with standard Gumbel margins and for $S(\tilde x,\tilde y) = \exp\{-(\tilde x + \tilde y)\ A(\tilde y/(\tilde x+\tilde y))\}$, $\tilde x,\tilde y \ge 0$ to be the survival function of bivariate minima with standard exponential margins. Let us recall that, for $A(u)$, $u \in [0,1]$, we have $A(0) = A(1) = 1$, $0 \le -A'(0), A'(1) \le 1$ and $A(u)$ convex.

We will call an <u>intrinsic estimator</u> of $A(u)$ (of $k(w)$) an estimator (i.e., a statistic depending on u and converging in probability to the function $A(u)$, for every $u \in [0,1]$) that satisfies the necessary and sufficient conditions to be $A(u)$ ($k(w)$), conditions that are restrictions on any development of the dependence functions, much stronger than the non-parametric analysis.

A solution, given by Pickands (1981) for a sample of n bivariate minima with exponential standard margins, is the following. Let $\tilde Z_i = \tilde Z_i(u) = \min(\tilde X_i/(1-u), \tilde Y_i/u))$ be a (dependent on u) random variable. It is immediate that its survival function is $\text{Prob}\{\tilde Z_i(u) > Z\} = e^{-Z\,A(u)}$, the exponential distribution with scale parameter $1/A(u)$. Its maximum likelihood estimator is, evidently, $\hat A_n(u) = 1/((1/n).\ \Sigma_1^n\ \tilde Z_i(u))$; $1/\hat A_n(u)$ is an unbiased estimator of $1/A(u)$. $\hat A_n(u)$ is <u>not</u> an intrinsic estimator of $A(u)$ chiefly because is not a convex function, but by Khintchine theorem, $\hat A_n(u) \overset{a.s.}{\to} A(u)$. Then Pickands (1981) sugests the computation of $\hat A_n(u)$ of the "turning" or "crossover" points $\tilde u_i = \tilde y_i/(\tilde x_i+\tilde y_i)$ and the numerical search of the greatest convex minorant in the plane (u,A) passing by the points (0,1), $(\tilde u_i, \hat A_n(\tilde u_i))$ and (1,1). It is very easy to see that if we take $n = 1$ as $\tilde u = \tilde y/(\tilde x+\tilde y)$ are have $\hat A(u) < \min(u, 1-u)$ iff $1 < \min(\tilde X, \tilde Y)$ and a basic relation is not verified; we have also $\hat A(u) > 1$ - not verifying another relation - when $\tilde X + \tilde Y > 1$ but in that case we could "consider" the greatest convex minorant as $A(u) = 1$. In fact, the law of large numbers "acts", overcoming this counter-example, and in practice the technique used did give good results in the cases it was applied.

REFERENCES

B.V. Finkelshteyn (1953) - Limiting distribution of extremes of a variational series of a two-dimensional random variable, Dokl. Ak. Nauk. S.S.S.R., vol. 91 (in russian), 209-211.

R.A. Fisher and L.H.C. Tippet (1928) - Limiting forms of frequency distribution of largest and smallest members of a sample, Proc. Cambr. Phil. Soc., vol. 24, pt. 2, 180-190.

M. Fréchet (1927) - Sur la loi de probabilité de l'écart maximum, Ann. Soc. Polon. Math. (Krakov), vol. 6, 93-116.

J. Galambos (1968)-The Asymptotic Theory of Extreme Order Statistics, J. Wiley, N. Y.

J. Geffroy (1958-59) - Contributions à la théorie des valeurs extrêmes, Thése de doctorat, Publ. Inst. Stat. Paris, vol 7/8, 37-185.

B.V. Gnedenko (1943) - Sur la distribution limite du terme maximum d'une série aléatoire, An. Math., vol. 44, 423-433.

E.J. Gumbel (1935) - Les valeurs extrêmes des distributions statistiques, Ann. Inst. H.Poincaré, vol. V, 115-158.

---------(1958) - Statistics of Extremes, Columbia University Press.

---------(1961) - Multivariate extremal distributions, Bull. Int. Stat. Inst., 33e sess., 2e livr., Paris, 191-193.

---------(1961') - Bivariate logistic distribution, J. Amer Stat. Assoc., vol. 56, # 307, 194-816.

E. J. Gumbel and Neil Goldstein (1964) - Analysis of empirical bivariate extremal distributions, J. Amer. Stat. Assoc., vol. 59, 794-816.

K. Joag - Dev (1983) - Independence via uncorrelatedness under certain dependence conditions, Ann. Prob., vol. 11, 1037-1041.

Albert W. Marshall and Ingram Olkin (1967) - A multivariate exponential distribution, J. Amer. Stat. Assoc., vol. 62, 30-44.

------------------------- (1983) - Domains of attraction of multivariate extreme value distributions, Ann. Prob., vol. 11, 168-177.

R. von Mises(1935)-La distribution de la plus grande de n valeurs, Rev. Math. Union Interbalkanique, vol. I, 141-160.

J. Pickands III (1981) - Multivariate extreme value theory, Bull. Int. Stat. Inst., 49th session I.S.I., Buenos Aires, 859-878.

M. Sibuya (1960) - Bivariate extremal statistics, I, Ann. Inst. Stat. Math. vol. XI, 195-210.

J.A. Tawn (1987) - Bivariate extreme value theory - models and estimation, Techn. Rep. nº 57, Dep. of Math., University of Surrey.

J. Tiago de Oliveira (1958) - Extremal distributions, Rev. Fac. Ciências Lisboa, 2 ser., A, Mat., vol. VIII, 299-310.

-------------- (1962-63) - Structure theory of bivariate extremes; extensions, Estudos Mat., e Econom., vol. VII, 165-195.

-------------- (1965) - Statistical decision for bivariate extremes, Portug. Math., vol. 24, 145-154.

-------------- (1968) - Extremal processes; definition and properties, Publ. Inst. Stat. Univ. Paris, vol. XVII, 25-36.

-------------- (1970) - Biextremal distributions: statistical decision, Trab. Estad. y Inv. Oper., vol. XXI, Madrid, 107-117.

-------------- (1971) - A new model of bivariate extremes: statistical decision, Studi di Probabilitá, Statistica e Ricerca Operativa in Onore di Giuseppe Pompili, Tip. Oderisi, Gubbio, Italia, 1-13.

-------------- (1974) - Regression in the non-differentiable bivariate models, J. Amer. Stat. Assoc., vol. 69, 816-818.

-------------- (1975) - Bivariate and multivariate extreme distributions Statistical Distributions in Scientific Work, G.P. Patil, S. Kotz and J.K. Ord. eds., vol. 1, D. Reidell Publ. Co, 355-361.

-------------- (1975') - Statistical decision for extremes, Trab. Estad. y Inv.Oper., vol. XXVI, Madrid, 453-471.

-------------- (1980) - Bivariate extremes: foundations and statistics, Proc. Int. Symp. Multiv. Analysis, P.R. Krishnaiah, ed., North Holland, 349-366.

-------------- (1982) - Decision and modelling for extremes, Some Recent Advances in Statistics, J. Tiago de Oliveira and B. Epstein eds., Academic Press, London, 101-110.

-------------- (1984) - Bivariate models for extremes; statistical decision, Statistical Extremes and Applications, J. Tiago de Oliveira ed., D. Reidel and Co, 131-153.

-------------- (1985) - Statistical Choice of non-separate models, Trab. Estad. y Inv. Oper. (Madrid), vol. 36, 136-152.

-------------- (1987) - Comparaison entre les modéles bivariées logistique et naturel pour les maxima et extensions, C.R. Acad. Sc. Paris, t.305, ser. I, 481-484.

MULTIVARIATE NEGATIVE EXPONENTIAL AND EXTREME VALUE DISTRIBUTIONS

James Pickands III

Department of Statistics, University of Pennsylvania

Philadelphia, Pa. 19104-6302 U.S.A.

Abstract. We survey multivariate extreme value distributions. These are limiting distributions of maxima and/or minima, componentwise, after suitable normalization. A distribution is extreme value stable if and only if its margins are stable and its dependence function is stable. Thus it is possible, without loss of generality, to choose any stable marginal distribution deemed convenient. We use the negative exponential one. The class of multivariate stable negative exponential distributions is characterized by the fact that weighted minima of components have negative exponential distributions. We examine several representations and the relationships among them and we consider, in terms of them, joint densities and scalar measures of dependence. We also consider estimation.

1 Introduction.

Let $\mathbf{X}$ be a random vector with components X_j. We say that $\mathbf{X}$ has a (multivariate) negative exponential distribution if and only if for any set of "weights" $a_j \in [0, \infty)$, with at least one $a_j > 0$, $\min(X_1/a_1, X_2/a_2, ..., X_J/a_J)$ has a (univariate) negative exponential distribution. The dimensionality J may be finite or infinite. Notice that if $a_j = 0$, $X_j/a_j = \infty$, and so the j th component plays no role in the minimization. This is why we divide by a_j instead of multiplying by it. Esary and Marshall [2] give five classes, ordered by inclusion, of multivariate negative exponential distributions. Their most sparse class is the class of distibutions with mutually independent negative exponential margins. Their widest class is the class of all distributions having negative exponential margins. The class we have just defined is their third.

Among other things, our family forms a basis for the study of multivariate extreme value distributions. A random vector $\mathbf{X}$, and equivalently, its distribution, are said to be "max-stable" iff $\mathbf{X}$ is the limit in distribution of $(\mathbf{M}_n - \mathbf{b}_n)/\mathbf{a}_n$, componentwise, as $n \to \infty$, where $\mathbf{M}_n$ is the maximum, componentwise, of n mutually independent identically distributed random vectors and $\mathbf{a}_n$ and $\mathbf{b}_n$ are nonrandom vectors. Notice that it is not enough that the margins be stable. We are requiring joint stability. With appropriate but obvious modifications, we define "min-stable" distributions. If the limiting joint distribution is formed, taking minima or maxima, possibly minima for some components and maxima for others we have what we call a multivariate "extreme-value" distribution. For stability in the multivariate case it is necessary and sufficient that the margins be stable and that the dependence function D be stable, where the joint cumulative distribution function $F = D(F_1, F_2, ..., F_J)$ and the F_j are the marginal cumulative distribution functions. This is Theorem 5.2.3, Galambos [3], page 294. Thus, there is a complete conceptual separation between the margins and the dependence function. To study the latter, we can choose any extreme-value margins we find most convenient. Tiago de Oliveira [16] uses the standard Gumbel distribution: $F_X(x) := \exp -e^{-x}$. de Haan and Resnick [9] use the Frechet distribution with shape parameter 1: $F_X(x) := \exp -x^{-1}$. Pickands [13] uses the standard negative exponential distribution: $F_X(x) := 1 - e^{-x}$. The first two are max-stable while the last is min-stable. Our multivariate negative exponential family is exactly the class of min-stable distributions with negative exponential margins. This is a consequence, after a simple transformation, of a result of de Haan [5]. It follows that most results for multivariate negative exponential distributions translate directly to corresponding ones for multivariate extreme-value distributions. The first book on extreme value theory was by Gumbel [4]. It emphasized applications. Two recent books are the comprehensive book by Galambos[3] and the book by Leadbetter, Lindgren and Rootzen [10], which emphasizes stochastic processes.

2 Representations and measures of dependence.

Using results of de Haan [6], it is shown in de Haan and Pickands [8] that a spectral representation exists for any multivariate negative exponential distribution as follows: Let $\{U_l, Y_l\}_{l=1,2,...\infty}$ be the points of a homogeneous Poisson process with unit intensity on the strip $[0,1] \times R_+$. Each component variable X_j can be represented:

$$X_j = \min_{l=1}^{\infty} \left\{ \frac{Y_l}{f_j(U_l)} \right\} \tag{1}$$

where $\{f_j(u), j = 1, 2, ...\infty\}$ are L^1_+ functions from $[0,1]$ to R_+. Such representations are non-unique. Equivalences are examined in de Haan and Pickands [8]. A class of examples is the family of shock models considered by Marshall and Olkin [11]. For some other examples, see Pickands [13]. One unique functional representation is given and examined by Tiago de Oliveira [16].

In passage from the bivariate to the trivariate case, the difficulties are increased enormously. Jointly max- or min- stable random variables are always non-negatively associated. Marshall and Olkin [12] proved that they are always "associated" in a very strong sense, that is, that the covariance between any increasing functions of the components must be non-negative. A measure of dependence $\delta = \delta(X, Y)$ must, in order to be reasonable, satisfy the following:

1. $0 \leq \delta \leq 1$, for all X, Y,
2. $\delta = 0$ iff X, Y are independent,
3. $\delta(X/a, Y/b) = \delta(X, Y)$, for all $a, b \in (0, \infty)$,
4. δ is continuous with respect to the joint distribution of X and Y,
5. $\delta = 1$ iff $p\{X = aY\} = 1$ for some $a \in (0, \infty)$.

The first measure of dependence, applicable to bivariate negative exponential distributions, was introduced by Tiago de Oliveira [15]. We will call it δ_T, where

$$\delta_T := 2 - \frac{1}{E \min\left\{\frac{X}{EX}, \frac{Y}{EY}\right\}}.$$

Let X, Y have de Haan-Pickands spectral functions, respectively, $f_X(u)$ and $f_Y(u)$. Now

$$\delta_T = 2 - \int_0^1 \max\left\{\frac{f_X(u)}{||f_X||}, \frac{f_Y(u)}{||f_Y||}\right\} du$$

where

$$||f|| := \int_0^1 f(u) du. \tag{2}$$

The second bivariate measure of dependence was suggested by the author. See de Haan [7]. It is the probability that both X and Y are generated by the same point of the Poisson process on the strip. We call it δ_p. The subscript stands for "probability". In spectral terms:

$$\delta_p = \int_0^1 [\int_0^1 \max\left\{\frac{f_X(v)}{f_X(u)}, \frac{f_Y(v)}{f_Y(u)}\right\} dv]^{-1} du.$$

Both of these are "piston invariant" in the sense of de Haan and Pickands [8], as must be the case. That is, different spectral representations of the same bivariate negative exponential distribution will yield the same values of δ_T and δ_p. Both also satisfy 1. through 5. above. Using the simplex representation discussed below, we can show that

$$\delta_T/2 \leq \delta_p \leq \delta_T.$$

In some non-trivial cases $\delta_T = \delta_p$, but not all. But both are continuous in appropriate senses and so they are not "equivalent". That is, they don't provide the same dependence ordering.

We characterize the multivariate negative exponential family as follows:

$$-\ln p\{\mathbf{X} > \mathbf{x}\} = cE_{\mathbf{Q}} \max\{\mathbf{Qx}\}$$

where $0 < c < \infty$, operations are componentwise, the maximum is taken over all components and $\mathbf{Q}$, with elements Q_j, lies on the J-dimensional simplex. That is,

$$p_{\mathbf{Q}}\{\sum_{j=1}^{J} Q_j = 1\} = 1.$$

This representation is unique. See Galambos[3], pages 309-312 and Pickands [13]. Without loss of generality we can normalize so that all margins have mean 1. Then $c = J$ and $E_{\mathbf{Q}}Q_j = 1/J$ for all j. In the bivariate case (with $EX = EY = 1$),

$$-\ln p\{X > x, Y > y\} = 2E_Q \max(Qx, (1 - Q)y)$$

where Q is a $[0, 1]$ random variable with mean 1/2. The random variables X and Y are independent iff $p_Q\{Q = 0\} = p_Q\{Q = 1\} = 1/2$. They are completely dependent iff $p_Q\{Q = 1/2\} = 1$.

This representation can be related to the Poisson-spectral representation as follows: Let $f_X(u)$ and $f_Y(u)$ be the spectral functions for X, Y. We are assuming that

$$1/EX = 1/EY = \int_0^1 f_X(u)du = ||f_X|| = ||f_Y|| = 1.$$

We let

$$Q := \frac{f_X(U)}{f_X(U) + f_Y(U)}$$

where

$$U \sim (f_X(u) + f_Y(u))/2.$$

Conversely, we can let

$$f_X(u) := 2F_Q^{\leftarrow}(u)$$

and

$$f_Y(u) := 2(1 - F_Q^{\leftarrow}(u)).$$

In Pickands [13] we introduce a unique functional representation which is symmetric in the components. Let the vector $\mathbf{\Omega}$, with elements ω_j, be any point on the J-dimensional simplex: $\sum_{j=1}^{J} \omega_j = 1$. Then a unique dependence function is $A(\mathbf{\Omega})$, where

$$A(\mathbf{\Omega}) := \frac{1}{E \min\left(\frac{X_j}{\omega_j}\right)}.$$

In the bivariate case we write

$$A(\omega) = \frac{1}{E \min\left(\frac{X}{\omega}, \frac{Y}{1-\omega}\right)}. \tag{3}$$

We consider the relationship between $A(\omega)$ and the cumulative distribution function F_Q. Notice that for all ω,

$$\begin{aligned}
A(\omega) &= -\ln p\{\min\left(\frac{X}{\omega}, \frac{Y}{1-\omega}\right) > 1\} \\
&= -\ln p\{X > \omega, Y > 1-\omega\} \\
&= 2E_Q(\max[Q\omega, (1-Q)(1-\omega)]) \\
&= 2\omega E_Q Q 1_{\{Q>1-\omega\}} + 2(1-\omega) E_Q (1-Q) 1_{\{Q \le 1-\omega\}}.
\end{aligned}$$

It follows that

$$\begin{aligned}
dA(\omega) = {} & 2d\omega E_Q Q 1_{\{Q>1-\omega\}} - 2\omega(1-\omega) dF_Q(1-\omega) \\
& -2d\omega E_Q(1-Q) 1_{\{Q \le 1-\omega\}} + 2\omega(1-\omega) dF_Q(1-\omega)
\end{aligned}$$

$$= 2d\omega E_Q Q - 2d\omega F_Q(1-\omega)$$

$$= d\omega(1 - 2F_Q(1-\omega)).$$

That is the derivative

$$A'(\omega) = (1 - 2F_Q(1-\omega))$$

and so

$$F_Q(\omega) = (1 - A'(1-\omega))/2.$$

But

$$A(0) = -\ln p\{Y > 1\} = 1/EY = 1$$

and, similarly,

$$A(1) = -\ln p\{X > 1\} = 1/EX = 1.$$

Thus

$$A(\omega) = 1 + \int_0^\omega (1 - 2F_Q(1-z))dz.$$

By (3) an obvious choice of estimator is $\hat{A}(\omega)$, where

$$\hat{A}(\omega) := \frac{n}{\sum_{i=1}^n \min\left(\frac{X_i}{\omega}, \frac{Y_i}{(1-\omega)}\right)}.$$

With probability 1, the estimated function fails to be a dependence function because F_Q is not everywhere non-decreasing. As our estimator we use the convex hull of $\hat{A}(\omega)$, which is a dependence function, with probability 1. Unfortunately, two bivariate distributions are not continuous with respect to one another if the corresponding distribution functions F_Q have different jumps, except at 0 and 1. Smith [14] suggests a kind of (density estimation type) smoothing. In an interesting recent paper Emoto and Matthews [1] consider bivariate Weibull estimation under a form of censoring.

One can verify that

$$\delta_T = 2(1 - E_Q \max\{Q, 1 - Q\})$$

and

$$\delta_p = E_Q[1/E_{Q'} \max\left\{\frac{Q'}{Q}, \frac{1 - Q'}{1 - Q}\right\}]$$

where Q' is an independent replicate of Q. Some other possibilities are:

$$\delta_1 := 4\int_0^1 (1 - A(\omega))d\omega,$$

$$\delta_2 := 1 - 4\text{var}_Q(Q)$$

and, equivalently,

$$\delta_3 := 1 - 2\text{sd}_Q(Q).$$

Notice that the dependence functions constitute a convex set. That is, for all $\alpha \in [0, 1], \alpha\delta_A + (1 - \alpha)\delta_B$ satisfies 1.-5. if δ_A and δ_B do. Since all "spectral" probability measures are on $[0, 1]$ with means $1/2$, we can define a partial order for dependence functions, "more dependent" being equated with "p_Q less spread".

We can define linear combinations of joint distributions. Let $\alpha \in [0, 1]$, let $X, Y \sim e(F_Q)$ independently of $X', Y' \sim e(F_{Q'})$. Now let

$$X'', Y'' := \min\left(\frac{X}{\alpha}, \frac{X'}{1 - \alpha}\right), \min(\frac{Y}{\alpha}, \frac{Y'}{1 - \alpha}).$$

Now

$$X'', Y'' \sim e(\alpha F_Q + (1 - \alpha)F_{Q'}).$$

If

$$\delta'' = \alpha\delta + (1 - \alpha)\delta',$$

we say that δ is "linear". Notice that δ_T and δ_1, above, are linear, while the others are not.

We can define a "canonical association" between two vectors. Let

$$\delta(\mathbf{X}, \mathbf{Y}) := \sup \delta(\min \frac{\mathbf{X}}{\mathbf{a}}, \min \frac{\mathbf{Y}}{\mathbf{b}}),$$

where division is componentwise, the minima are taken over all of the components of $\mathbf{X}/\mathbf{a}$ and of $\mathbf{Y}/\mathbf{b}$ and the supremum is over the choice of vectors $\mathbf{a}$ and $\mathbf{b}$. Notice that by scale invariance (condition 3. above) we can let the components of a and b be so normalized that

$$E \min \frac{\mathbf{X}}{\mathbf{a}} = E \min \frac{\mathbf{Y}}{\mathbf{b}} = 1,$$

without loss of generality, the minimum being taken over all components.

3 The case of more than two dimensions.

Without further restriction the complexity is increased enormously when the number of dimensions $J \geq 3$. In the trivariate case, the normalized joint distributions are indexed by the measures on the three dimensional simplex. This is too rich a class to be statistically useful. We can simplify somewhat by assuming exchangeability or stationarity for example.

We prefer a different approach. We call the joint distribution "ratio-regular" if there exists a spectral representation such that each of the spectral functions has at most a countable number of discontinuities. This is, apparently, always true if $J = 2$, but it is not generally true if $J \geq 3$. Consider p_Q in the case $J = 3$. Ratio-regularity means that if Q_2/Q_1 has a positive density, but not a positive probability at its value, then, conditionally, Q_3/Q_2, hence also Q_3/Q_1, is constant. Ratio-regularity holds if p_Q is discrete. Thus the set of ratio-regular measures is dense in the set of all measures. If the spectral functions are powers, p_Q is, again, ratio-regular. We consider an example where it is not. Let $u = e_1 e_2 \ldots e_k \ldots$, the binary expansion. Now let

$$f_j(u) := e_j e_{j+3} \ldots e_{j+3k} \ldots,\ j = 1, 2, 3.$$

At present we are seeking a continuous non-negative functional $h(p_Q)$ which is finite in the case of ratio-regularity and infinite otherwise. We are also considering non-parametric estimation in some special cases, in particular where p_Q is a finite mixture of "ratio-monotone" measures.

For forecasting we recommend the conditional median which minimizes the L_1 error. This is the only L_p error whose minimizer is invariant under monotone transformation of the margins. At present we don't have a way of computing this.

References

[1] Emoto S. E. and Matthews P. C. (1987) A Weibull model for dependent censoring. *Research Report* 87-14. Dept. of Math. Univ. Maryland, Catonsville Md.

[2] Esary J. D. and Marshall A. W. (1974) Multivariate distributions with exponential minimums. *Ann. Statist.* 2, 84-93.

[3] Galambos J. (1987) *The asymptotic theory of extreme order statistics.* 2nd ed. Robert E. Krieger Publishing Co. Malabar Fla.

[4] Gumbel E. J. (1958) *Statistics of Extremes.* Columbia Univ. Press, New York.

[5] de Haan l. (1978) A characterization of multivariate extreme- value distributions. *Sankhya Series A,* 40, 85-88.

[6] de Haan L. (1984) A spectral representation for max-stable processes. *Ann. Probab.* 12, 1194-1204.

[7] de Haan L. (1985) Extremes in higher dimensions: the model and some statistics. *45th Session ISI,* Amsterdam.

[8] de Haan L. and Pickands J. (1986) Stationary Min-Stable Stochastic Processes. *Probab. Th. Rel. Fields* 72, 477-492.

[9] de Haan L. and Resnick S. I. (1977) Limit theory for multivariate sample extremes. *Z. Wahrsch.* 40, 317-337.

[10] Leadbetter M. R., Lindgren G. and Rootzen H. (1983) *Extremes and related properties of random sequences and processes.* Springer, New York.

[11] Marshall A. W. and Olkin I. (1967) A generalized bivariate exponential distribution. *J. Appl. Probab.* 4, 291-302.

[12] Marshall A. W. and Olkin I. (1983) Domains of attraction of multivariate extreme value distributions. *Ann. Probab.* 11, 168-177.

[13] Pickands J. III (1981) Multivariate extreme value distributions. *43rd Session, ISI,* Buenos Aires, 859-878.

[14] Smith R. L. (1985) Statistics of extreme values. *45th Session, ISI,* Amsterdam.

[15] Tiago de Oliveira J. (1962/63) Structure theory of bivariate extremes: extensions. *Estudos de Math. Estat. Econom.* 7, 165-195.

[16] Tiago de Oliveira J. (1975) Bivariate extremes. Extensions. *40th Session, ISI,* Warsaw.

Author-Index

Subject-Index